Vegetation and Erosion

Processes and Environments

Edited by
J. B. Thornes
Department of Geography
University of Bristol

John Wiley & Sons
CHICHESTER · NEW YORK · BRISBANE · TORONTO · SINGAPORE

Copyright © 1990 by John Wiley & Sons Ltd
            Baffins Lane, Chichester
            West Sussex PO19 1UD, England

Other Wiley Editorial Offices

John Wiley & Sons, Inc., 605 Third Avenue,
New York, NY 10158-0012, USA

Jacaranda Wiley Ltd, G.P.O. Box 859, Brisbane,
Queensland 4001, Australia

John Wiley & Sons (Canada) Ltd, 22 Worcester Road,
Rexdale, Ontario M9W 1L1, Canada

John Wiley & Sons (SEA) Pte Ltd, 37 Jalan Pemimpin 05-04,
Block B, Union Industrial Building, Singapore 2057

Library of Congress Cataloging-in-Publication Data:
Vegetation and erosion : processes and environments / edited by
   J. B. Thornes.
      p.   cm.—(British Geomorphological Research Group symposia
series)
   Includes bibliographical references.
   ISBN 0 471 92630 2
   1. Geomorphology. 2. Erosion. I. Thornes, John B.   II. Series.
   GB401.5.G46   1990   89–25010
   551.3′02—dc20   CIP

British Library Cataloguing in Publication Data:
Vegetation and erosion.
   1. Geomorphology. Biological aspects
   I. Thornes, John B. (John Barrie), *1940–*   II. Series
   551.4

   ISBN 0 471 92630 2

Phototypeset by Dobbie Typesetting Limited, Plymouth, Devon
Printed and bound in Great Britain by Biddles Ltd, Guildford, Surrey

8.6,92

To the memory of Ted Culling

Contents

Series Preface

The British Geomorphological Research Group (BGRG) is a national multi-disciplinary Society whose object is 'the advancement of research and education in geomorphology'. Today, the BGRG enjoys an international reputation and has a strong membership from both Britain and overseas. Indeed, the Group has been actively involved in stimulating the development of geomorphology and geomorphological societies in several countries. The BGRG was constituted in 1961 but its beginnings lie in a meeting held in Sheffield under the chairmanship of Professor D. L. Linton in 1958. Throughout its development the Group has sustained important links with both the Institute of British Geographers and the Geological Society of London.

Over the past three decades the BGRG has been highly successful and productive. This is reflected not least by BGRG publications. Following its launch in 1976 the Group's journal, *Earth Surface Processes* (since 1981 *Earth Surface Processes and Landforms*) has become acclaimed internationally as a leader in its field, and to a large extent the Journal has been responsible for advancing the reputation of the BGRG. In addition to an impressive list of other publications on technical and educational issues, BGRG symposia have led to the production of a number of important works including *Nearshore Sediment Dynamics and Sedimentation* edited by J. R. Hails and A. P. Carr; *Geomorphology and Climate* edited by E. Derbyshire; *River Channel Changes* edited by E. Derbyshire; *River Channel Changes* edited by K. J. Gregory, and *Timescales in Geomorphology* edited by R. Cullingford, D. Davidson and J. Lewin. This sequence of books culminated in 1987 with a publication of the *Proceedings of the First International Geomorphology Conference* edited by Vince Gardiner. This international meeting, arguably the most important in the history of geomorphology, provided the foundation for the development of geomorphology into the next century.

This open-ended BGRG Symposia Series has been founded and is now being fostered to help maintain the research momentum generated during the past three decades, as well as to further the widening of knowledge in component fields of geomorphological endeavour. The series consists of authoritative volumes based on the themes of BGRG meetings, incorporating, where appropriate, invited contributions to complement chapters selected from presentations at these meetings under the guidance and editorship of one or more suitable specialists. Whilst maintaining a strong emphasis on pure geomorphological research, BGRG meetings are diversifying, in a very positive way, to consider links between geomorphology *per se* and other disciplines such as ecology, agriculture, engineering and planning.

The first volume in the series was published in 1988. *Geomorphology in Environmental Planning*, edited by Janet Hooke, reflects the trend towards applied studies. The second volume, edited by Keith Beven and Paul Carling, *Floods—Hydrological, Sedimentological and Geomorphological Implications,* focuses on a traditional research theme. *Soil Erosion on Agricultural Land* reflects the international importance of this topic for researchers during the 1980s. The volume, edited by John Boardman, Ian Foster and John Dearing, forms the third in the series.

The role of vegetation in geomorphology is a traditional research theme, recently revitalized with the move towards interdisciplinary studies. *Vegetation and Erosion* edited by John Thornes, reflects this development in geomorphological endeavour, and raises several research issues for the next decade.

The BGRG Symposia Series will contribute to advancing geomorphological research and we look forward to the effective participation of geomorphologists and other scientists concerned with earth surface processes and landforms, their relation to Man, and their interaction with the other components of the Biosphere.

Geoffrey Petts
BGRG Publications

September 1989

Preface

Despite the implicit acceptance of the importance of vegetation by geomorphologists, we have been strangely reluctant to enter this arena in a serious fashion. There are of course some notable exceptions, such as J. T. Hack, and we can count in the hundreds the papers which 'explain' the impact of climatic change through the mechanisms of plant cover or use pollen analysis to infer climatic change and thereby claim to account for geomorphological change. Yet even in the excellent *Geomorphology* by R. J. Chorley, S. A. Schumm and D. Sugden, vegetation gets very short shrift.

In conversations with Karol Rotnicki in 1983 I became convinced that something should be done about this state of affairs and tried to make a start by shifting my own semi-arid work in this direction, and encouraging others. In the event Heather Viles has produced a much needed review of what has been done in *Biogeomorphology* published by Blackwell in 1988 in which the chapter by Stan Trimble provides a useful backcloth to this book. The conference of the British Geomorphological Research Group in 1988 attempted to bring together the work in progress or recently completed whilst specifically excepting the historical dimension. The idea was to see what is known about the processes and their interactions rather than producing more of the ill-informed speculations about plant and geomorphological interactions which have characterized the last forty years.

The results, as presented in this book, are encouraging but only a start. They indicate a preponderance of work (including some of my own) which bring in vegetation only as an extrinsic variable, parameter or even constant rather than a highly dynamic and vital component which affects virtually all processes and therefore all geomorphological histories. Nor is there much to show on the other side of the coin; how the geomorphic processes affect the plant cover. In the Introduction I try to raise some of these issues and create an agenda in the hope that some future work will move into this area.

The success of the Conference was largely attributable to the astonishing and continuing vigour of the British Geomorphological Research Group. It was also due in no mean part to the energy and resourcefulness of the Bristol Geography Department and my secretary Sarah Howell whose help is gratefully acknowledged. In producing this book I am also pleased to acknowledge the help of Simon Godden and Tony Philpott of the Department which came at a time of unprecedented tumult in the Department and to the many referees who made the book better than it would otherwise have been.   ·

The Conference was marred only by the death of W. E. H. Culling a few weeks before. The loss of this brilliant, ingenious, sympathetic and completely likeable geomorphologist will leave a serious gap in theoretical geomorphology of which he was a world leader. This book is dedicated to his memory.

List of Contributors

Roy Alexander, Department of Geography, Chester College, Chester CH1 4BJ, UK

Ken Atkinson, Geography Department, University of Leeds, Leeds LS2 9JT, UK

Yvan Biot, School of Development Studies, University of East Anglia, Norwich NR4 7TJ, UK

Jackie Birnie (née Watts), Geography Department, Cricklade College, Andover, Hants, UK

Tony G. Brown, Geography Department, University of Leicester, Leicester LE1 7RH, UK

T. P. Burt, Geography Department, University of Oxford, Oxford OX1 3TB, UK

Adolfo Calvo, Department of Geography, University of Valencia, Valencia, Spain

Angela Clark, Geography Department, Birkbeck College, London WC1 7HX, UK

Ian Douglas, School of Geography, University of Manchester, Manchester M13 9PL, UK

P. J. Edwards, Biology Department, University of Southampton, Southampton SO9 5NH, UK

Walter Erdelen, Department of Biogeography, University of Saarland, D-6600, Saarbrucken, Federal Republic of Germany

Hazel Faulkner, Department of Geography, Polytechnic of North London, London N7 ORN, UK

Carolyn F. Francis, Geography Department, University of Bristol, Bristol BS8 1SS, UK

J. Carlos Gonzalez Hidalgo, Department of Geography, University of Zaragoza, Zaragoza 50009, Spain

Tony Greer, School of Geography, University of Manchester, Manchester M13 9PL, UK

Angela M. Gurnell, Geography Department, University of Southampton, Southampton SO9 5NH, UK

Louise Heathwaite, Geography Department, University of Sheffield, Sheffield S10 2TN, UK

P. Anne Hughes, Geography and Biology Departments, University of Southampton, Southampton SO9 5NH, UK

Cliff R. Hupp, United States Geological Survey, 430 National Center, Reston, Virginia 22092, USA

L. Husain, University of Pertanian, Malaysia

Mike J. Kirkby, Geography Department, University of Leeds, Leeds LS2 9JT, UK

D. Leggett, Geography Department, University of Hull, Hull HU6 7RX, UK

John Lockwood, Geography Department, University of Leeds, Leeds LS2 9JT, UK

M. V. Lopez, Mediterranean Agronomic Institute, Zaragoza, Spain

Mark G. Macklin, Department of Geography, University of Newcastle-upon-Tyne, Newcastle-upon-Tyne NE1 7RU, UK

David Mitchell, School of Applied Sciences, The Polytechnic, Wolverhampton WV1 1SB, UK

Roy P. C. Morgan, Silsoe College, Silsoe, Bedford MK45 4DT, UK

Stephen Nortcliff, Department of Soil Science, University of Reading, Reading, UK

Antony R. Orme, Geography Department, University of California, Los Angeles, California 90024, USA

Robert Parkinson, Seale-Hayne College, Newton Abbott, Devon TQ12 6NQ, UK

Francisco Pellicer, Department of Geography, University of Zaragoza, Zaragoza 50009, Spain

John Pethick, Geography Department, University of Hull, Hull HU6 7RX, UK

Christoph Preu, Department of Geography, University of Augsburg, Augsburg, Federal Republic of Germany

Ian Reid, Geography Department, Birkbeck College, London WC1E 7HX, UK

R. Jane Rickson, Department of Rural Land Use, Silsoe College, Silsoe, Bedfordshire MK45 4DT, UK

Sheila M. Ross, Geography Department, University of Bristol, Bristol, BS8 1SS, UK

Maria Sala, Department of Physical Geography, University of Barcelona, Barcelona, Spain

Leonel Sierralta, Ecology Department, Catholic University of Santiago, Santiago, Chile

Andrew Simon, United States Geological Survey, Nashville, Tennessee 37203, USA

Roger S. Smith, Department of Agricultural and Environmental Science, University of Newcastle-upon-Tyne, Newcastle-upon-Tyne, NE1 7RU, UK

Tom Spencer, Department of Geography, University of Cambridge, Cambridge, CB2 3EN, UK

David S. G. Thomas, Geography Department, University of Sheffield, Sheffield, S10 2TN, UK

Colin R. Thorne, Department of Geography, University of Nottingham, Nottingham, UK

John B. Thornes, Geography Department, University of Bristol, Bristol, BS8 1SS, UK

Stanley W. Trimble, Department of Geography, University of California, Los Angeles, California 90024, USA

Stephen T. Trudgill, Geography Department, University of Sheffield, Sheffield, S10 2TN, UK

Haim Tsoar, Geography Department, Ben Gurion University, Beer Sheva, 84120, Israel

Stephen Twomlow, Seale-Hayne Faculty of Agriculture, Food and Land Use of Polytechnic Southwest, Newton Abbott, Devon, TQ12 6NQ, UK

Heather Viles, Geography School, University of Oxford, Oxford, OX1 3TB, UK

G. D. Watts, School of Biological Sciences, University of East Anglia, Norwich, UK

Waidi Sinun, Danum Valley Field Centre, Yayasan Sabah, 91108, Lahad Datu, Sabah, Malaysia

1 Introduction

J. B. THORNES

This book appears at a time of unprecedented interest and concern in the environment and at a time of rapid change in earth and environmental sciences. The enormous political pressure to understand the impact of anthropogenic activity, coupled with an injection of financial resources, could achieve more in a few years than has been achieved in decades of scientific endeavour in the poorly funded and hitherto unglamorous backwaters of science. Two of the most pressing issues are transformation of the troposphere and stratosphere through so-called 'greenhouse gases' and transformation of the surface and its agricultural productivity through so-called 'desertification'. The 'greenhouse' issue puts the emphasis on the interactions between the climate, the biosphere, oceans and ice sheets and human activity. The 'desertification' issue puts the emphasis on the interaction between climate, earth surface processes, especially as they are affected by the vegetation cover, and human activity.

Two major research questions are involved. The first seeks to establish the processes by which such change is being brought about. The second asks the extent to which such changes are part of the natural state of affairs as revealed by the history of past changes. Both the issues and both of the research questions have been traditional concerns of geomorphologists. Climate provides the driving agents of geomorphic processes and climatic change results in changes in the magnitude and intensity of the earth surface processes. This critical symbiosis between process and change, change and process is mediated almost everywhere on earth through the hydrological cycle and the vegetation cover.

In the last three decades through the separation of studies of process and evolution, through the concentration on hydrologically driven processes and through the concern for local rather than regional scale phenomena and short and medium rather than long timescales, geomorphologists have lost sight of the essential linkages between process, form and evolution. Above all there has been a major neglect of the vegetative cover at almost all scales despite the recognition of its importance by countless students of Quaternary geomorphology in the 1950s and 1960s.

It is not our purpose to review the field. This is taken care of by Heather Viles in Chapter 2 and, for the wider context, in her recently published book

Vegetation and Erosion
Edited by J. B. Thornes
©1990 John Wiley & Sons Ltd

on *Biogeomorphology*. Rather it is to indicate the kind of issues which should appear in the future research agendas of both geomorphologists and ecologists and so set the context for the papers which follow. Three major areas can generally be recognized: (i) processes and mechanisms, (ii) interactions and environments, and (iii) evolution. They are neither mutually exclusive nor can they only be tackled in that order, though the order is one of increasing complexity. Most of the chapters in this book fall into the first group and there is a certain logic in this. However, both the quantitative and qualitative modellers at least, with their usual impatience, feel a need to press on even though the base may be rather shaky. This is acceptable because neither science nor human needs ever proceed in the linear fashion advocated in classical scientific methodology.

At the simplest level vegetation is important in geomorphology because it controls the nature and rate of operation of the processes. For example, it provides an element of roughness in stream channels and on hillslopes or determines the rate of interception of rain in erosional processes. There is a very considerable literature here, much of it characterized by treatment of vegetation as a relatively static phenomenon. The role of vegetation is then often considered as an extrinsic variable which appears in regression analyses, for example, as density, cover, or leaf area index. Even at this level there are evident lacunae. The emphasis so far has been mainly on characteristics which are hydrologically important and engineers have been left to worry about root strength, chemists to think about dissolved organic matter and the biologists about seed banks. Moreover, although cover has received plenty of attention, for example in agricultural erosion studies, life form, composition and pattern have attracted little interest even at this static level.

In the short term and when considered only as an extrinsic variable, the vegetative cover is dynamic rather than static. The phenology of plants results in strong seasonal contrasts which may be geomorphologically very important. This phenology is essentially controlled by moisture, temperature and genetics and although its relationship to erosional processes is well known for certain crops, much less is known about phenology and process in natural and semi-natural vegetation which is critical for questions of longer-term change. This is especially important when dealing with annuals which, in the drier regions, often provide vegetation cover at the most critical times but the whole nutrient cycling and weathering processes are intimately related to the uptake and production of minerals and organic matter which itself is a dynamic phenomenon.

At the second level of complexity it is the interaction between plants and geomorphic processes which needs more attention, both by geomorphologists and ecologists. On the one hand physical processes are materially accelerated or retarded by vegetation dynamics. On the other the structure and composition of the cover may be controlled by geomorphic processes. At the simplest level this may involve wholesale destruction of the cover, for example under the Quaternary ice sheets, by marine transgression or by volcanic eruption. More usually the processes rates determine the stability and suitability of the plant growth environment. This is most notable in the weathering–soil–plant complex but extends to many other areas ranging from soil creep to the establishment of vegetation on river banks or lichens on badland slopes. Sometimes the interactions are competitive, at other times they are symbiotic. In addition these interactions are played out in space as well as time and there is an evident need for research into vegetationally induced geomorphological patterns, such as the early work in periglacial and arid environments. There is also a need for explanation of geomorphic processes in primary colonization or after fire.

The third and most complex level is that of landform history and evolution. Although there are still a few who believe that the relationships between climate, vegetation and landform evolution are a relatively simple and straightforward matter, for the most part geomorphologists appreciate that the links between climate and vegetation cover and between cover and geomorphology are both complex. In the past, discussion of these relationships often led to circular argument whereby pollen analysis was used to infer climate and these climates were used to account for plant cover. Only with the development of dating and independent estimates of temperature from oxygen isotope ratios have we reached a position when these earlier hypotheses relating climate to vegetation cover can be tested and apparently accepted. In the past two decades the sensitivity of both vegetation to climatic change and of geomorphology to vegetation change, and the dynamic behaviour of response systems through lags and thresholds have been recognized as major factors. We have to accept moreover that important effects occur not only at equilibria but also in the transitional states between them and to acknowledge that the plant species and communities are evolving at rates different from the those of the landforms. It does not follow that in modelling geomorphic change we must necessarily completely resolve the detailed structural, compositional and spatial response of the vegetation cover. For some purposes, such as channel palaeohydrology, a crude identification into dominant vegetation cover types may be enough. Unfortunately our understanding of vegetational impact on processes is still insufficient to know just what is required though we can and should make a good first approximation.

Finally it is worth pointing out that earth scientists have much to learn from the methodology of the ecologists through a common conceptual and technical base. Both subjects are evolving rapidly but the fundamental interest in process and change linked to climate should bring them closer together. The chapters in this book are written largely by and for geomorphologists but the next stage of development will have to rely heavily on cooperation with ecologists if meaningful progress is to be achieved.

2 'The Agency of Organic Beings': A Selective Review of Recent Work in Biogeomorphology

H. A. VILES
University of Oxford

SUMMARY

The term 'biogeomorphology' may be applied to studies of the influence of plants and animals on earth surface processes and landform development. A preliminary classification of biological effects on geomorphological processes distinguishes between active and passive, direct and indirect effects of individuals and ecosystems, and the effects of changing ecosystems. All of these types can be subsumed within a framework of ecosystem dynamics and denudation. A review is presented in this paper of such effects, specifically involving the action of plants in weathering, erosion and sedimentation. A consideration of the overall role of plants in the development of landform assemblages is also made.

INTRODUCTION

'I shall offer a few remarks on the superficial modifications caused directly by the agency of organic beings, as when the growth of certain plants covers the slope of a mountain with peat, or converts a swamp into dryland, or when vegetation prevents the soil in certain localities from being washed away by running water.'

(Lyell, 1835, p. 175)

Enquiries into the natural world in the nineteenth century recognized many links between inorganic and organic phenomena. Charles Darwin, for example, made detailed investigations into the role of earthworms as tillers of the soil, concluding that:

Vegetation and Erosion
Edited by J. B. Thornes
©1990 John Wiley & Sons Ltd

'It may be doubted whether there are many other animals which have played so important a part in the history of the world, as have these lowly organized creatures.'

(Darwin, 1881, p. 316)

As scientific perspectives changed and disciplinary boundaries became entrenched, so landform studies in the Davisian tradition in Britain and America became somewhat divorced from studies of vegetation and animals. Within Europe, however, the morphoclimatic tradition in geomorphology has long recognized the importance of organisms to morphogenic processes (e.g. Tricart and Cailleux, 1972, pp. 65–86; Büdel, 1982, p. 31). Recent intellectual and technological advances have encouraged a greater *rapprochement* between geomorphology and ecology within the Anglo-American tradition. One manifestation of this has been the proliferation of papers on 'bio-geomorphology', i.e. the influence of animals and plants on earth surface processes and landform development (Viles, 1988).

There have been several reviews of biological influences on geomorphology in the past, including the excellent summary of Pitty (1971, pp. 147–68). The aims of this chapter are to provide a survey of some important areas of 'biogeomorphology' and to suggest a possible theoretical framework for future investigations.

TYPES OF ORGANIC EFFECTS ON GEOMORPHOLOGY

Plant influences on earth surface processes can be classified in several ways. Here, we aim to split organic effects in a geomorphologically useful fashion. Firstly, they can be divided according to whether they involve individuals (or small communities); ecosystems; or changing ecosystems. This is an arbitrary division according to scale. Secondly, active and passive effects can be recognized, where active effects involve a biological process and passive effects involve the organism(s) simply as structures. Finally, organic effects can be split into those that have a direct effect upon earth surface processes and those that operate indirectly. In reality, there is a similar set of influences operating in reverse whereby geomorphological processes and features affect vegetation, leading to a complex, dynamic set of interactions.

An ecosystem approach can be used to set these different types of organic influences in a coherent framework. Bormann and Likens (1979) in their work on the Hubbard Brook drainage basin provide a useful method of integrating ecosystem dynamics and degradation. Under 'normal' conditions within an aggrading temperate forest ecosystem, biological regulation of the hydrological cycle, export of particulate material and streamflow chemistry can be recognized involving processes and mechanisms as documented in Table 2.1. These 'normal'

Table 2.1   Ecosystem dynamics and denudation within temperate forests

1. *Regulation of hydrological cycle*

   Transpiration: affects streamflow and soil moisture; diminishes summertime peak storm discharge rates
   Leaves and litter: intercept and disperse raindrop energy
   Dead biomass in soil: aids water storage in soil; aids infiltration and percolation
   Log and debris dams: regulate streamflow

2. *Control of particulate matter export*

   (a) Through biological control of hydrology
       Transpiration: Reduces storm peaks during growing season when erodibility is greatest
       Organic matter: enhances infiltration and percolation, thereby reducing erosional surface flow

   (b) Through biological regulation of erodibility
       Roots: bind soil, reducing erodibility
       Canopy and litter layer: protect soil from raindrop impact
       Log and debris dams: regulate streamflow and retain particulate matter in ecosystem
       Tree throw: provides exposed soil for erosion

3. *Regulation of streamflow chemistry* (dissolved load)

   Canopy: alters chemistry of throughfall and stemflow
   Soil organisms: involved in biogeochemical alteration of soil water composition
   Roots: uptake of elements from soil and bedrock

(Adapted from Bormann and Likens, 1979)

conditions involve the active and passive, direct and indirect effects of individuals and the total ecosystem as identified above. Bormann and Likens (1979) also investigated the breakdown, and subsequent re-establishment of such biotic regulation due to disturbance of the forest vegetation (analogous to the effects of changing ecosystems recognized above).

Forest geomorphology may be viewed as a sequence of 'normal' biologically regulated states alternating with disturbances when 'inorganic' processes may dominate. This framework can also be used to investigate the biological impact on geomorphology in other environments, where biotic regulation under 'normal' conditions may not be so complete, and 'disturbances' may occur far more frequently (e.g. tundra and desert environments). As ecosystems and geomorphological systems are not necessarily completely congruent (i.e. do not occupy exactly the same area) the links between vegetation and earth surface processes may be more complicated.

Active direct effects (e.g. animal bioturbation) have received most attention from geomorphologists as they are the easiest to quantify. Much progress has also been made in studying passive, direct effects such as the influence of vegetation on erosion, which are amenable to laboratory and field experimentation and modelling. The task that awaits geomorphology, however, involves a more complete consideration of the whole suite of organic influences in geomorphological explanation within a framework of ecosystem dynamics.

PLANTS AND PROCESSES

In most environments the fundamental earth surface processes of weathering, erosion and deposition occur under non-sterile conditions (cf. Cawley, Burruss and Holland, 1969). Communities of plants, animals and micro-organisms are widely distributed over the earth's surface. Even those environments experiencing extreme conditions, such as hot springs and Antarctic cold deserts, are capable of supporting a rich flora of micro-organisms and lower plants (Staley, Palmer and Adams, 1982; Hale, 1987). Newly exposed rock and sediment surfaces are quickly colonized and a plant community succession begins (Fritsch, 1907; Brock, 1973). Because of the diversity and complexity of plant life it is impossible to document all the ways in which plants affect earth surface processes, but in the following sections some of the more frequently studied interactions are reviewed.

Weathering

Processes of *in situ* breakdown of rocks and minerals are commonly split into physical (disintegration) and chemical (decomposition) forms. At the micro scale it is apparent that such a classification is hard to uphold because chemical and physical mechanisms are often inextricably linked in real weathering situations and both of these are regularly mediated by organisms. Two major areas of interest in biological weathering studies have developed recently: the impact of higher plants/ecosystems upon weathering and the impact of lower plants upon weathering at the micro scale. An excellent, comprehensive survey of organic weathering is presented in Yatsu (1988).

Most textbooks on geomorphology mention the impact of plant roots on weathering, often including a photograph of roots involved in joint widening. Beyond this, however, there has been little serious study of whether plant roots *do* instigate rock weathering and what processes may be involved, although for example, root effects have been shown to produce some brecciation in calcrete (Klappa, 1980a). In part, this lack of research is due to the difficulty of studying such processes (especially where large, slow-growing trees are involved). On some rocks, e.g. limestone, small-scale morphological features ('root karst') have been ascribed to root action although there has been no real investigation of mechanisms (Wall and Wilford, 1966; Esteban and Klappa, 1983).

Investigations have now moved away from such interesting oddities to a more general consideration of overall ecosystem impacts upon weathering.

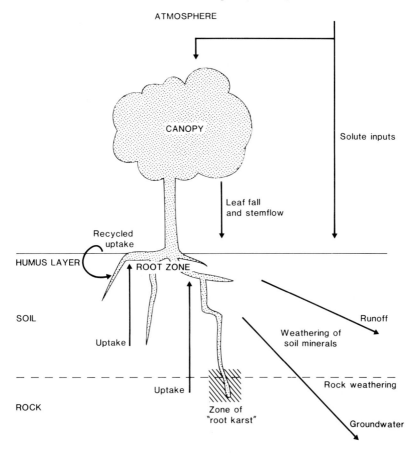

Figure 2.1 Processes and pathways of solute movements within an ecosystem

This includes the chemical and physical effects of roots on weathering of rocks and of minerals within the soil profile as shown in Figure 2.1. Such studies attempt to quantify the interdependence of plant productivity, nutrient cycling, soil leaching and rock weathering within particular environments. So, for example, Lelong and Wedraogo-Dumazet (1987) studied the impact of forest vegetation on weathering within similar small granite drainage basins in France. In this study water chemistry was sampled at various points within the drainage basins. Elemental concentrations in river water were found to be low, due to high rates of elemental uptake by plant roots.

Crowther (1987) carried out a similar study involving detailed water chemistry investigations within tropical forest-covered karst terrain in Malaysia. He found that most chemical weathering there occurs in the soil-vegetation zone and that tree roots remove large amounts of calcium and magnesium directly from the soil and bedrock. This appears to be very different from the situation in temperate areas, where biological uptake is limited relative to inorganic weathering (Trudgill, 1977). This root 'weathering' removes cations from the geomorphological system and is therefore 'invisible' when weathering is estimated from groundwater and runoff chemistry. Here is one example of how standard geomorphological techniques may fail to recognize biological contributions. As Trudgill (1986, p. 4) indicates, however, the complexity of cycling processes within the ecosystem makes it difficult to specify how much plant uptake of elements is actually due to geomorphologically important biological weathering and not simply recycling within the soil. Root processes are only one aspect of ecosystem functioning and other organisms within these ecosystems may also contribute to weathering, e.g. micro-organisms in soils (Berthelin *et al.*, 1985) and snails on bare rock (Crowther, 1979; Stanton, 1984).

Table 2.2 Rock weathering by micro-organisms and lower plants

Rock type	Location	Organism	Process	Reference
Sandstone	Antarctic/ subaerial	Endolithic lichens	Iron mobilization	Friedmann & Weed, 1987
Sandstone	South Africa/ subaerial	Lichens & algae	?	Cooks & Pretorius, 1987
Limestone	Aldabra Atoll/ subaerial	Blue-green algae	Boring, etc.	Viles, 1987
Limestone	Negev Desert/ subaerial	Endolithic lichens/algae	Surface weakening	Shachak *et al.*, 1987, Danin & Garty, 1983
Limestone/ dolomite	Scotland/ coastal	Lichens/blue-green algae	Surface weakening	Kelletat, 1986, 1988
Limestone/ dolomite	Spain/ subaerial	Lichens	Oxalic acid attack	Ascaso *et al.*, 1982
Calcareous aeolianite	Madagascar/ coastal	Blue-green algae/bacteria	Increase corrosion	Battistini, 1981
Recent lava	Hawaii/ subaerial	Lichens	Increase chemical weathering	Jackson & Keller, 1970
Granite	Scotland/ subaerial	Lichens	Etching of felspars, etc.	Jones *et al.*, 1981
Basalt	Scotland/ subaerial	Lichens	Etching of minerals	Jones *et al.*, 1981
Schist/shale gneiss	N. Wales/ subaerial	Lichens	Physical weathering	Fry, 1927
Obsidian	Yellowstone Park USA/ subaerial	Lichens	Physical	Fry, 1927

Table 2.3 Mineral weathering by micro-organisms and lower plants

Mineral	Process	References
Calcite	Algal and fungal etching	Jones, 1987
Calcite	Blue-green algal boring	Le Campion-Alsumard, 1979
Calcite and dolomite	Lichen etching	Muxart & Blanc, 1979
Calcite and pyrite	Bacterial etching	Kieft & Caldwell, 1984
Amorphous and crystalline silicates	Fungal etching	Callot et al., 1987
Glass	Algal boring	Jones & Goodbody, 1982
Felspars, mica, biotite, olivine, quartz	Lichen etching	Wilson & Jones, 1983

The impact of whole ecosystems on weathering as described above is very sensitive to human and natural disturbance of the vegetation, which affects nutrient cycling and plant uptake.

Studies of the role of lower plants and micro-organisms in rock and mineral weathering have focused especially on the micro-scale results of such activity. Table 2.2 presents some examples of organic weathering on different rocks. Some species of lichens, algae, fungi and bacteria have been found to be capable of weathering common rock-forming minerals such as felspars and calcite, as shown in Table 2.3. There are several possible interactions between micro-organisms and minerals involving chemical and/or physical processes. Micro-organisms may protect the surface of rocks and minerals, thereby retarding weathering; alternatively they may weather directly through acid attack and other processes, or indirectly through their encouragement of other weathering processes. In most cases all three situations occur. At this scale, and due to the complexity of the system, it is difficult to assess rates of biological weathering and compare them with inorganic process rates.

The importance of plants to weathering does not directly relate to the size of the biomass. In some cases relatively limited organic cover can be the dominant influence on weathering, where other weathering processes are limited and where nutrient uptake from rocks/minerals is large, i.e. 'stressed' environments where nutrient cycles are 'leaky' or conservation mechanisms have developed (Jordan, 1988, p. 151). Under semi-arid conditions in Jerusalem, for example, blue-green algae have been found to be the dominant weathering agent on limestone walls (Danin, 1983).

Some workers regard biological and inorganic weathering processes as acting in competition (Friedmann and Weed, 1987); others regard them as acting in some kind of synergistic association. Certainly weathering is highly related to ecosystem functioning and the complexities are difficult to unravel. Lower plants and micro-organisms are particularly important to weathering processes.

Weathering is also influenced by animals with, for example, earthworms and termites aiding comminution of minerals within the soil zone; organisms such as chitons weathering rocky limestone coasts; and urine from rock wallabies and other mammals aiding terrestrial limestone weathering.

Erosion

Plants influence erosion in all areas except the glacial zone. Subaerial erosion, for example, is influenced by plants both directly (e.g. tree fall causing mass movements of soil) and indirectly (e.g. through canopy influences on the erosivity of rainfall). Also, the removal or change of vegetation can alter erosion rates. A general overview of man, vegetation and erosion is presented by Thornes (1987) and several attempts have been made to include vegetation in mathematical models of soil erosion (Thornes, 1985; Kirkby and Neale, 1987).

There has been a number of interesting studies recently investigating the role of tree fall in both soil erosion and the linked question of soil profile development (Mills, 1984; Mitchell, 1988). In England, the potential importance of tree fall to erosion was indicated forcefully in the 'Great Storm' of 16 October 1987 when some 15 million trees were brought down over a large area of southeastern England. A large amount of soil may be disturbed, depending upon the interplay of soil conditions (wetness, etc.), tree characteristics (root network, etc.) and the external conditions (wind direction, etc.) as described more fully in Table 2.4. Mounds of soil are often produced by tree fall (Hamann, 1984) and these, as well as the pits created, will be subject to subsequent erosion.

The phenomenon of tree fall illustrates well some of the complexities of biogeomorphology. Initially, tree fall causes a mass movement of soil, thereby

Table 2.4 Factors influencing the geomorphological importance of tree fall

Factor	Effects
External	
Nature of storm	Number of trees felled; frequency of tree fall events
Nature of fire	Number of trees felled; frequency of tree fall events
Wind direction	Direction of tree fall and soil movement
Forest	
State of trees	Ease with which trees fall
Density of trees	Nature and density of tree falls
Nature of root systems	Amount and shape of soil moved
Soil	
Soil structure	Amount and shape of soil moved
Soil moisture	Amount and shape of soil moved

directly influencing erosion. Tree fall also exposes soil to further erosion, thus indirectly increasing erosion and potentially influencing soil profile development (Schaetzel, 1986). Tree fall may also lead to the production of small-scale landforms (Bucklewiesen; see Hamann, 1984) which are also of geomorphological interest and will influence the nature of future soil movements. Feedbacks occur as tree fall leads to changes in the soil which then may affect vegetation development. Finally, where tree falls are caused by some external factor, such as a storm, other types of erosion may also be triggered at the same time. Overall, therefore, the example of tree fall illustrates that there is no direct, simple correlation between a biological event (tree fall) and a geomorphological process (erosion).

There has been considerable recent work on the role of vegetation in slope stability (see, for example, the reviews of Selby, 1982, and Greenway, 1987). Vegetation may both encourage and discourage slope instability. For example, de Ploey and Cruz (1979) illustrate how on steep slopes with shallow soils tree root channels may encourage rapid interflow and landsliding. The stabilizing effect of vegetation may last for several years after clearance (Crozier, 1986, pp. 156–9).

Vegetation indirectly influences erosion through its impact on climate and hydrology as shown in Table 2.5. As can be seen from this table, vegetation may influence rainfall routes, runoff and subsurface flow, temperature and wind characteristics. The size, shape, growth characteristics and density of vegetation are all important factors conditioning its influence on erosion; dense, tall forest will have different effects to low, sparse grassland. At the ground surface percentage cover, stem density, surface litter and soil organic matter content are the relevant parameters (Thornes, 1987).

Vegetation communities are not static and changes to vegetation (especially caused by human interference) often lead to spectacular changes in erosion rates. Heede (1987), however, found erosion only increased markedly in harvested forest areas in Arizona where disturbance had been severe. Vegetation changes occur because of either biological factors (e.g. succession, disease, fire, human planting) or external factors (e.g. climatic change, volcanic eruptions). In the first case geomorphological processes are affected through vegetation change alone; in the second case they are also directly influenced by the external event itself. Recently, various palaeoecological studies have revealed the history of vegetation/geomorphological interactions with lake deposits providing particularly useful evidence. Stott (1987), for example, uses mineral magnetic studies of reservoir sediments to infer that in a northwest England catchment increased erosion occurred through tree fall and gully development when canopy closure occurred in a managed forest.

Fire is an important mechanism of change in many vegetation communities, including boreal forest, tropical rain forest and savannas. Although fires can be started naturally, through volcanic eruptions and lightning, human activity

Table 2.5 Indirect effects of vegetation on erosion

Factor	Effects	Example
Stemflow impact on rainfall input	Leads to infiltration excess runoff—increasing soil erosion	TRF, Queensland (Herwitz, 1986)
Canopy layer/ rainfall input	Increases kinetic energy of raindrops—increasing soil erosion	Beech forest, New Zealand (Mosley, 1982); TRF, Colombia (Vis, 1986)
Canopy layer/rainfall input	Intercepts rainfall— reducing soil erosion	
Litter layer/runoff	Increases infiltration— reducing soil erosion	TRF, Kalimantan Timur (Besler, 1987)
Roots/subsurface flow	Bind soil—reducing soil erosion	Coastal marshes, Netherlands, (van Eerdt, 1987)
Roots/subsurface flow	Produce piping—increasing soil erosion	
Crop residue/ soil temperature	Reduces soil frost— reducing soil erosion	Oregon wheat belt, (Pikul et al., 1986)
Tree shading/ soil temperature	Reduces soil temperature— leading to permafrost development	Black spruce, sporadic permafrost zone (Williams, 1988)
Trees/wind	Act as windbreaks—reducing soil erosion	
Crops/wind	Increase drag velocity— increasing soil erosion	Onions and sugar beet, (Morgan & Finney, 1987)
Vegetation cover/wind	Reduces wind velocity— decreasing sand transport	Australian dunes, (Ash & Wasson, 1983)

is now responsible for much burning. In most environments fire leads to an increase in erosion as protective vegetation cover is removed, and chemical and physical changes are made to the soil (e.g. Chinen, 1987). In some areas, for example the chaparral vegetation in the USA hydrophobic soils may develop (Trimble, 1988; Brown, Chapter 18 of this volume). In a rather different environment peat fires have recently been found to play a role in cyclic channel change in the Okavango Delta, Botswana (McCarthy et al., 1988). In serious fires the indirect links between vegetation and erosion shown in Table 2.5 are destroyed as the vegetation itself is killed and different plants may grow back afterwards. Extensive tree fall may also occur during and after fires with concomitant effects on soil movement. Obviously the extent of fire impact on erosion rates in any one area will depend upon the nature of fires (intensity, extent, duration, frequency). In most fires animal communities will also be affected, thereby upsetting any zoogenic bioturbation. In tundra and boreal forest areas fires often cause thawing of permafrost, producing thermokarst terrain and causing slumps and slides on highly saturated soils (Williams, 1988).

In terms of subaerial erosion, therefore, vegetation communities play a range of roles in reducing and increasing erosion rates. Direct and indirect effects occur together and vegetational change will disturb both of these. Substantial progress has been made in measuring some vegetation effects, but an assessment of the overall situation is more difficult. Ecologists have made particular progress in this field and the framework of Bormann and Likens (1979) presented above has much to offer. Bioturbation by animals is also an extremely important direct influence on erosion which has received much recent attention, and is obviously not unrelated to vegetation effects.

Studies of fluvial and coastal erosion have shown some interesting organic effects. Log and debris jams, for example, have been found to have often spectacular effects upon channel processes, including bank and bed erosion (Duijsings, 1987; Graf, 1988; Gregory and Gurnell, 1988; Marston, 1982). Again, influences may be both direct and indirect and are affected by vegetation change. In coastal and shelf situations work has focused on the direct effect of animals: whales, walruses, tile fish, geese, crabs and gastropods, for example, are bioturbators or bioeroders (Nelson, Johnson and Barber, 1987; Dionne, 1985; Twichell et al., 1985). The indirect effect of vegetation on sand dune erosion processes has been studied by Carter and Stone (1989).

Deposition

Both physical and chemical deposition may have a biological component. Where plants reduce erosion they often also act to aid sedimentation. Key areas for such sedimentation processes are dunes, bogs, swamps and marshes. Vegetation plays two main roles in promoting physical sedimentation, namely, trapping mineral sediment in both root networks and above-ground structures, and producing sediment through the build-up of dead plant material. Eriksson et al. (1989) report on a rather more complex relationship between vegetation and animals and parabolic dune patch development in a stabilized dune field in the southern Kalahari. Algae also aid sedimentation, for example in salt marshes (Coles, 1979) and dunes (Van den Ancker and Jungerius, 1985), and animal faecal pellets are important in some coastal situations (Pryor, 1975). Cycles of build-up and erosion may occur as vegetation communities grow and are then killed off by an excess of sediment. In some cases distinctive landforms may develop as a result of vegetation influencing sedimentation. For example, nebkha dunes develop where bushes interrupt wind flow, causing scour at the sides and deposition behind.

Recent work has focused on clarifying and quantifying the relationship between deposition and biological processes. So, for example, studies have been made of the effect of vegetation on sand movements over dunes (Buckley, 1987; Wasson and Nanninga, 1986) to produce quantitative information. In coastal areas, there has been much debate over the relative importance of physical and biological sedimentation in salt marshes and mangrove swamps (Frey and Basan, 1985; Bird, 1986). In this case, factors such as local coastal geometry determine whether vegetation plays an important role or not. De Laune, Smith and Patrick (1986), for example, show how organic material accounts for most marsh deposition under normal conditions in a backbarrier marsh, with inorganic sediment deposited primarily during major storms. In California, Collins, Collins and Leopold (1987) suggest that biogenic sedimentation dominates away from creek bank areas.

Table 2.6   Influence of plants and micro-organisms on chemical sedimentation

Sediment	Organism	Process	Reference
Tufa/travertine	Bacteria	Bacterial precipitation of calcite	Folk & Chafetz, 1983; Chafetz & Folk, 1984
	Blue-green algae	Precipitation of calcite around algae	Love & Chafetz, 1988
	Blue-green algae	Algal trapping of calcite particles	Pentecost, 1978
	Mosses	Inorganic precipitation of calcite onto moss surface	Irion & Muller, 1968
Calcretes	Plant Roots	Precipitation of calcite around roots	Semeniuk & Meagher, 1981; Klappa, 1980b
	Mycorrhizae	Calcification which produces Microcodium	Klappa, 1978
	Lichens	Lichen stromatolite formation	Klappa, 1979
Laterites	Vegetation cover	Aids accumulation of laterite	McFarlane, 1983
Cave deposits	Bacteria	Bacterial precipitation of calcite	Danielli & Edington, 1983
	Algae	Aid precipitation of micrite	Jones & Motyka, 1987
	Fungi	Crystallization of calcite around fungal hyphae	Went, 1969
Desert varnish	Bacteria	Fixation of manganese	Dorn & Oberlander, 1981
	Blue-green algae/fungi	Fixation of manganese	Krumbein & Jens, 1981
Beach-rock	Blue-green algae/bacteria	Precipitation of carbonates	Krumbein, 1979
Coastal algal platforms	Coralline algae	Algal precipitation of calcium carbonate	Focke, 1978; Spencer, 1985; Kelletat, 1985

Chemical sedimentation may be seen as the converse to chemical weathering. Most chemical sediments have been formed by a combination of weathering and sedimentational processes (Goudie and Pye, 1983). Detailed studies on the role of organisms in such processes are referenced in Table 2.6. Animals do not seem to play any great role in chemical sedimentation, except of course in the production of skeletal carbonates.

Plants can trap and bind chemically precipitated material; act as nuclei for precipitation; and directly precipitate minerals themselves. Algae, fungi, bacteria, lichens and mosses are often involved, although higher plants may also play a role through their root network (e.g. Semeniuk and Meagher, 1981). Assessing the contribution of organisms to the formation of chemical sediments can be difficult, not least because of the range of potential organic influences as documented above. In the case of tufas and travertines, for example, evidence has been collected of mosses, bacterial and algal involvement in trapping, binding and precipitation of calcite (Schneider, 1977). Conversely, tufa deposits are equally well developed without any biological input (Golubic, 1969) and algae have been observed to be eroding tufa deposits through boring processes (Pentecost, 1978). Within a tufa deposit cycles of algal growth/sedimentation may result as the algae-induced sedimentation kills off the algae by reducing light levels followed by recolonization. Quantitative estimates have been made, from *in vitro* and *in vivo* studies, of the contribution made by organisms to tufa accretion, but these normally consider only active organic precipitation.

Plants aid both physical and chemical sedimentation, as we have seen above. Higher plants are dominantly involved in physical processes, usually playing a passive, trapping role. Lower plants dominate chemical processes, playing a range of active and passive roles.

LANDSCAPE IMPLICATIONS

The review presented above indicates that any general consideration of the relations between ecosystem dynamics and denudation has to consider a whole suite of impacts on weathering, erosion and sedimentation. Tying this information in with landform development poses further complications. A useful concept to employ in an attempt to unite different biological and inorganic processes in an overview of geomorphology is that of process domains or process associations (Brunsden and Thornes, 1979).

Brunsden and Thornes suggest that characteristic associations of processes develop in an area according to climatic and other factors. Process associations occur at a number of scales, so for example, within an arid area certain microenvironments might favour a process association characteristic of more humid areas. Process associations, if operative over a long enough period of time, will lead to the development of characteristic landforms which can be relatively easily explained. This occurs mainly for small-scale landforms, formed over a period of about 100 years. For larger and more resistant landforms, process associations will have changed many times over the history of the landform, and so contemporary process–form links will be difficult to establish.

It is evident from the work discussed in this chapter that biological processes, and biological mediation in other processes, are important and neglected aspects of process associations in many landscapes. Three different scales of interest may be identified. At the smallest scale one can consider the role of individual organisms and small communities within process associations. Such organisms may dominate these process associations, leading to the production of recognizable, small-scale landforms or 'biokarst' (Viles, 1984). At the medium scale complete ecosystems become important in general process associations, as they contribute to overall rates of denudation within the landscape. At the largest scale the links between changes in vegetation and changes in process associations are of interest. These three levels may be compared with the position of vegetation as a drainage basin variable over steady, graded and cyclic time in the often-quoted diagram in Schumm and Lichty (1965, p. 112).

Several landforms have been recognized as being produced by organic processes, e.g. the 'phytokarst' of Folk, Roberts and Moore (1973). In the situation envisaged by Folk *et al.* an active organic process (i.e. algal boring) is proposed to have a direct and dominant influence on limestone weathering processes, producing a recognizable landform. This is, if true, a simple, congruent association of the biological and geomorphological worlds. However, as indicated by Viles (1987) and others, the situation is often more complicated. Passive and indirect organic effects on weathering also occur when limestone surfaces are covered in algal growths, complicating the picture. Undoubtedly, algae and other lower plants play a role in the solution and precipitation of calcium carbonate, and such activity is probably most important in wetter parts of the landscape (springs, hollows, etc.) where organic growths are most luxuriant.

At the medium scale, ecosystem processes are one component of process associations acting upon the landscape, contributing to overall denudation rates. In karst areas one important manifestation of vegetation at this scale is its contribution to soil carbon dioxide levels. Variation in soil carbon dioxide levels has been recognized as a major factor in explaining differences in dissolution rates between zones. Jakucs and Barany (1987, p. 389) show the variation in the contribution of organically produced carbon dioxide to dissolution over the major climatic zones which suggests that in tropical areas over 90 per cent of dissolution is due to organic acids and organically produced carbon dioxide. Pham Khang (1985), however, indicates that in the tropical environment of Vietnam organic influences on dissolution are very limited. In temperate areas the figure is about 70 per cent, whereas in high mountain areas it is about 40 per cent. Finally, in arid areas biogenic carbon dioxide apparently makes no contribution to dissolution. More quantitative information is needed to back up these suggestions.

On the largest scale, changes to ecosystems may be involved in changes in process associations which themselves lead to a complex response in the landscape. Vegetational change may be the cause of such process association changes, or may be only a linked factor. In the Burren in western Ireland, for example, Neolithic–Bronze Age forest clearance led to soil erosion with concomitant effects on limestone pavement development (Drew, 1983). Here vegetation change has led to geomorphological change. On the longer time scale changes in climate are presumed to affect karst landscapes through changes in runoff and also in biogenic carbon dioxide. In this case, vegetational change and geomorphological change are more complexly interrelated and there is insufficient information on plant/process links.

CONCLUSIONS

Enough evidence has been presented in this rather selective review to show that biogeomorphology is in a healthy state, at least in terms of the numbers of investigations into different vegetation–geomorphology interactions. Quantitative information is being produced on many of the plant/land interactions noted by Lyell and other nineteenth-century workers. Such information provides more detail on the relationship of ecosystem dynamics to denudation in different environments, as sketched by Bormann and Likens (1979) for temperate forests (Table 2.1).

Further work is required by geomorphologists and ecologists to relate all the different biological influences to overall denudation rates and to landform development within some coherent theoretical framework. The temperate forest environment seems to provide the best place to attempt such a synthesis, although work in other areas such as coastal wetlands would also be of great interest.

REFERENCES

Ascaso, C., J. Galvin and C. Rodriguez-Pascual (1982). The weathering of calcareous rocks by lichens. *Pedobiologia*, **24**, 219–29.

Ash, J. E. and R. J. Wasson (1983). Vegetation and sand mobility in the Australian desert dunefield. *Zeitschrift für Geomorphologie, SupplementBand*, **45**, 7–25.

Battistini, R. (1981). La morphogénèse des plateformes de corrosion littorale dans les gres calcaires (plateforme superieure et plateforme à vasques) et le problème des vasques, d'aprés des observations faites a Madagascar. *Revue de Géomorphologie Dynamique*, **30**, 81–94.

Berthelin, J., M. Bonne, G. Belgy, and F. X. Wedraogo (1985). A major role for nitrification in the weathering of minerals of brown acid forest soils. *Geomicrobiological Journal*, **4**, 175–90.

Besler, H. (1987). Slope properties, slope processes and soil erosion risk in the tropical rain forest of Kalimantan Timur (Indonesian Borneo). *Earth Surface Processes and Landforms*, **12**, 195–204.

Bird, E. C. F. (1986). Mangroves and intertidal morphology in Westernport Bay, Victoria, Australia. *Marine Geology*, **69**, 251–71.

Bormann, F. H. and G. E. Likens (1979). *Pattern and Process in a Forested Ecosystem*, Springer-Verlag.

Brock, T. D. (1973). Primary colonisation of Surtsey, with special reference to the blue-green algae. *Oikos*, **24**, 239–43.

Brunsden, D. and J. B. Thornes (1979). Landscape sensitivity and change. *Transactions, Institute of British Geographers*, **4**, 463–84.

Buckley, R. (1987). The effect of sparse vegetation on the transport of dune sand by wind. *Nature*, **325**, 426–8.

Büdel, J. (1982). *Climatic Geomorphology*, Princeton University Press.

Callot, G., M. Maurette, L. Pottier, and P. Dubois (1987). Biogenic etching of microfractures in amorphous and crystalline silicates. *Nature*, **328**, 147–9.

Carter, R. W. G. and G. W. Stone (1989). Mechanisms associated with the erosion of sand dune cliffs, Magilligan, Northern Ireland. *Earth Surface Processes and Landforms*, **14**, 1–10.

Cawley, J. L., R. C. Burruss and H. D. Holland (1969). Chemical weathering in central Iceland: An analog of pre-Silurian weathering. *Science*, **165**, 391–2.

Chafetz, H. S. and R. L. Folk, (1984). Travertines: Depositional morphology and the bacterially controlled constituents. *Journal of Sedimentary Petrology*, **54**, 289–316.

Chinen, T. (1987). Hillslope erosion after a forest fire in Etajima Island, Southwest Japan. In A. Godard and A. Rapp (eds), *Processus et mesure de l'érosion*, CRNS, pp. 199–210.

Coles, S. M. (1979). Benthic microalgal populations on intertidal sediments and their role as precursors to salt marsh development. In R. L. Jefferies and A. J. Davy (eds), *Ecological Processes in Coastal Environments*, Blackwell Scientific Publications, pp. 25–42.

Collins, L. M., J. L. Collins and L. B. Leopold (1987). Geomorphic processes of an estuarine marsh: Preliminary results and hypotheses. In V. Gardiner (ed.), *International Geomorphology 1986* I, Wiley, 1049–72.

Cooks, J. and J. R. Pretorius (1987). Weathering basins in the Clarens formation sandstone, South Africa. *South African Journal of Geology*, **90**, 147–54.

Crowther, J. (1979). Limestone solution on exposed rock outcrops in West Malaysia. In A. F. Pitty (ed.), *Geographical Approaches to Fluvial Processes*, GeoBooks, pp. 31–50.

Crowther, J. (1987). Ecological observations in tropical karst terrain, West Malaysia. III: Dynamics of the vegetation–soil–bedrock system. *Journal of Biogeography*, **14**, 157–64.

Crozier, M. J. (1986). *Landslides*, Croom Helm.

Danielli, H. M. C. and M. A. Edington (1983). Bacterial calcification in limestone caves. *Geomicrobiological Journal*, **3**, 1–16.

Danin, A. (1983). Weathering of limestone in Jerusalem by cyanobacteria. *Zeitschrift für Geomorphologie*, **27**, 413–21.

Danin, A. and J. Garty (1983). Distribution of cyanobacteria and lichens on hillsides of the Negev Highlands and their impact on biogenic weathering. *Zeitschrift für Geomorphologie*, **27**, 423–44.

Darwin, C. (1881). *Vegetable Mould and Earthworms*, John Murray.

de Laune, R. D., C. J. Smith and W. H. Patrick (1986). Sedimentation patterns in a Gulf Coast backbarrier marsh: response to increasing submergence. *Earth Surface Processes and Landforms*, **11**, 485–90.

de Ploey, J. and O. Cruz (1979). Landslides in the Serra do Mar, Brazil. *Catena*, **6**, 111–22.

Dionne, J.-C. (1985). Tidal marsh erosion by geese, St Lawrence estuary, Quebec. *Geographie, Physical et Quaternarie*, **39**, 99–105.

Dorn, R. J. and T. M. Oberlander (1981). Rock varnish origin, characteristics and usage. *Zeitschrift für Geomorphologie*, **25**, 420–36.

Drew, D. P. (1983). Accelerated soil erosion in a karst area: the Burren, Western Ireland. *Journal of Hydrology*, **61**, 113–26.

Duijsings, J. J. H. M. (1987). A sediment budget for a forested catchment in Luxembourg and its implications for channel development. *Earth Surface Processes and Landforms*, **12**, 173–84.

Eriksson, P. G., N. Nixon, C. P. Snyman and J. duP. Bothma (1989). Ellipsoidal parabolic dune patches in the southern Kalahari desert. *Journal of Arid Environments*, **16**, 111–24.

Esteban, M. and C. F. Klappa (1983). Subaerial exposure. In P. A. Scholle, D. G. Bebout and C. H. Moore (eds), *Carbonate Depositional Environments*, AAPG, pp. 1–54.

Focke, J. W. (1978). Limestone cliff morphology on Curaçao (Netherlands Antilles) with special attention to the origin of notches and vermetid/coralline algal surf benches ('cornices', 'trottoirs'). *Zeitschrift für Geomorphologie*, **22**, 329–49.

Folk, R. L. and H. S. Chafetz (1983). Pisoliths (Pisoids) in Quaternary travertines of Tivoli, Italy. In T. M. Peryt (ed.), *Coated Grains*, Springer-Verlag, pp. 474–87.

Folk, R. L., H. H. Roberts and C. H. Moore (1973). Black phytokarst from Hell, Cayman Islands, British West Indies. *Bulletin, Geological Society of America*, **84**, 2351–60.

Frey, R. W. and P. B. Basan (1985). Coastal salt marshes. In R. A. Davis (ed.), *Coastal Sedimentary Environments*, Springer-Verlag, pp. 225–301.

Friedmann, E. I. and R. Weed (1987). Microbial trace-fossil formation, biogenous and abiotic weathering in the Antarctic cold desert. *Science*, **236**, 703–4.

Fritsch, F. E. (1907). The role of algal growth in the colonisation of new ground and in the determination of scenery. *Geographical Journal*, **30**, 531–47.

Fry, E. J. (1927). The mechanical action of crustaceous lichens on substrata of shale, schist, gneiss, limestone and obsidian. *Annals of Botany*, **XLI**, 437–60.

Golubic, S. (1969). Cyclic and non-cyclic mechanisms in the formation of travertine. *Verhandlungen der Internationale Vereinigung für Limnologie*, **19**, 2315–23.

Goudie, A. S. and K. Pye (eds) (1983). *Chemical Sediments and Geomorphology*, Academic Press.

Graf, W. L. (1988). *Fluvial Processes in Dryland Rivers*, Springer-Verlag.

Greenway, D. R. (1987). Vegetation and slope stability. In M. G. Anderson and K. S. Richards (eds), *Slope Stability*, Wiley, 187–230.

Gregory, K. J. and A. M. Gurnell (1988). Vegetation and river channel form and process. In H. A. Viles (ed.), *Biogeomorphology*, Basil Blackwell, pp. 11–42.

Hale, M. E. (1987). Epilithic lichens in the Beacon sandstone formation, Victoria Land, Antarctica. *The Lichenologist*, **19**, 269–88.

Hamann, C. (1984). Windwurf als Ursache der Bodenbuckelung an Sudrand des Tennengebirges, ein Beitrag zur Genese der Buckelwiesen. *Berliner Geographische Abhandlung*, **36**, 69–76.

Heede, B. H. (1987). Overland flow and sediment delivery 5 years after timber harvest in a mixed conifer forest. *Journal of Hydrology*, **91**, 205–16.

Herwitz, S. R. (1986). Infiltration-excess caused by stemflow in a cyclone-prone tropical rainforest. *Earth Surface Processes and Landforms*, **11**, 401–12.

Irion, G. and G. Muller (1968). Mineralogy, petrology and chemical composition of some calcareous tufa from the Schwabische Alb, Germany. In G. Muller and G. M. Friedman (eds), *Recent Developments in Carbonate Sedimentology in Central Europe*, Springer-Verlag, pp. 157–71.

Jackson, T. A. and W. D. Keller (1970). A comparative study of the role of lichens and 'inorganic' processes in the chemical weathering of recent Hawaiian lava flows. *American Journal of Science*, **269**, 446–66.

Jakucs, L. and I. Barany (1987). Ecological factors playing part in karst denudation dynamism for different geographical zones. In A. Godard and A. Rapp (eds), *Processus et mesure de l'érosion*, CRNS, pp. 387–92.

Jones, B. (1987). The alteration of sparry calcite crystals in a vadose setting, Grand Cayman Island. *Canadian Journal of Earth Sciences*, **24**, 2292–314.

Jones, B. and Q. H. Goodbody (1982). The geological significance of endolithic algae in glass. *Canadian Journal of Earth Sciences*, **19**, 671–8.

Jones, B. and A. Motyka (1987). Biogenic structures and micrite in stalactites from Grand Cayman Island, British West Indies. *Canadian Journal of Earth Sciences*, **24**, 1402–11.

Jones, D., M. J. Wilson and W. J. McHardy (1981). Lichen weathering of rock-forming minerals: application of scanning electron microscopy and microprobe analysis. *Journal of Microscopy*, **124**, 95–104.

Jordan, C. F. (1988). The tropical rain forest landscape. In H. A. Viles (ed.), *Biogeomorphology*, Basil Blackwell, pp. 145–65.

Kelletat, D. (1985). Bio-destruktive und bio-konstruktive Formelemente an der spanischen Mittelmeerkuste. *Geookodyamik*, **6**, 1–20.

Kelletat, D. (1986). Die Bedeutung biogener Formung in Felslittoral Nord-Schottland. *Essener Geographische Arbeiten*, **14**, 1–83.

Kelletat, D. (1988). Quantitative investigations on coastal bioerosion in higher latitudes: An example from Northern Scotland. *Geookodynamik*, **9**, 41–51.

Kieft, T. L. and D. E. Caldwell (1984). Weathering of calcite, pyrite and sulphur by *Thermothrix thiopara* in a thermal spring. *Geomicrobiological Journal*, **3**, 201–16.

Kirkby, M. J. and R. H. Neale (1987). A soil erosion model incorporating seasonal factors. In V. Gardiner (ed.), *International Geomorphology 1986* II, Wiley, pp. 189–210.

Klappa, C. F. (1978). Biolithogenesis of Microcodium: elucidation. *Sedimentology*, **25**, 489–522.

Klappa, C. F. (1979). Lichen stromatolites: Criterion for subaerial exposure and a mechanism for the formation of laminar calcretes (caliche). *Journal of Sedimentary Petrology*, **49**, 387–400.

Klappa, C. F. (1980a). Brecciation textures and tepee structures in Quaternary calcrete (caliche) profiles from eastern Spain; the plant factor in their formation. *Geological Journal*, **15**, 81–9.

Klappa, C. F. (1980). Rhizoliths in terrestrial carbonates: Classification, recognition, genesis and significance. *Sedimentology*, **27**, 613–29.

Krumbein, W. E. (1979). Photolithotropic and chemoorganotropic activity of bacteria and algae as related to beachrock formation and degradation (Gulf of Aqaba, Sinai). *Geomicrobiological Journal*, **1**, 139–203.

Krumbein, W. E. and K. Jens (1981). Biogenic rock varnishes in the Negev Desert (Israel): An ecological study of iron and manganese transformation by cyanobacteria and fungi. *Oecologia*, **50**, 25–8.

Le Campion-Alsumard, T. (1979). Les cyanophycées endolithes marines. Systematique, ultrastructure, écologie et biodestruction. *Oceanologica Acta*, **2**, 143–56.

Lelong, F. and B. Wedraogo-Dumazet (1987). Influence de la végétation sur la mobilisation chimique des elements dans les bassins versants granitiques du Mont-Lozere (France). In A. Godard and A. Rapp (eds), *Processus et mesure de l'érosion*, CRNS, pp. 299–311.

Love, K. M. and H. S. Chafetz (1988). Diagenesis of laminated travertine crusts, Arbuckle Mountains, Oklahoma. *Journal of Sedimentary Petrology*, **58**, 441–5.

Lyell, C. (1835). *Principles of Geology* (4th edn), John Murray.

McCarthy, T. S., I. G. Stanistreet, B. Cairncross, W. N. Ellery, K. Ellery, R. Oelofse and T. S. A. Grobicki (1988). Incremental aggradation on the Okavango Delta-fan, Botswana. *Geomorphology*, **1**, 267–78.

McFarlane, M. J. (1983). Laterites. In A. S. Goudie and K. Pye (eds), *Chemical sediments in Geomorphology*, Academic Press, pp. 7–58.

Marston, R. A. (1982). The geomorphic significance of log steps in forest streams. *Annals, Association American Geographers*, **72**, 99–108.

Mills, H. M. (1984). Effect of hillslope angle and substrate on tree tilt and denudation of hillslopes by tree fall. *Physical Geography*, **5**, 253–61.

Mitchell, P. B. (1988). The influences of vegetation, animals and micro-organisms on soil processes. In H. A. Viles (ed.), *Biogeomorphology*, Basil Blackwell, pp. 43–82.

Morgan, R. P. C. and H. J. Finney (1987). Drag coefficients of single crop rows and their implications for wind erosion control. In V. Gardiner (ed.), *International Geomorphology 1986* II, Wiley, pp. 449–58.

Mosley, M. P. (1982). The effect of a New Zealand beech forest canopy on the kinetic energy of water drops and on surface erosion. *Earth Surface Processes and Landforms*, **7**, 103–7.

Muxart, T. and P. Blanc (1979). Contribution à l'étude de l'alteration differentielle de la calcite et de la dolomie dans les dolomies sous l'action des lichens. Premières observations au microscope optique et au M. E. B. *Proceedings, International Symposium on Karst Erosion*. Union Internationale de Spéléologie, Aix en Provence-Marseille-Nîmes, pp. 165–174.

Nelson, C. H., K. R. Johnson and J. H. Barber (1987). Grey whale and walrus feeding excavation on the Bering shelf, Alaska. *Journal of Sedimentary Petrology*, **57**, 419–30.

Pentecost, A. (1978). Blue-green algae and freshwater carbonate deposits. *Proceedings, Royal Society of London*, B, **200**, 43–61.

Pham Khang (1985). The development of karst landscapes in Vietnam. *Acta Geologica Polonica*, **35**, 305–19.

Pikul, J. L., J. F. Zuzel and R. N. Greenwalt (1986). Formation of soil frost as influenced by tillage and residue management. *Journal of Soil and Water Conservation*, **41**, 196–9.

Pitty, A. F. (1971). *Introduction to Geomorphology*, Methuen.

Pryor, W. A. (1975). Biogenic sedimentation and alteration of argillaceous sediments in shallow marine environments. *Geological Society of America, Bulletin*, **86**, 1244–54.

Schaetzel, R. J. (1986). Complete soil profile inversion by tree uprooting. *Physical Geography*, **7**, 181–8.

Schneider, J. (1977). Carbonate construction and decomposition by epilithic and endolithic microorganisms in salt- and fresh-water. In E. Flugel (ed.), *Fossil Algae*, Springer-Verlag, pp. 248–60.

Schumm, S. A. and R. W. Lichty (1965). Time, space and causality in geomorphology. *American Journal of Science*, **263**, 110–19.

Selby, M. J. (1982). *Hillslope Materials and Processes*, Oxford University Press.

Semeniuk, V. and T. D. Meagher (1981). Calcrete in Quaternary coastal dunes in southwest Australia: A capillary-rise phenomenon associated with plants. *Journal of Sedimentary Petrology*, **51**, 217–68.

Shachak, M., C. G. Jones and Y. Granot (1987). Herbivory in rocks and the weathering of a desert. *Science*, **236**, 1098–9.

Spencer, T. (1985). Marine erosion rates and coastal morphology of reef limestones on Grand Cayman Island, West Indies. *Coral Reefs*, **4**, 59–70.

Staley, J. T., E. Palmer and J. B. Adams (1982). Microcolonial fungi: Common inhabitants on desert rocks. *Science*, **215**, 1093–5.

Stanton, W. I. (1984). Snail holes in Mendip limestones. *Proceedings, Bristol Naturalists' Society*, **44**, 15–18.

Stott, A. P. (1987). Medium-term effects of afforestation on sediment dynamics in a water supply catchment: A mineral magnetic interpretation of reservoir deposits in the Macclesfield forest, NW England. *Earth Surface Processes and Landforms*, **12**, 619–30.

Thornes, J. B. (1985). The ecology of erosion. *Geography*, **70**, 222–34.

Thornes, J. B. (1987). The palaeoecology of erosion. In J. M. Wagstaff (ed.), *Landscapes and Culture*, Basil Blackwell, pp. 37–55.

Tricart, J. and A. Cailleux (1972). *Introduction to Climatic Geomorphology*, Longman.

Trimble, S. W. (1988). The impact of organisms on overall erosion rates within catchments in temperate regions. In H. A. Viles (ed.), *Biogeomorphology*, Basil Blackwell, pp. 83–142.

Trudgill, S. T. (1977). *Soil and Vegetation Systems*, Clarendon Press.

Trudgill, S. T. (ed.) (1986). *Solute Processes*, Wiley.

Twichell, D. C., C. B. Grimes, R. S. Jones and K. W. Able (1985). The role of erosion by fish in shaping topography around Hudson submarine canyon. *Journal of Sedimentary Petrology*, **55**, 712–19.

Van den Ancker, J. A. M. and P. D. Jungerius (1985). The role of algae in the stabilisation of coastal dune blowouts. *Earth Surface Processes and Landforms*, **10**, 189–92.

van Eerdt, M. M. (1987). The influence of basic soil and vegetation parameters on salt marsh cliff strength. In V. Gardiner (ed.), *International Geomorphology 1986* I, Wiley, pp. 1073–86.

Viles, H. A. (1984). Biokarst: Review and prospect. *Progress in Physical Geography*, **8**, 523–42.

Viles, H. A. (1987). Blue-green algae and terrestrial limestone weathering on Aldabra Atoll: An SEM and light microscope study. *Earth Surface Processes and Landforms*, **12**, 319–30.

Viles, H. A. (ed.) (1988). *Biogeomorphology*, Basil Blackwell.

Vis, M. (1986). Interception, drop size distributions and rainfall kinetic energy in four Colombian forest ecosystems. *Earth Surface Processes and Landforms*, **11**, 591–603.

Wall, J. D. R. and G. E. Wilford (1966). Two small-scale solution features in limestone outcrops in Sarawak, Malaysia. *Zeitschrift für Geomorphologie*, **10**, 90–4.

Wasson, R. J. and P. M. Nanninga (1986). Estimating wind transport of sand on vegetated surfaces. *Earth Surface Processes and Landforms*, **11**, 505–14.

Went, F. W. (1969). Fungi associated with stalactite growth. *Science*, **16**, 385–6.

Williams, R. B. G. (1988). The biogeomorphology of periglacial environments. In H. A. Viles (ed.), *Biogeomorphology*, Basil Blackwell, pp. 222–52.

Wilson, M. J. and D. Jones (1983). Lichen weathering of minerals: Implications for pedogenesis. In R. C. L. Wilson (ed.), *Residual Deposits*, Geological Society Special Publication 11, pp. 5–12.

Yatsu, E. (1988). *The Nature of Weathering*, Sozosha.

3 Aspect, Vegetation Cover and Erosion on Semi-arid Hillslopes

MIKE J. KIRKBY, KEN ATKINSON and JOHN LOCKWOOD
School of Geography,
University of Leeds

SUMMARY

The integrated effect of aspect on clear-sky radiation can be calculated directly from latitude, local time and topographic data. Computed values are used to drive a simplified model for actual evapotranspiration and vegetation growth in semi-arid areas. It shows the contrasts which may be expected in vegetation biomass and cover, as they vary through an average year. Large north–south contrasts are forecast in a steep valley for vegetation biomass, overland flow and erosion. They reflect both the local topography and the overall climate of the area. Combining this model for water balance with an overland flow model driven by the natural or man-modified vegetation cover, forecasts are also made about rates of sediment yield, and the resulting asymmetry in slope form which may develop. Results are compared with observed distributions for southeast Spain.

INTRODUCTION

The important role of aspect in controlling the distribution of plants and vegetation communities has long been recognized. Shreve (1922) was one of the first to quantify the influence of differences in solar radiation due to slope and aspect. He recognized the effects, acting through soil moisture and temperature, on the vegetation of the western USA. Tansley and Chipp (1926) found that the most important effect of radiation contrasts, for low-latitude

Vegetation and Erosion
Edited by J. B. Thornes
©1990 John Wiley & Sons Ltd

semi-arid areas, was through differences in soil water balance which influence plant germination. For north–south contrasts in southwest Texas, Cottle (1932) found that soils on south-facing slopes were 5–11 °C warmer (at 5 cm depth), had 5–15 per cent lower moisture content and 24–44 per cent greater evaporation than soils on the north-facing slopes. Boyko (1947) emphasized differences in vegetation structure for slopes in Israel, finding greater diversity on north-facing and shaded slopes.

Since 1960 there has been improved data on site microclimates and improved analysis of statistical differences between vegetation communities. Ayyad and Dix (1964) found the greatest contrasts between NNE- and SSW-facing slopes in a study of Saskatchewan grasslands. The best statistical predictors of vegetation were found to be soil moisture and soil temperature. Position on the slope, identified with runoff, was also found to be an important determinant of soil organic matter and pH. Mayland (1972) used a measure of incident solar radiation, but found that it only accounted for 22 per cent of the variation in plant frequency. He concluded that soil moisture and other climatic factors were also very important. Geiger (1973) noted the principles governing the daily and seasonal variation in solar radiation received on a sloping surface. He noted that vegetation contrasts persisted into low latitudes where solar radiation contrasts are relatively weak. He attributed the effect to the timing of maximum radiation during the course of the day. On SW-facing slopes, maximum radiation occurs in the afternoon, when surface soil moisture has already evaporated, so that the moisture is less available to plants than on NE-facing slopes with a morning radiation maximum.

Dargie (1984, 1987) has used multivariate ordination techniques (detrended correspondence analysis: DECORANA) to analyse vegetation contrasts in Murcia, southeast Spain. He concludes that temperatures are the main determinant of contrasts in floristic composition, and soil moisture of contrasts in vegetation cover and plant biomass. He also notes a significant offset of the maximum contrast from the north–south axis in a NE–SW direction. Offsets vary from 7°–19°, and are attributed to the asymmetry in the daily temperature, and consequently moisture cycles. Hackett (1985) examined presence and density of indicator species in the Corbières region, southern France. He found that vegetation contrasts were associated with reduced levels of humification, and consequently of exchangeable magnesium and potassium, on the drier south-facing slopes.

These studies agree on the underlying importance of differences in solar radiation produced by aspect contrasts, but show a range of possible intermediates which influence the vegetation type, cover and biomass. The present study provides some basis for extending the discussion, by modelling some of the relevant processes explicitly. Because aspect contrasts are thought to be greatest in semi-arid mid-latitude areas, with minimum cloud cover, the study focuses on the Almeria province of southeast Spain, where annual rainfall is about

200–250 mm, falling on 40–50 days. An explicit summation of direct beam solar radiation, with allowance for diffuse and long-wave radiation, has been used to drive a model for soil moisture. Temperatures have not been modelled explicitly, but the observed delay of maximum temperatures after solar noon has been included to give some measure of possible offset from the north–south axis.

Vegetation and organic soil have been 'grown' in the model explicitly, but there has been no allowance for consequent changes in nutrient status. That is to say that nutrients have not been considered to be limiting for the vegetation. Vegetation types are not explicit in this type of model, but can be partially inferred from differences in seasonal persistence and biomass.

METHODOLOGY

The intensity of direct beam solar radiation is proportional to the cosine of the angle between the sun's beam and a perpendicular to the local slope surface. Intensity is greatest when the sun shines squarely on a slope facet, and diminishes rapidly when the sun is shining at a low angle to the surface. Allowance should also be made for attenuation of the beam as it passes through the atmosphere. This loss is least when the sun is vertically overhead and the beam therefore passes through the least thickness of atmosphere. Even with clear skies, there is some diffuse radiation from the blue sky. When the sun's beam is shaded by surrounding hills, this diffuse radiation becomes the only short-wave contribution.

The radiation received at any point varies continuously throughout the day and throughout the year at every site. Instantaneous measurements can provide local values for the proportion of diffuse radiation and of atmospheric attenuation, but knowledge of the regular movement of the sun through the day and the seasons allows calculation of the integrated total short-wave radiation for any site, and its distribution. This analysis provides only an estimate of short-wave radiation from a cloudless sky, and radiation contrasts are greatest under these conditions. The present study has therefore concentrated on southeast Spain, an area with only 200 mm rainfall a year and generally clear skies, where aspect contrasts may be expected to be high.

In areas with extensive cloud cover, radiation contrasts are low, and the dominant effect may be that due to local shading which limits the solid angle of sky that delivers diffuse radiation to a site. In a latitude transect from the north pole to the equator, aspect effects should also provide greater warming on south-facing slopes, but the effect on vegetation and landforms is not consistent in direction. In high arctic latitudes, south-facing slopes may thaw more frequently than north-facing slopes, leading to greater solifluction. At slightly lower latitudes, north-facing slopes may freeze more often, so that

solifluction is greater on north-facing slopes. Within the semi-arid zone, greater radiation may limit vegetation and promote soil erosion on south-facing slopes, while in the humid tropics, vegetation may grow better on the slightly dryer south-facing slopes. At the same time, there is a decrease in radiation contrasts towards the equator due to the increasing symmetry of solar elevation over the year. Southeast Spain is at latitude 38° N, where radiation contrasts are still strong, and the vegetation is exposed to considerable moisture stress. It is therefore an area which is expected to show strong responses to aspect differences.

Computed solar radiation levels are one important input to any model of slope evapotranspiration and hydrology. In this chapter a model based on the use of Bowen ratios (Priestley and Taylor, 1972) has been used to estimate potential evapotranspiration, using average air temperatures, with a superimposed daily cycle to give the value for the ratio, and using an average albedo for the surface. This potential value has then been used to estimate actual evapotranspiration and other components of the hillslope water balance, using an explicit hillslope flow model, based on TOPMODEL (Beven and Kirkby, 1978). The details of this procedure are set out in Kirkby and Neale (1986). Actual evapotranspiration is constrained by the estimated potential value and by available soil water. Forecast net runoff of soil water is incorporated into a monthly water budget to update soil water deficit below saturation. Overland flow is obtained by integrating over the frequency distribution of daily storms which exceed the forecast deficit.

Natural vegetation growth is estimated from actual evapotranspiration and size-dependent leaf fall. Accumulation of soil organic matter is estimated from leaf fall and a temperature-dependent rate of decomposition, assuming aerobic conditions. The organic matter content is used in turn to estimate the parameters of soil water storage which control subsurface flow. The biological parameters of this model have been derived from empirical data, much of it collected as part of the International Biological Programme (e.g. Lieth and Whittaker, 1975), within a global rather than a Spanish local framework. There is therefore some scope for improvement at the local scale, and no account has, at present, been taken of land use which interferes with the natural vegetation, notably grazing by sheep and goats for the area of immediate interest.

The hydrological and biological model has been run for a sufficient number of years to stabilize the biomass and runoff values, using average monthly values for rainfall. This has been found to occur in between 10 and 50 years, depending mainly on the equilibrium age of the dominant vegetation. This equilibrium approach has been adopted on the reasonable premise that changes through erosion occur slowly enough to allow vegetation and organic soil to keep pace with them.

For a slope profile, overland flow may be summed downslope to give an estimate of overland flow discharge, which may in turn be used to estimate sediment discharge associated with soil erosion. In principle, flow estimates may be incorporated into a fully interactive model for slope evolution. Changes in topography due to erosion could then be used to update the radiation calculations, evapotranspiration and overland flow, to give erosion values at the next modelled time increment. At present a simpler approach has been adopted. The contrast in rates of overland flow production on opposing slopes has been used to parameterize a simpler existing slope evolution model (Kirkby, 1986) for a valley cross-section to give an estimate of the expected rate of growth of valley asymmetry. The radiation parameters have not been continuously updated, but merely checked at the end of the slope model run to confirm that there have been no gross changes in hydrological conditions.

MODEL SPECIFICATION

The Radiation Model

Regional data required to estimate the direct solar beam radiation consist of latitude, overhead atmospheric attenuation and proportion of diffuse blue-sky radiation. For each site, local values are needed for the slope gradient and direction, and for the elevation of the local skyline at sufficient points of the compass to estimate when the site is shaded. These data may be generated quickly for a field site, or computed for points on a slope profile, with the assumption that the valley is of indefinite extent. Similar data could also be computed for every point in a digital terrain model, and some work has been done on this although without taking account of shading effects.

The relevant equations used are as follows (Robinson, 1966):

Solar declination, $d = -23.5 \cos(30m)$
Solar elevation, $\alpha = \sin^{-1}[\sin\theta \sin d - \cos\theta \cos d \cos(15h)]$
Solar azimuth, $\phi = \tan^{-1}\{[\cos d \sin(15h)]/[\sin\theta \cos d \cos(15h) + \cos\theta \sin d]\}$
taking due care about the appropriate quadrant for ϕ

where angles are taken as degrees,
h is the local time in hours and fractions of an hour (0–24)
m is the month (0 on 1st Jan to 12 on 31st Dec)
and θ is the latitude (positive in northern hemisphere).

The direct solar beam radiation at the top of the atmosphere is about 1360 W m^{-2}, with a small correction for the closer approach to the sun in winter. Ignoring diffuse radiation, the direct beam radiation on a slope surface is:

$$r = r_0 \, p^{\mathrm{cosec}\,\alpha} \, [\cos\beta \, \sin\alpha + \sin\beta \, \cos\alpha \, \cos(\theta - \psi)]$$
if $\alpha >$ local horizon elevation in direction θ.

Taking account of diffuse radiation, the total short-wave radiation is:

$$R = r(1-\lambda) + pr_0\lambda[1 - \exp(-\alpha/30)]$$

without the second term if sun is below global horizon
where λ is the proportion of diffuse radiation,
$\quad\quad p$ is the proportional attenuation for overhead sun,
$\quad\quad \beta$ is the slope angle
and $\quad \psi$ is the slope azimuth (down line of local slope).

The Potential Evapotranspiration Model

Hourly mean temperatures have been estimated from mean monthly temperatures and ranges, with the daily variation treated as a sine wave, with maximum temperatures at 3 p.m. local solar time to allow for the delay between radiation and temperature maxima. The Bowen ratio is calculated for a given temperature, following the method of Priestley and Taylor (1972), originally calculated for an extensive area of moist pasture. Here it is approximated as:

$$B = 0.778 - 0.0396T + 4.714 \times 10^{-4} \ T^2$$

where T is the hourly mean temperature.
Potential evapotranspiration is then calculated as:

$$E^P = [(1 - A)R - \mu\sigma \ (T+273)^4] / [L/1 + B)]$$

where L is the latent heat of vaporization of water,
$\quad\quad A$ is the albedo of the surface,
$\quad\quad \mu$ is the proportion of long-wave radiation escaping (ca. 25%)
and $\quad \sigma$ is the Stefan–Boltzmann constant.

In a real situation, even with cloudless skies, the proportion of long-wave radiation lost will depend on the limits of the solid angle through which the slope views the sky. This is because the effective radiating temperature of the atmosphere is usually lowest directly overhead, and the net outgoing long-wave radiation is greatest in this direction and decreases with increasing zenith distance. For example, the net outgoing long-wave radiation from a deep basin whose rim extends 40° above the horizon will be 64–71 per cent of that from a flat plain.

Summation of the above expression over the day provides an estimate of monthly potential evapotranspiration, for use within the slope hydrology and vegetation growth model. The details of this model are not specified here, but are taken directly from Kirkby and Neale (1986, Figure 2).

The assumptions of the model used are appropriate for short vegetation rather than for forests. The nature of the vegetation cover can have a considerable effect on the importance of aspect effects. The Penman–Monteith equation (Monteith, 1965) may be presented in the form:

$$LE_A = \Omega[s/(s+\gamma)]R_N + (1-\Omega)[\varrho o_p D/(\gamma r_c)]$$

where $\Omega = [1 + \gamma/(s+\gamma)\, r_c/r_a)]^{-1}$

s is the slope of the local slope of the saturated humidity–temperature curve

R_N is the net radiation

r_c, r_a are the canopy and aerodynamic resistances

γ is the psychometric constant

ϱ is the density and c_P the specific heat of air

D is the saturation deficit

L is the latent heat of vaporization

and E_A is the rate of actual evapotranspiration.

Where Ω is large the surface may be considered to be isolated from conditions in the atmosphere overhead, so that the radiation effects modelled here are predominant. In contrast, small values of Ω give evapotranspiration rates which depend mainly on heat exchanges with the atmosphere. Typically extensive areas of freely transpiring short vegetation give Ω values of about 0.8, while forests give $\Omega = 0.2$.

For the conditions modelled in this chapter, potential evapotranspiration in the summer is so much higher than available moisture that it has little effect on the water balance of the slope. In winter, actual evapotranspiration approaches potential values, so that the model should be most reliable under these conditions. Thus the approximations which assume a high value of Ω, and high radiative control of evapotranspiration rates appear to be suitable.

The Slope Model

The possible erosional impact of aspect differences is illustrated for a symmetrical sinusoidal valley, which is assumed to be 200 m wide, 50 m deep and of indefinite length. The local horizon elevations may then be calculated for any chosen valley orientation. The aspect contrasts in sediment yield and the long-term evolution of the slope form may then be simulated. Average rate of overland flow production from the opposite valley sides have been used to parameterize wash erosion rates in a slope model for a valley cross-section. Asymmetric development is allowed by considering the valley form as repeated, so that divides and streams may migrate laterally if there is a process imbalance. This model has been written to include the slope processes of creep, splash, wash and landslides, but the rate of sliding has been set to zero to emphasize the effect of aspect, acting through the wash erosion. The model is described in full in Kirkby (1986).

Sample Results

Figure 3.1 shows an example of the computed local horizon for a site near the bottom of an east–west trending sine wave shaped valley (shown in Figure 3.2a) with a southerly aspect near Almeria, at 38° N. In directions due east and west, the horizon is level, looking along the length of the valley. To north and south, the valley sides partially hide the sun at certain times of day and year. The figure shows the sun's path through the sky month by month. It may be seen that, for these rather moderate slopes, shading is mainly in the morning and evening and that, even in January, the midday sun is visible over the hilltop to the south.

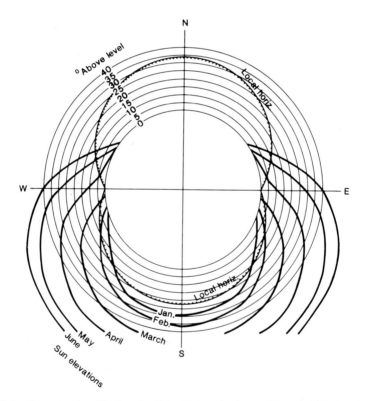

Figure 3.1   Computed local horizon for Point 8, near the base of the south-facing slope, and solar elevation/azimuth tracks for January to June, showing the effect of local shading, particularly in the winter months

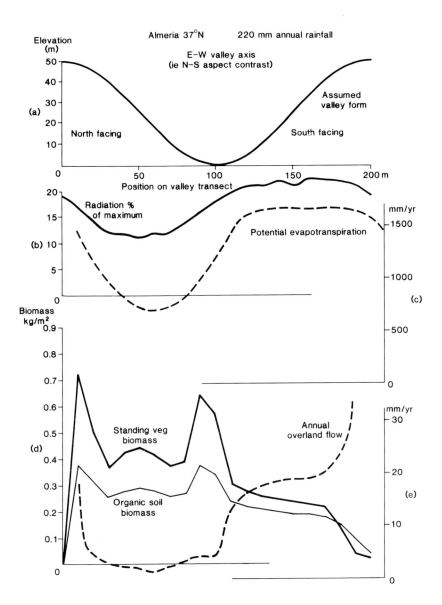

Figure 3.2 Forecast variations in climatic and vegetation values on a symmetrical sinusoidal slope with directly north- and south-facing slopes at 37° N. (a) The assumed slope form. (b) Percent of maximum possible radiation received over the year. (c) Estimated potential annual evapotranspiration. (d) Forecast mean standing biomass and organic soil biomass (kg. m^{-1}). (e) Forecast annual overland flow (mm)

Figure 3.2 shows the computed pattern, for each point of the same valley profile, of annual averages for radiation (Figure 3.2b) and for simulated potential evapotranspiration (Figure 3.2c). The only asymmetry about the north–south axis is in the delay of maximum temperature after radiation maximum, which has been set at three hours in this example. Making use of the water balance model, the equilibrium pattern of living vegetation and organic soil (Figure 3.2d) is simulated to show a double peak at moderate gradients on the north-facing slope, and a clear difference in average biomass between the two valley sides. There is a similar high contrast in the rates of overland flow production (Figure 3.2e) from the two sides, with average values of about 3 mm and 19 mm respectively from the north- and south-facing slopes. Only the divides are exceptional, with a forecast of high overland flow and sparse vegetation cover which merits further examination.

The effect of changing valley orientation is shown in Figure 3.3. As might be expected, aspect contrasts are weakest when the valley slopes face east and west. If the more cooler slope faces between northeast and northwest lines, vegetation on the warmer (south-facing) slope is forecast to be very insensitive to aspect. On the north-facing slope, however, there appears to be a critical slope gradient of about 10–20° on which vegetation is lushest, although the exact gradient varies. The largest biomass in both plants and soil is forecast to occur on slopes facing either northwest or northeast, whereas true north-facing slopes show a less marked contrast with the south-facing slopes opposite.

The long-term effects of the vegetation contrast on hillslope forms have been modelled, as briefly described above. The overland flow contrast has been used to parameterize a model including splash and wash transport. This approach may underestimate the difference because it makes no allowance for the greater flow resistance offered by the thicker vegetation on the north-facing slopes. On the other hand, overland flow discharge has been assumed to increase linearly with distance from the divide, whereas storms may be brief enough to limit the increase of discharge. If this is the case, wash erosion would be less effective than forecast on both slopes, and splash erosion would be proportionately more significant so that asymmetry would develop more slowly.

Figure 3.4 illustrates some example simulations which show the evolution of valley asymmetry from an initially symmetrical valley. At the valley bottom, the increment in stream transporting capacity is assumed to be directly proportional to height above a base level of zero. This allows a dynamic interaction between slope and channel processes. As stream behaviour is outside the scope of the present discussion, two different values of stream sediment increment have been chosen to illustrate its influence over the slope development. In Figure 3.4a a low rate of stream activity is assumed, so that valley asymmetry develops with aggradation of the valley bottom. In Figure 3.4b, stream activity is ten times greater, allowing overall degradation, and an even greater asymmetry. Because landslides have not been included in this model run, the maximum slopes

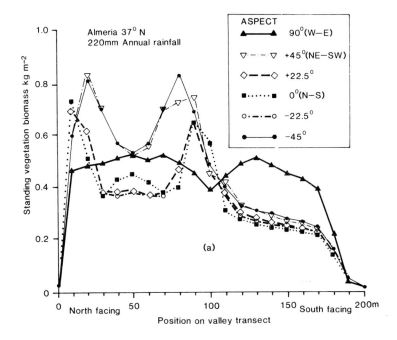

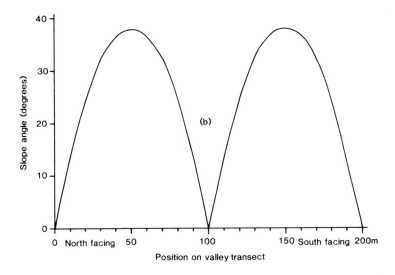

Figure 3.3 (a) Forecast variations in mean standing biomass along the slope transect shown in Figure 3.2, for different orientations of the valley. The aspect quoted is for the more northerly of the two opposite slopes. The 90° slope goes from west to east. (b) Slope gradient angles across the transect for comparison

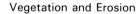

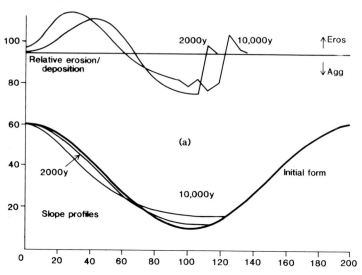

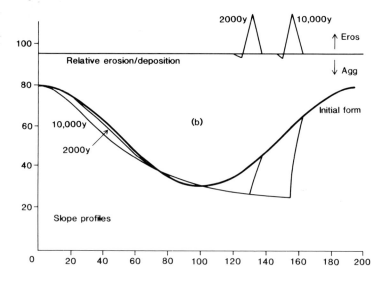

Figure 3.4 Forecast evolution of a north–south slope profile pair, assuming overland flow contrasts similar to those forecast above. In each case, evolution begins from the sinusoidal form assumed above. Each figure shows the relative rates of erosion and deposition (above) and the form of the slope profile (below) after 2000 and 10 000 years. (a) Stream sediment increment $\propto 10^{-6} \times$ Stream elevation. (b) Stream sediment increment $\propto\, ^{-5} \times$ Stream elevation

are unrealistically steep on the north-facing (right-hand) slope. Inclusion of landslides reduces the extent of asymmetry and the maximum slope. At this stage, it is perhaps rash to go beyond the conclusion that aspect differences offer the potential for developing evident valley asymmetry over any period long enough for appreciable erosion. Although there are other possible causes of valley asymmetry, aspect appears to be the only one which is able to provide a consistent north–south contrast irrespective of river behaviour and lithology.

COMPARISON WITH FIELD VEGETATION DATA

Preliminary field observations have been made at several sites in Almeria province. Results show strong aspect contrasts in plant diversity, cover and biomass. At a series of sites near Turre, NW- to NE-facing slopes showed a strong increase in species diversity and biomass downslope, which was not forecast in the model, suggesting that the model has underestimated the role of lateral subsurface flow on the moister slopes. The drier south-facing slopes show no comparable trend.

The most direct comparison with the model is for January biomass. Figure 3.5 shows simulated and measured values. The values are comparable, which provides some validation for the model, which has not been fitted to the local

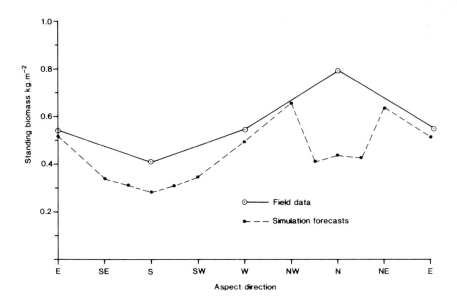

Figure 3.5 Mid-slope standing biomass (kg. m^{-2}) for January. (a) Measured in the field near Mojacar, Almeria province. (b) Forecast from the simulation model for a symmetrical sinusiodal slope form

data at this stage. Both field and model values show an aspect contrast, although the data are not sufficiently detailed to test for small offsets in the axis of maximum asymmetry, as reported by Dargie (1987). The model appears to underestimate biomass on the north-facing slopes, particularly close to north–south. Some of this discrepancy is thought to be due to underestimation of subsurface flow, but a part is also due to the assumed geometry of infinitely extending valleys, with shading effects which imperfectly reproduce the three-dimensional geometry of the actual valleys. The assumed geometry is also a symmetrical one, whereas many actual slopes show at least some degree of asymmetry. This increases the shading of north-facing slopes, and hence the expected moisture and vegetation contrast.

The field data so far support one of Dargie's (1984) main conclusions, that soil moisture differences are of importance in determining vegetation cover and biomass. The model makes use of a water balance as an essential component in estimating plant growth, and so provides evidence that solar radiation differences are sufficient to drive biomass differences using a hydrological mechanism. Air temperatures are assumed the same, and soil temperatures are not explicitly forecast by the model, so that there is no basis for testing the strength of Dargie's second conclusion, that differences in floristic composition are driven mainly by temperature contrasts.

Although there remain many questions to be answered about the way in which plants adapt to moisture and temperature stress in semi-arid environments, the model is seen to provide a methodology for discussing the effects of aspect in a coherent physical framework, which can be related to climatology, hydrology and geomorphology. There is plainly scope for investigating the effects of greater subsurface flow on the moister slopes, and a need to relate vegetation growth models to the overall structure of the plant community, rather than simply to total biomass.

REFERENCES

Ayyad, M. A. G. and R. L. Dix (1964). An analysis of a vegetation–micro-environment complex on prairie slopes in Saskatchewan. *Ecological Monographs*, **34**, 421–42.

Beven, K. J. and M. J. Kirkby (1978). A physically based, variable contributing area model of basin hydrology. Hydrological Sciences Bulletin, **24**(1), 43–69.

Boyko, H. (1947). On the role of plants as quantitative climate indicators and the geo-ecological law of distributions. *J. Ecology*, **35**, 138–57.

Cottle, H. J. (1932). Vegetation on north and south slopes of mountains in S.E. Texas. *Ecology*, **13**, 121–34.

Dargie, T. C. D. (1984). On the integrated interpretation of indirect site ordinations: a case study using semi-arid vegetation in southeastern Spain. *Vegetatio*, **55**, 37–55.

Dargie, T. C. D. (1987). An ordination analysis of vegetation patterns on topoclimate gradients in south-east Spain. *J. Biogeography*, **14**, 197–211.

Geiger, E. (1973). *The Climate Near the Ground* (2nd edn), Harvard Univ. Press.

Hackett, D. J. (1985). *Factors affecting the distribution of plant and animal communities of dolomitised limestone in the Eastern Corbières, France.* Unpublished Ph.D. thesis, University of Leeds.

Kirkby, M. J. (1986). *Conditions for valley asymmetry derived from a slope evolution model.* Working paper 477, Leeds University, School of Geography.

Kirkby, M. J. and R. H. Neale (1986). A soil erosion model incorporating seasonal factors. In V. Gardiner (ed.), *International Geomorphology 1986*, Part II, John Wiley, pp. 189–210.

Lieth H. and R. H. Whittaker (eds) (1975). *The Primary Productivity of the Biosphere*, Ecological Studies, 14, Springer-Verlag, New York.

Mayland, H. E. (1972). Correlation of exposure and potential solar radiation to plant frequency of *Agropyron desetorum*. *Ecology*, **53**, 1204–6.

Monteith, J. L. (1965). Evaporation and environment. In *The State and Movement of Water in Living Organisms*, 19th Symposium, Society for Experimental Biology, pp. 205–34.

Priestley, C. H. B. and R. J. Taylor (1972). On the assessment of surface heat flux and evaporation using large scale parameters. *Monthly Weather Review*, **100**, 81–92.

Robinson, N. (1966). *Solar Radiation*, Elsevier, Amsterdam.

Shreve, F. (1922). Conditions indirectly affecting vertical distribution on mountains. *Ecology*, **3**, 269–74.

Tansley, A. G. and T. F. Chipp (1926). *Aims and Methods of Vegetation Study*, The British Empire Vegetation Committee, London, 383 pp.

4 The Interaction of Erosional and Vegetational Dynamics in Land Degradation: Spatial Outcomes

JOHN B. THORNES
Bristol University

SUMMARY

The model starts from the basic assumption that there is competition between erosion and vegetation and seeks to establish the conditions under which one or other will dominate. Both are modelled logistically, the essential constraints being soil and climate characteristics and interplant competition. The differential equations for plant and erosion growth are solved digitally and resolved graphically to provide stability conditions for the model. Finally some examples are given of the spatial interactions between species and erosion for two species, two different soil erodibility conditions and two different wetness conditions. All of the submodels are simple and relatively crude but the basic principles are considered to be correct.

INTRODUCTION

The investigation of water constraints on vegetation productivity is well documented but much less is known about the limiting effects of erosion. In this chapter we focus on the competitive interaction between erosion and the vegetation cover, which was sketched in an earlier paper (Thornes, 1985) and developed in an archaeological context (Thornes, 1988). Here we attempt to incorporate more realistic hydrological and boundary conditions as well as the impact of organic matter distribution throughout the soil profile. The resulting model is then examined for spatial varations at the scale of a hillslope.

Vegetation and Erosion
Edited by J. B. Thornes
©1990 John Wiley & Sons Ltd

BASIC MODEL — SOIL EROSION

One approach to the vegetation–erosion problem is to budget inputs and outputs to the soil store from weathering, erosion, organic matter production and decomposition. This can be done either analytically (Biot, 1988) or using digital simulation (Kirkby and Neale, 1987). In this chapter the problem is approached by considering the equilibrium conditions for vegetation and erosional change when taken singly or in combination to reveal the stability of the system and therefore follows the strategy adopted in ecological modelling of non-linear systems.

In considering erosion we start with a simple model of wash and of excess runoff production and throughout the development the wash model assumes that the change in soil depth normal to the surface is given by equation (1) in which erosion is a power function of surface runoff and slope:

$$dz/dt = k_1 q^m s^n \qquad (1)$$

in which k_1, m and n are parameters and q, the surface runoff per unit width, is obtained as the difference between intensity and infiltration rate for an average storm magnitude recurring N times/year and assumed to be of average duration T. This yields, for a unit width,

$$q = (I - F_z) * x * T * N * 0.01$$

with I = intensity and x the slope length in metres. F_z is the infiltration rate which exists if the soil profile has been truncated to a depth z from the original soil surface and is described by

$$F_z = F_0 * e^{-Rz} + a * V * e^{-bz} \qquad (2)$$

with R an infiltration decay coefficient due to vertical increase in bulk density. The first term is then the vertical decrease in surface bare soil infiltrability (F_0). In the second term b is a decay coefficient, V the vegetation cover (in %) and a a scaling factor relating increased surface infiltration to vegetation cover at the surface. This equation produces an exponential decay of infiltration rate which is a function of the decrease in organic matter and bulk density down from the surface and b can be regarded as a mixing factor reflecting the importance of translocation of organic matter by physical and biotic processes. By contrast, the a term essentially represents the joint effects of humification and mineralization on the available organic matter which is assumed to be directly proportional to vegetation cover. V is assumed to be a function of erosion and the local V_{max} which itself is climatically and for edaphically determined. This can be compared with the organic profile derived in Kirkby's (1985) model where the mixing factor was shown to be a major determinant of the resulting profile.

At the other end of the profile the essential control is determined by the reduction in erodibility due to increasing bulk density and stone content. This is expressed by

$$(1 - (z/z_1)^R)$$

and R is to be calibrated by the relationship between erosion rate and bulk density and for stoniness on various soils (Thornes *et al.*, in press) and z_1 is profile depth. As R increases, the effect of stoniness on erosion diminishes. Erosion is considered to be reduced exponentially in relation to the bare soil value by increased vegetation cover as follows:

$$dz/dt = -k_1 * s^n * q_b^m * (1 - e^{-BV}) \qquad (3)$$

with B a decay constant and q_b the bare soil runoff (equal to $(I - F_z)$ when $V = 0$). This equation follows the work of Elwell and Stocking (1976) confirmed more recently by Dunne, Dietrich and Brunengo (1978), Lee and Skogerboe (1985) and by Francis and Thornes (Chapter 22, this volume). Finally we may assume that there is a positive effect of grazing proportional to grazing intensity G reflecting the compaction effect of animals in relation to erosion. As yet, despite some recent studies, we have a poor understanding of this phenomenon.

When these effects are combined, the differential equation for soil loss becomes

$$dz/dt = k_1 * s^n * [q^m * (1 - (z/z_1)^R - q_b^m (1 - e^{-BV})] + k_2 G \qquad (4)$$

The effects of inhibiting depth and varying vegetation cover are shown in Figures 4.1(a) and (b). As soil losses increase, erosion rate increases up to some limit which is a function of the value R and the total soil depth z_1. As R becomes large (relatively stone-free soil) the effects of the lower boundary become very sharp. When there is no soil left, there can be no erosion. The diagram also shows that for lower mixing rates ($b = 0.07$) the overall erosion rates are higher. Figure 4.1(b) shows for fixed R and b the general effect of increased vegetation cover on the reduction of erosion.

If equation (4) is set to zero and solved for V and z (the erosional loss from the surface down) we obtain the isocline for erosion shown in Figure 4.1(c). In this diagram the isoclines for different values of R represent the vegetation required to keep erosion rates constant at different soil losses. For stonier soils (low values of R) the minimum vegetation required to keep erosion in check is less. For a relatively stone-free soil the minimum vegetation required for equilibrium is an increasing function of the amount of soil removed.

There are two other important aspects to Figure 4.1(c). First, using equation (4), the value of dz/dt can be calculated for any point in the domains of V and z.

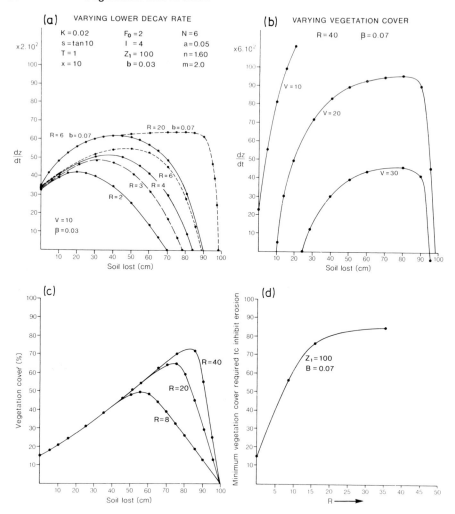

Figure 4.1 (a) Erosion control by stoniness (R), organic matter decay coefficient and amount of soil already lost (z). (b) Rates of erosion for different vegetation cover amounts. (c) Isoclines for wash erosion with varying stoniness. (d) Effects of R on minimum amount of vegetation required to keep erosion in check

When this is done for the area beneath the isocline, the horizontal lines of change in z (for any constant V) are towards maximum erosion. Outside the curve they are towards z = 0. These vectors represent the dynamics of change. Second, the net effects of this is that the equilibria to the left of the maximum values of V are unstable (vectors flow away from them) whereas to the right the equilibria are stable (vectors flow towards them). As R decreases, soils are more stony and the range of stable values increases, reflecting the induced stability due to slope armouring. As a result the minimum vegetation cover needed to provide equilibrium is less (Figure 4.1(d)).

VEGETATION COVER

The differential equation for vegetation growth is assumed to be:

$$dV/dt = P_1 * V(1 - V/V_r) - P_2 z \qquad (5)$$

with P_1 and P_2 as constants. V is now the vegetation biomass and V_r the maximum possible biomass as constrained by regional climatic factors and local edaphic controls. The first term is a logistic growth equation limited by V_r and this is widely used for vegetation growth reflecting density effects under conditions of growth limited by an exhaustible water supply. If the latter is considered to be the water available at the beginning of the growing season then it appears appropriate for semi-arid environments (cf. Noy-Meir, 1978). This is the very short-term or seasonal dynamic. In the long term V_r increases as more vegetation retains more water, so that V_r itself can change in the manner of Eagleson's (1978) dynamic optimum. The value of V_r will also be species dependent since it depends on the water use efficienty of the plant.

The second term represents the reduction in plant growth due to nutrient losses in the soil. At the moment this is assumed to be a linear function of depth so that P_2 is the proportion of nutrients per unit thickness of soil, but a more elaborate inorganic profile associated with nutrient cycling can readily be envisaged, especially in arid regions (cf. Kirkby, 1985). Two typical curves for dV/dt are shown in Figure 4.2(a) with a common maximum vegetation cover (V^r). They are characteristic of logistic curves with low rates of increase at limiting amounts of biomass.

In the original model (Thornes, 1988) the maximum vegetation cover was assumed to be a fixed value relating to rainfall. It seems more realistic, however, to consider it a function of soil moisture (theta) and available water storage capacity ($z_1 - z$). To this end we have followed Biot (1988) in using the following equation:

$$V_{max} = V_r / [1 + P_5 \exp(-P_6 * \theta * (z_1 - z))] \qquad (6)$$

and assuming, thereby, that available soil storage capacity is an exponential function of soil depth. We assume moreover that moisture is

$$F_z * N * T / (z_1 - z)$$

which is itself a function of V.

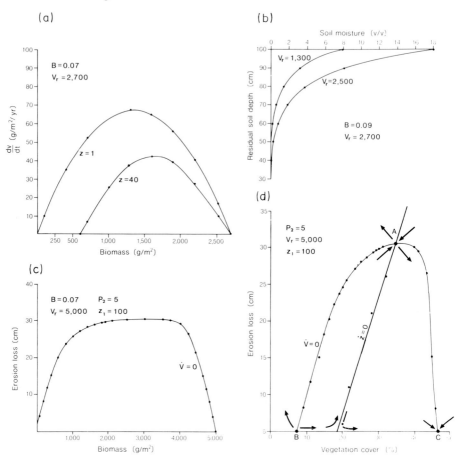

Figure 4.2   (a) Vegetation growth rates for different amounts of prior erosion. (b) Soil moisture content for a given rainfall according to vegetation cover and residual soil depth. (c) Isocline for vegetation biomass. (d) Phase diagram with isoclines for wash erosion and vegetation cover superimposed

The effects of this are shown in Figure 4.2(b) where available soil moisture is shown as a function of vegetation cover and residual soil depth. These are not profiles of moisture against depth because equation (6) assumes that moisture is uniformly distributed with depth. Rather they show for a given residual depth of soil and a given vegetation cover how soil moisture will vary. Bringing together these various terms and incorporating them into equation (5) we can again solve the equation for equilibrium; in this case when $dV/dt = 0$. A typical result is shown in Figure 4.2(c). This is quite different from the isoclines obtained in earlier models, reflecting the more critical profile-controlled interaction between soil thickness and vegetation missing from the earlier model

(Thornes, 1988), where organic matter loss through erosion was the sole constraint on vegetation growth. It also throws into question models which use available water *capacity* rather than available moisture as the driving mechanism for the increase in vegetation cover. Soil moisture (θ) in this model still rests on the rather primitive assumption of the average storm infiltration input used in the erosion model. Clearly a more eleborate model is ultimately required but this will need to accommodate a more realistic mixture of surface types than existing water balance models. For the moment it is driven to provide soil moistures which are typical of late spring values in semi-arid Mediterranean areas. The isocline shown in Figure 4.2(c) is characterized by a quite flat upper limit to the isocline over biomass values of between about 1500 and 4500 g/m^2.

The implication of this is that over this range and with these particular parameters, vegetation cover is kept sharply in check by erosion and that a slight increase in erosion will result in a complete loss of vegetation cover. This is because above the isocline growth rates are negative, whereas below it they are positive. Again the equilibria on the left-hand limb are unstable, while those on the right are stable. Provided soil loss does not exceed about 30 cm, once biomass exceeds the equilibrium value of the left-hand limb it will increase inexorably to the highest possible values on the right-hand limb along horizontal vectors.

VEGETATION–EROSION INTERACTION

Vegetation–erosion interactions could be viewed as integrations of the two basic equations over time to provide analytical or numerical solutions. In Figure 4.2(d) we choose again to use the phase-space representation. Here the erosion and vegetation isoclines are plotted together. To achieve this the biomass has to be converted to cover and this is done using the expression

$$V_c = 16.6(V/0.625)^{0.5}$$

with V now expressed in kg/ha and V_c as percentage of total area, after Whittaker and Marks (1975). In so doing we note that the vegetation isocline ($dV/dt = 0$) is now much steeper and narrower. In the feasible area of overlap, the vegetation cover isocline intersects the erosion isocline on the lower limb of the latter. In the example shown, a rise from 16 to 28 per cent in vegetation cover is enough to constrain erosion rates from increasing over a soil loss depth of 0–30 cm. This reflects the strong effects of the vegetation damping in equation (3). Above 32 cm erosion, cover decreases to zero as erosion dominates.

By examining the direction and magnitude of the resultant of the two growth equations we can determine the behaviour of any point on Figure 4.2(d), the phase plot. Generally, to the left of the $dz/dt = 0$ isocline, soil erosion is increasing (vectors of dz/dt are vertical) whereas to the right it is decreasing so that all the equilibria along the isocline are unstable. As before, outside the $dV/dt = 0$ isocline (vectors horizontal) the vegetation is decreasing while within it the vegetation cover is increasing. The directions and magnitudes of the resultant vectors determine the ultimate endpoint of these changes. When the vegetation change vectors are strong and erosional organic matter losses are small, the dynamics are dominated almost entirely by dV/dt. As coefficient P_2 in equation (5) increases, the erosional effects produce a saddle point at A, a repellor at B and an attractor at C. The sector to the right of the $dz/dt = 0$ isocline and beneath the $dV/dt = 0$ is thus the 'safe' sector since if the system enters that section of phase space it will move to the right-hand limb of the $dV/dt = 0$ isocline and more particularly to point C of full cover with little or no erosion.

In the example shown, a cover of at least 8 per cent is required to have any effect. Above this there are two zones: (i) a semi-stable zone taking the system to the saddle point at A, and a (ii) fully stable zone taking the cover to C. With the parameter values used here, on soils 1 m thick, with small stone content, vegetation cover would never exceed about 47 per cent but erosion would be firmly in check at that value.

If this model can be validated and adequately parameterized the implications for management of erosion–vegetation interactions are obvious.

GRAZING

As a first approximation to grazing, consider the inverted exponential model of Noy-Meir (1978). Here the slope of consumption by grazing animals decreases continuously with increasing V, approaching satiation asymptotically according to the equation:

$$dV/dt = k_2 G(1 - \exp(-V/V_r)$$

i.e. the biomass is decreased at a decreasing rate as V_r (or satiation, whichever is the lesser) is approached, reflecting the assumption that the animal adjusts its marginal grazing effort in proportion to the food available. We shall assume that V_r equals the satiated value because in Mediterranean areas there is an adjustment by herders to what the vegetation can stand. Under this assumption the total consumption never exceeds about 63 per cent of the available biomass. As the mean consumption rate (k_2) multiplied by the number of animals (G) increases, the safe sector is diminished accordingly and colonization becomes impossible even with only a limited soil loss (10–25 cm), though the stable limb for vegetation sets in at progressively lower cover values.

SENSITIVITY

Some idea of the sensitivity of the wash erosion and vegetation change models can be obtained from Figure 4.3. Here the change in dz/dt is shown for an n-fold increase in the parameters about typical values for the operation of the model. For the wash model (4.3a) the equation is highly responsive to slope length (L) and mean storm intensity (I). Considered for a single location and a single climatic type, the vertical erosion decay coefficient (R), reflecting increased stoniness or bulk density and decreased erodibility with depth, is most critical, followed by the exponents of the wash model. The sensitivity to R is greater the thinner the soil cover and the case shown here is much diminished. The coefficient of infiltration (a) plays a lower role than expected in overall magnitude and varies over a nine-fold range. There is least sensitivity to the vegetation decay constant (B) at the level of vegetation here ($V = 15\%$, $z = 50$ cm).

For the vegetation submodel (Figure 4.3b), the model is most sensitive to the scaling factor for the control of infiltrability through the vegetation cover (and hence moisture and productivity). This is followed by the parameter controlling

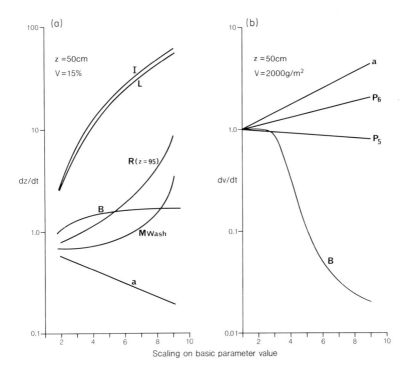

Figure 4.3 Sensitivity of parameters

productivity in relation to available water (P_6) which is species dependent and conceptually identical to the water use factor. By contrast P_5 plays a small role. For values of z and V used here (50 cm and 2000 g/m², B again plays a small role. Both of these results suggest that after z and V themselves, the most significant roles are played by mean intensity and slope length on the one hand and the impact of vegetation cover on soil moisture and the productivity change of residual soil in relation to moisture content (water use factor), on the other. A fuller and more complex analysis of the model sensitivity awaits more realistic parameters.

SPATIAL EFFECTS ON THE HILLSLOPE SCALE

At the hillslope scale we are especially interested in two outcomes:

1. If we vary slope parameters such as length and shape for a given set of climatic conditions, how does the stability of the plant cover change as a response to erosion?
2. If we have plants with different water use factors (P_6) how are they likely to vary in their competitive ability with erosion and, given the erosion, what patterns of plants of different water use factors are likely to emerge?

Assume for the purposes of illustrating the model that we have a rectilinear slope on which the soil increases in a linear fashion in the downslope direction. The effect of this condition, which is certainly common in arid and semi-arid environments, is to offset the effects of increased runoff in the downslope direction by increasing infiltration, soil moisture and hence plant growth. The result in this simple case is shown in Figure 4.4. on bare slopes, erosion extends over the entire slope increasing downslope but at a strongly decreasing rate.

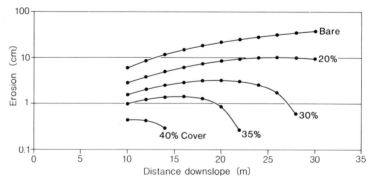

Figure 4.4   Downslope variations in slope erosion with a fixed vegetation cover of various percentages

At 30 per cent plant cover erosion falls very rapidly at the foot of the slope and by 35 per cent is only occurring on the upper 15 m or so. This is converse to the belt of no erosion envisaged by Horton (1945) and illustrates the strong interaction between erosion and cover. In the upper part of the slope erosion is still able to compete with vegetation. This is common in semi-arid environments, where the Horton model may be inverted.

Of course this does not address the more important questions as to whether the vegetation or the erosion is stable or whether the transition is sharp or gentle. For this the isoclines for erosion and vegetation have to be constructed in a spatial context. This is illustrated in Figure 4.5. Consider the case in which the climatically limiting biomass is at $2500\,g/m^2$, typical of the drier parts of Europe. If we assume the same slope conditions as before, and a fixed value of P_6, which controls water use efficiency (as in species B), we can compute the positions on the slope where plant growth drops to zero for given initial conditions. We can also construct the erosional spatial isoclines. On Figure 4.5 two are shown cutting obliquely across the plot. Outside the spatial vegetation isocline plant productivity is decreasing; inside it productivity is increasing. To the left of and above the erosional isoclines erosion is increasing; to the right and below it is decreasing. As a consequence, using the same arguments as in Figure 4.2(d), we can define conditions (area 1) for successful competition between plants and erosion for species B.

The curve shows that there is a sharp upper boundary to successful plant growth, regardless of the initial vegetation cover, at about one-third of the distance upslope. At the highest vegetation densities (in zone 2) the plants compete with each other for water, reducing their individual effectiveness in erosion competition.

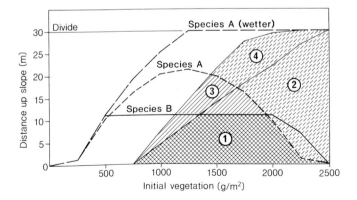

Figure 4.5 Spatial isoclines for vegetative growth for species of different water use efficiency (A and B) and under different annual moisture amount and distribution

A plant with more effective water use will be able to compete more effectively with erosion at higher points upslope (species A) and is thus able to occupy up to two-thirds of the slope under optimal conditions. This is area 3 in the diagram. Finally if conditions changed to become wetter, species A can colonize the entire slope over the area marked 4 in the figure. The cover will only be stable to the right of the erosional isocline (area 2 for less resistant plants and areas 2 and 4 for more resistant plants).

LIST OF SYMBOLS

N number of storm events per year
T average duration of storms
I average storm intensity
k_1 coefficient of wash erosion equation (taken as 0.2)
q discharge per unit width of slope (m/hr)
q_b discharge per unit width of bare slope
s slope
z erosion from surface downwards (cm)
m,n exponents in erosion equation
F_0 surface infiltration rate (cm/hr)
V vegetation cover (% or g/m^2)
R decay coefficient for effects of stoniness and bulk density on erosion
b decay coefficient for organic matter at depth
a scaling factor relating vegetation cover to surface infiltration rate
z_1 profile depth to bedrock
B vegetation decay coefficient
K_2 grazing mean consumption rate (g/m^2 per animal)
G grazing, number of animals
V_{max} climatically determined regional vegetation maximum
V_r edaphically and climatically determined vegetation cover
P_1 constant in logistic growth equation
P_2 coefficient reflecting reduction in plant growth per centimetre of soil due to erosional nutrient and organic matter losses
P_5 scaling factor for vegetation growth
P_6 plant water consumption coefficient
x distance from divide
F_z Infiltration rate at depth z
Θ Soil moisture content

REFERENCES

Biot, Y. (1988). *Forecasting productivity losses by sheet and rill erosion in semi-arid rangeland*. Unpublished PhD Thesis, University of East Anglia, Norwich.

Dunne, T., W. Dietrich and Brunengo, M. J. (1978). Recent and past erosion rates in semi-arid Kenya. *Z. Geomorph, Supplbd.*, **29**, 130–40.

Eagleson, P. S. (1978). Climate, vegetation and soils. *Water Resources Res.*, **14**(5), 705–76.

Elwell, H. A. and M. A. Stocking (1976). Vegetative cover to estimate soil erosion hazard in Rhosesia. *Geoderma*, **15**, 61–70.

Horton, R. E. (1945). Erosional development of streams and their drainage basins: hydrophysical approach to quantitative morphology. *Bulletin of the Geological Society of America*, **56**, 275–370.

Kirkby, M. J. (1985). A basis for soil profile modelling in a geomorphic context. *Soil Science*, **36**, 97–122.

Kirkby, M. J. and Neale, R. H. (1987). A soil erosion model incorporating seasonal factors. In V. Gardiner (ed.), *International Geomorphology 1986*, vol. 2, Wiley, pp. 189–210.

Lee, C. R. and J. G. Skogerboe, 1985. Quantification of erosion control by vegetation on problem soils. In S. A. El-Swaify, W. C. Moldenhauer and A. Lo (eds.), *Soil Erosion and Conservation*, Soil Conserv. Soc. Am., Ankenny, pp. 437–44.

Noy-Meir, I. (1978). Stability in simple grazing models: Effects of explicit functions. *J. Theor. Biol.*, **71**, 347–80.

Thornes, J. B. (1985). The ecology of erosion. *Geography*, **70**(3), 222–36.

Thornes, J. B. (1988). Erosional equilibria under grazing. In J. Bintliff, D. Davidson and E. Grant (eds.), *Conceptual Issues in Environmental Archaeology*, Edinburgh University Press, pp. 193–210.

Thornes, J. B., C. F. Francis, F. Lopez-Bermudez and A. Romero-Diaz (in press). Reticular overland flow with course particles and vegetation roughness under Mediterranean conditions. In J. Luis Rubio (ed.), *Proceedings of the Valencia Conference on Mediterranean Desertification*, European Community.

Whittaker, R. H. and P. L. Marks (1975). Methods of assessing terrestrial productivity. In H. L. Leith and R. H. Whittaker (eds.), *Primary Productivity of the Biosphere*, New York, Springer-Verlag.

5 Geomorphic Effects of Vegetation Cover and Management: Some Time and Space Considerations in Prediction of Erosion and Sediment Yield

STANLEY W. TRIMBLE

Department of Geography, University of California, Los Angeles

SUMMARY

Predictions of erosion and sedimentation based on vegetative cover and management are subject to error from variance in time and space. Soils may be in a deteriorating or improving phase which will significantly alter rates of erosion and sedimentation over time. Relict rills and gullies can enhance runoff while continuous fields may increase erosion potential. The downstream cascade and storage of sediment presents great variance with time and basin area. Good vegetation cover may actually enhance downstream sedimentation, at least temporarily. All of these topics require much additional research.

INTRODUCTION

Vegetative cover is the variable controlling erosional activity which is most subject to human manipulation and is therefore an important component of any predictive model. However, such models may not always give satisfactory predictions because relatively unknown long-term factors may be operative. Prediction becomes even more difficult as one moves downstream because the variable of storage flux is increasingly introduced to the sediment cascade. Some

Vegetation and Erosion
Edited by J. B. Thornes
©1990 John Wiley & Sons Ltd

longer-term considerations, all of which are time and/or space-dependent, are discussed in the following sections.

SOIL CONDITION

Vegetational cover may be altered radically within a short time, but biophysical changes *within* the soil which also affect erosion rates may take long periods of time. Soil under the conditions of good plant cover and stable conditions may improve with time by accumulating organic material, increasing floral and faunal activity with possible macropore formation, undergoing structural improvement with corresponding decreases in bulk density, increases of infiltration capacity and water-transmission capacity, and decreases of erosion potential (Trimble, 1976, 1988). However, such improvements (termed 'residual effects' by Wischmeier and Smith, 1978) take time and there is little research showing the rates and extent of such improvements. Although several primary hydrologic plot studies demonstrating the residual effect are cited in this chapter, those studies were *not specifically designed* to evaluate residual effects. Thus, the data and conclusions reported here are ancillary to the goals of these studies

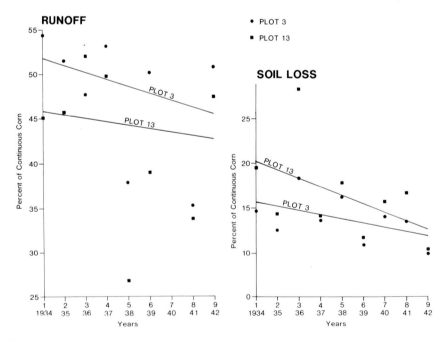

Figure 5.1 Annual runoff and soil loss from conservation rotation as percentage of that from continuous corn (two plots, 3 and 13), 1934–42, Zanesville, Ohio. Lines fitted by eye. (Data from Borst *et al.* 1945.)

and the fact that we have even such poor information appears to be simply fortuitous.

Borst, McCall and Bell (1945) studied the effects of crop rotations on soil erosion and runoff in Ohio, USA. Although residual time effects were not part of their research objectives, some of their data for a nine-year period are useful for this purpose (Figure 5.1). The trends of their data indicate that the runoff (overland flow) from the conservation rotation averaged about 49 per cent of that from continuous corn during the first year, but decreased to about 44–45 per cent by the ninth year. The trend of soil loss from the conservation rotation as compared to continuous corn decreased from about 18 per cent to about 12 per cent during the nine-year period. The work of Harrold *et al.* (1962) indicated similar results. The poor statistical fit illustrates part of the problem (Figure 5.1). So many variables enter the process that isolating 'residual effects' is difficult and it appears that most researchers have not chosen to deal with it.

Hendrickson *et al.* (1963) compared a conservation rotation to continuous corn over a six-year period in Georgia, USA. They found that runoff (overland flow) from soil placed under conservation practices decreased from 100 per cent to about 40 per cent of that from continuous cotton while soil loss decreased from 100 per cent to about 10 per cent of that for continuous cotton during the period (Figure 5.2). For both runoff and soil loss, the decreases took the form of a negative exponential decay function. Unfortunately, these studies included few measurements or speculations on edaphic changes which may have caused these gradual improvements, but Hole (1981) and Trimble (1988) have reviewed many possible factors. For the condition where land reverts from tillage to forest, Dissmeyer and Foster (1980, 1981) point out that soil 'reconsolidates'

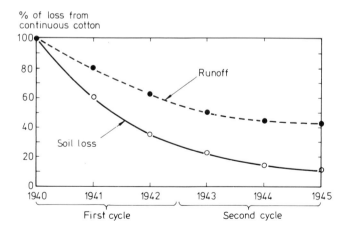

Figure 5.2 Annual runoff and soil loss from cotton plots under a three-year rotation of (1) oats–Kobe lepedeza, (2) volunteer lespedeza, and (3) cotton, compared with those from continuous cotton, 1940–45, Watkinsville, Georgia. (From Hendrickson *et al.*, 1963.)

(particles regain cohesion) over a period of six years or so, so that erosion is reduced.

Conversely, changes to poorer vegetative cover and management may lead to a deterioration of the soil and its associated biophysical attributes. For example, as nutrients are depleted and organic material is oxidized, the biochemical structural bonds are weakened. Plowing, heavy traffic, and grazing continue to break down the peds, compacting the soil, decreasing infiltration, increasing overland flow, and increasing erosion. As loamy, well-structured topsoil is eroded, infiltration capacity is further reduced so that runoff and erosion are accelerated. But this all takes considerable time.

The deterioration phase has happily received more attention. Dissmeyer and Foster (1981), for example, recognized that the erosional response of a soil depends in part upon the soil's recent history. Most workers examining this phenomenon have concentrated on the effects of organic material (OM) content in the soil, a content which generally decreases in the deterioration phase. Morgan (1985) cites studies which indicate decreases of 40–50 per cent of OM during periods of 35–60 years of cultivation. Greenland, Rimmer and Payne (1975) suggest that some soils with less than about 3.5 per cent OM can have unstable aggregates but this variable can be soil specific. Morgan (1985) points out that while there is some evidence of a linear relationship between soil erodibility and OM, the evidence is slight. However, a linear relationship is indicated by the commonly used soil erodibility nomograph (Wischmeier and Smith, 1978).

Luk (1979) supported the common contention (e.g. Bryan, 1974) that aggregate stability is a significant factor on splash and wash erosion and found that, for up to a content of about 7 per cent, OM was essential to the formation and maintenance of soil aggregates. Cultivation had reduced the OM in many samples of prairie soils and, for these, there was a good positive correlation of OM and aggregate stability (Luk, 1979). In Luxembourg, Imeson and Jungerius (1974) found that soils under forest cover had more aggregate stability than the soils of nearby areas which had been cultivated. For the farmland samples there was a good correlation between OM and resistance to splash erosion. The overall impression from the literature is that there is a good and generally linear relation between OM and erodibility. This seems to take us back the soil erodibility nomograph (Wischmeier and Smith, 1978).

With regard to infiltration alone, Wischmeier and Mannering (1965) found that incorporating crop residues into the soil over a 10–15 year period reduced runoff by almost one-half. This effect would presumably reverse with cropping practices which reduce OM. Wischmeier (1966) presented data from a large experiment in Indiana, USA, showing that percent OM explained about 45 per cent of runoff for a standardized storm of 6.35 cm. The predictive equation for runoff from the standardized storm is:

$$Y = 4.01 - 0.56X$$

where
$$Y = \text{runoff, cm}$$
$$X = \text{percent OM}$$

The equation indicates that less than half of that storm precipitation would be infiltrated with 1 per cent OM present in the soil, but over three-quarters would be infiltrated with 4 per cent OM. It is to be noted that OM is often associated with faunal activity such as that of earthworms, insects, and small mammals, and this activity often significantly enhances infiltration (see reviews in Hole, 1981 and Trimble, 1988) but such faunal development might itself lag behind the habitat enhancement of increasing OM. Also, OM may promote, and be a surrogate for, other soil qualities such as good structure and lower bulk density which themselves promote greater infiltration.

I have found few plot studies which demonstrate the deterioration phase. Copley *et al.* (1944) burned annually the surface litter of a forested plot in North Carolina, USA, for 9 years (1932–40). Runoff increased from less than 0.4 per cent of precipitation during the first 2 years to almost 20 per cent in

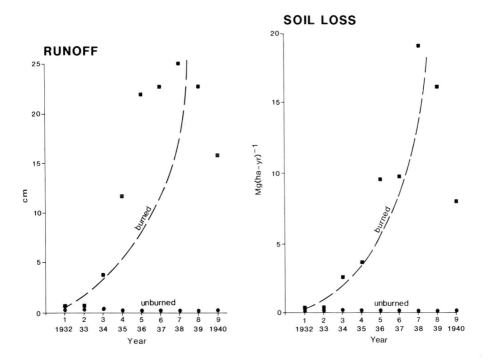

Figure 5.3 Time change of annual runoff and erosion, burned and unburned plots, 1932–40, Statesville, North Carolina. Curves fitted by eye. (Data from Copley *et al.*, 1944.)

the last 3 years, while soil loss increased from about 0.05 Mg/ha in the first 2 years to about 14 Mg/ha in the last 3 years (Figure 5.3). The authors did not comment on causes for the rapid increases and these need to be addressed.

In another plot study in North Carolina, Copley, *et al.* (1944) followed 7 years of sod with 2 years of continuous cotton. During the latter period, runoff and soil loss were only 1.0 per cent and 26.1 per cent respectively, of that from a nearby control plot which had been in continuous cotton for the entire experimental period of nine years. From a management viewpoint, the idea is to precede destructive land use (such as erosive crops) with one or more years of a soil-building land use, the best of which is sod or forest. Short-term effects of this type are incorporated in the Universal Soil Loss Equation (USLE) and the Erosion-Productivity Impact Calculator (EPIC) models. For example, when an established meadow is plowed and planted to maize, erosion approximately doubles by the third year of maize (Wischmeier and Smith, 1965) but further effects are not predicted after the third year for cropland. More recent revision (Wischmeier and Smith, 1978) make adjustments for only 2 years. However, Dissmeyer and Fostor (1980, 1981) believe that the time effect lasts up to 6 years under forest conditions because the organic material takes longer to decay.

For both the improving and the deterioration phases, it appears that there is a lag time of several years, but decades may be the proper time scale in some cases. More research is clearly needed.

HYDRAULIC EFFICIENCY

As increasing proportions of a basin are cleared, or are placed under poorer vegetative cover and/or management, surface runoff amounts and flow lengths become greater. As is well-known, rills and gullies can develop over time which more efficiently convey water, leading to increased erosion of uplands and possibly of downslope channels. Conversely and not as well known, such channels tend to persist as relicts long after an area is placed under better vegetative cover and management. They can continue to convey water even under conditions of partial-area runoff because they allow interflow to become open-channel flow. This is especially common in areas reclaimed by forest which were formerly rilled and gullied such as in the southeastern United States. Such forests may produce inordinate amounts of runoff for decades after vegetation becomes well established. The channels may themselves erode, but it is common for the channel to be completely vegetated and stable. In that case, the energy may be cascaded onto a more vulnerable downslope area such as a cultivated field. Thus recently created landforms can themselves affect biological factors, and measures of vegetation stand, condition, and management may not accurately assess the physical processes that could occur.

MANIFESTATIONS AT THE BASIN SCALE

The discussion thus far has concerned only the field or hillside scale, but the lag effects are also seen downstream at the basin scale. Working in southwestern Wisconsin, Potter and Faulkner (1987) have used catchment response (lag) times to predict flooding potential. Potter (personal communication, 1988) found that the increased stream response from the crop expansion of the 1970s did not become significant until several years later. In the region as a whole, however, William Krug (personal communication, 1988) has found evidence of increasingly greater base flow over the past 40 years which I would partially attribute to the conservation efforts of the 1930s, 1940s and 1950s. In the upper Mississippi River valley over the past 15 years, I have found increasing numbers of rejuvenated groundwater springs which had not heretofore flowed within memory, except of the oldest residents. This phenomenon may be, in part, the delayed result of conservation measures begun in the 1930s and which were generally in place by the early 1960s. The implication is that increasing amounts of infiltration, as evidenced by increasing base flow, is allowing progressively smaller stormflows. Thus, the potential for geomorphic work appears to be progressively decreasing.

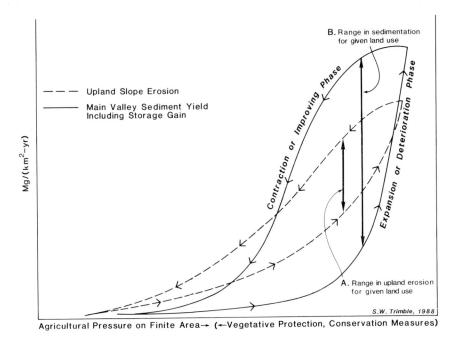

Figure 5.4 Differential relations of upland slope erosion and downstream sediment yield to agricultural pressure on finite areas (or vegetative protection, soil, conservation measures). Severe scenario, Coon Creek, Wisconsin, 1853–1975. (Based on a study by Trimble and Lund, 1982.)

Also working at the basin scale, Trimble and Lund (1982) found a pronounced lag in the response of upland erosion to erosive land use (vegetative cover weighted with conservation practices). As agriculture spread in the basin, it took years for the soil condition to deteriorate as already outlined. As fields were expanded, steeper slopes were cultivated so that the increasing surface flow had longer flow routes, topsoils were eroded, and there was a positive feedback so that erosion was accelerated. Conversely, as vegetative cover was expanded and improved in recent decades, it took time for soils to regain their good hydrologic condition and for upland waterways to heal so that, again, a lag was encountered. Thus a given vegetative cover during the expansion phase generally caused less erosion than in the contracting phase. The overall result is a hysteresis loop (Figure 5.4, loop A). Trimble (1988) has generalized this hysteresis model for other scenarios. Knox (1987) presented sedimentation data which also suggest a lag in sedimentation rates behind changes of land use, although no land use data were given. Knox (1972) shows a diagram which suggests that the lag phenomenon was important during the Holocene. Although partly speculative, it indicates that the rate of geomorphic work continued to change long after vegetation had attained equilibrium following an abrupt climatic change.

BASIN SIZE

The general model of unit sediment yield decreasing with basin size is well known if not well understood (Walling, 1983). One explanation of this widespread phenomenon is storage accretion. When vegetative cover is poor and upland erosion supplies streams with more sediment than they can transport, sediment is stored as colluvium and alluvium. Since storage opportunity generally increases as a power function of basin size, so may storage increase under the above conditions. Thus, the more easily measurable geomorphic effects of a vegetative cover may decrease downstream.

Conversely, the establishment of a good vegetative cover on disturbed land may lead to downstream sediment yields being augmented from storage. If vegetative cover curtails upland erosion without commensurate decreases of storm runoff, then downstream colluvium and alluvium may become unstable and be eroded, maintaining downstream sediment yields at high levels. This principle was identified as a potential problem by Happ, Rittenhouse and Dobson (1940) and shown to be operating on a regional basis by Trimble (1974, 1977).

Thus, the general unit sediment yield model mentioned above is not a universal model in time and space because it was based on data from disturbed basins where sediment storage was probably occurring (Trimble, 1977). As these basins are given improved vegetative cover, it may be that the rate of diminution of sediment yield with basin size is decreased and, in some of these basins, unit sediment yield may now actually be increasing with basin size (Trimble, 1977, 1983). Indeed, Church and Slaymaker (1989) show unit sediment yield in British Columbia to be increasing downstream with augmentation coming from Holocene deposits.

Schumm (1976) argues that sediment deposits, especially alluvium, can become unstable without changes of land use. Even with vegetative cover thus *held constant*, sediment yield can abruptly increase, at least in semi-arid regions.

Trimble and Lund (1982) have established the relationship of downstream sediment accumulation to land use in the Coon Creek basin of Wisconsin. They found that, with an increase of poor vegetative cover and management, there was a strong lag in downstream rates of sediment accumulation, which was partially a function of the lag in erosion rates and partially the result of upstream sediment accumulation. With improved vegetative cover, there was again a response lag of sedimentation, again due to a lag in erosion rates, but also to the release of sediment from alluvial storage. Thus, the overall process was again a hysteresis loop (Figure 5.4, loop B). The hysteresis effect was enhanced late in the deterioration phase by increased runoff from overgrazing which increased stormflow and helped transfer sediment downstream (Trimble and Lund, 1982). Despite these dynamic changes of storage, Trimble (1983) showed that there was little temporal change of downstream sediment yield notwithstanding the great changes of vegetative cover and management for the basin as a whole.

CONCLUSIONS

Geomorphic effects ultimately caused by changes of vegetative cover and management are not always directly linked to those vegetative changes in time and space. This makes uncertain predictions of erosion and sediment yield based on vegetation alone. The hysteresis models presented suggest that prediction presents a problem analogous to that of predicting soil water content from soil water capillary tension. Since that is also a hysteretic function, one must know (a) whether the soil is in a wetting or drying phase, and (b) the non-linear relationship of the two variables for each phase. Similarly, to predict geomorphic work from vegetative cover, one must know (a) the recent land use and geomorphic history of a field or basin (e.g., improving or deteriorating phase of soils, relative drainage density, gain or loss phase of sediment storage within the system), and (b) for whatever phase, the non-linear relationship between vegetative cover and geomorphic work. Published research is hardly more than adequate to outline the lags and transfers of mass and momentum that complicate predictions and is inadequate to describe all the physical and biological processes involved but it is clear that they are far more complicated than vegetation cover alone. Floral and faunal developments *within* the soil, for example, may sometimes prove to be as important as the vegetation *covering* the soil. Further research is needed at all scales: basin, hillslope, field, and soil. Biological scales range from micro to mega. This research frontier is a promising realm of collaboration for geographers, agronomists, agricultural engineers, and soil scientists.

ACKNOWLEDGEMENTS

I thank Chase Langford for the illustrations and Alesia Haymon for typing. George R. Foster and Francis D. Hole made important improvements, but any errors remain my own.

REFERENCES

Borst, H. L., A. G. McCall and F. G. Bell (1945). Investigations in erosion control and the reclamation of eroded land at the Northwest Appalachian Conservation Experiment Station, Zanesville, Ohio, 1934–1942. *USDA Technical Bulletin*, 888.

Bryan, R. B. (1974). Water erosion by splash and wash and the erodibility of Albertan soils. *Geografiska Annaler*, **56A**, 159–81.

Church, M. and O. Slaymaker (1989). Disequilibrium of Holocene sediment yield in glaciated British Columbia. *Nature*, **337**, 452–4.

Copley, T. L., L. A. Forrest, A. G. McCall and F. G. Bell (1944). Investigations in erosion control and reclamation of eroded land at the Central Piedmont Conservation Experiment Station. Statesville, NC, 1930–40. *USDA Technical Bulletin*, 873.

Dissmeyer, G. E. and G. R. Fostor (1980). A guide for predicting sheet and rill erosion on forest land. United States Department of Agriculture US Forest Service *Technical Publication*, SA-TP11.

Dissmeyer, G. E. and G. R. Fostor (1981). Estimating the cover-management factor (C) in the universal soil loss equation for forest conditions. *Journal of Soil and Water Conservation*, **36**, 235–40.

Greenland, D. J., D. Rimmer and D. Payne (1975). Determination of the structural stability class of English and Welsh soils using a water coherence test. *Journal of Soil Science*, **26**, 294–303.

Happ, S. C., G. Rittenhouse and G. C. Dobson (1940). Some principles of accelerated stream and valley sedimentation. *USDA Technical Bulletin*, 695.

Harrold, L. L., D. L. Brakensiek, J. L. McGuiness, C. R. Amerman and F. R. Dreibelbis (1962). Influences of land use and treatment on the hydrology of small watersheds at Cashocton, Ohio, 1938–1957. *USDA Technical Bulletin*, 1256.

Hendrickson, B. H., A. P. Barnett, J. R. Carreker and W. E. Adams (1963). Runoff and erosion control studies on Cecil soil in the Southern Piedmont. *USDA Technical Bulletin*, 1261.

Hole, F. D. (1981). Effects of animals on soil. *Geoderma*, **25**, 75–112.

Imeson, A. C. and P. D. Jungerius (1974). Landscape stability in the Luxembourg Ardennes as exemplified by hydrological and (micro)pedological investigations of a catena in an experimental watershed. *Catena*, **1**, 273–96.

Imeson, A. C. and P. D. Jungerius (1976). Aggregate stability and colluviation in the Luxembourg Ardennes: an experimental and micromorphological study. *Earth Surface Processes*, **1**, 259–71.

Knox, J. C. (1972). Valley alluviation in southwestern Wisconsin. *Annals AAG*, **62**, 401–10.

Knox, J. C. (1987). Historical valley floor sedimentation in the upper Mississippi Valley. *Annals AAG*, **77**, 224–44.

Luk, S. H. (1979). Effects of soil properties on erosion by wash and splash. *Earth Surface Processes*, **4**, 241–55.

Morgan, R. P. C. (1985). Soil degradation and erosion as a result of agricultural practice. In K. S. Richards, R. R. Arnett and S. Ellis (eds), *Geomorphology and Soils*, George Allen & Unwin, London, pp. 379–95.

Potter, K. W. and E. B. Faulkner (1987). Catchment response time as a predictor of flood quantities. *Water Resources Bulletin*, **23**, 857–61.

Schumm, S. A. (1976). Unsteady state denudation. *Science*, **191**, 871.

Trimble, S. W. (1974). *Man Induced Soil Erosion on the Southern Piedmont 1700–1970*. Ankeny, Iowa: Soil Conservation Society of America.

Trimble, S. W. (1976). Sedimentation rates in Coon Creek Valley, Wisconsin. Oral presentation at the Inter-Agency Sedimentation Conference, Denver, Colorado, March 1976.

Trimble, S. W. (1977). The fallacy of stream equilibrium in contemporary denudation studies. *American Journal of Science*, **277**, 876–87.

Trimble, S. W. (1983). A sediment budget for Coon Creek basin in the Driftless Area, Wisconsin, 1853–1977. *American Journal of Science*, **283**, 454–74.

Trimble, S. W. (1988). The impact of organisms on overall erosion rates within the catchments in temperate regions. In H. Viles (ed.), *Biogeomorphology*, Basil Blackwell, Oxford.

Trimble, S. W. and S. W. Lund (1982). Soil conservation and the reduction of erosion and sedimentation in the Coon Creek basin, Wisconsin. *USGS Professional Paper*, 1234.

US Department of Agriculture, Soil Conservation Service (1971). *Sedimentation in National Engineering Handbook*, Sec. 3, pp. 1–19.

Walling, D. E. (1983). The sediment delivery problem. *Journal of Hydrology*, **69**, 209–37.

Wischmeier, W. H. (1966). Relation of field plot runoff to management and physical factors. *Proceedings, Soil Science Society of America*, **30**, 272–77.

Wischmeier, W. H. and J. V. Mannering (1965). Effect of organic matter content of the soil on infiltration. *Journal of Soil and Water Conservation*, **20**, 150–2.

Wischmeier, W. H. and D. D. Smith (1965). Predicting rainfall-erosion losses from cropland east of the Rocky Mountains. *USDA Agricultural Handbook*, 282.

Wischmeier, W. H. and D. D. Smith (1978). Predicting rainfall losses—A guide to conservation planning. *USDA Agriculture Handbook* 537.

6 Recurrence of Debris Production under Coniferous Forest, Cascade Foothills, Northwest United States

ANTONY R. ORME
Department of Geography, University of California
Los Angeles

SUMMARY

Recent major debris avalanches and debris torrents in the Cascade foothills, northwest United States, have created a need to know more about the recurrence of such events under coniferous forest and the extent to which forest practices may have changed the frequency and magnitude of related debris production. Six approaches to the problem are pursued. Of these, historical records and climatic data are misleading. However, analyses of tree-rings, channel deposits, alluvial fan stratigraphy and lake sediments reveal a record of debris production over 3400 years. This record indicates that major debris avalanches and debris torrents are recurrent processes in steep drainage basins beneath coniferous forest. The magnitude of debris production is not significantly changed by timber harvesting alone, but may be locally accentuated by slope failures associated with abandoned logging roads.

INTRODUCTION

On January 9 and 10, 1983, a major winter storm moved off the Pacific Ocean and stalled over Whatcom and Skagit counties in northwest Washington. At the coast less than 5 cm of rain fell during these two days. Inland, however, in the forested foothills west of the Cascade Range, rainfall was both more intense

Vegetation and Erosion
Edited by J. B. Thornes

and more persistent, with valley stations recording over 15 cm in two days and more than $0.5\,\mathrm{cm\,hr^{-1}}$ during the middle of the period. At higher elevations, orographic enhancement and an antecedent snow cover greatly increased the flood potential of this storm.

The geomorphological impact of the storm was dramatic. Slope failures occurred over a broad area of forest terrain, mainly between the Nooksack and Skagit rivers. These triggered numerous debris avalanches which in turn generated debris torrents that moved swiftly downstream, discharging abundant sediment and organic debris onto alluvial fans and into lakes and major rivers. Widespread property damage occurred where alluvial fans had been developed for agricultural, residential and recreational purposes. Some property owners attributed the damage mainly to the effects of forest practices in upland drainage basins. Others understood the unusual intensity and persistence of the storm rainfall and the hazards of alluvial fan occupance.

From the scientific perspective, the geomorphological impact of the storm created a need to know more about the spatial and temporal aspects of debris avalanches and debris torrents. In a spatial context, it is important to understand the relation between debris production and vegetation type and age in forested drainage basins, particularly with respect to the patchwork of vegetation generated by timber harvesting. In a temporal context, it is important to know whether the nature and volume of debris produced during a major storm are reflections of the natural system or whether they are significantly changed by forest practices in the contributing basins. This chapter addresses the temporal aspect, namely the recurrence of debris production from forested terrain. A companion study will address the spatial context.

In northwest Washington, little is known about the recurrence of major debris-producing events and, prior to this study, no attempt has been made to evaluate a broad range of evidence in this context. There are no sediment-gauging stations in the numerous rugged basins producing the debris, and the few stations on major rivers have mostly short or discontinuous records which integrate sediment yields from very large areas. In the wider forested arena of the Pacific coastal slopes of North America, from northern California to Alaska, there have indeed been numerous studies of the relationship between erosion, sediment yields and forest practices. Many studies have involved monitoring drainage basins before, during and after timber harvesting, with comparable unlogged basins held as controls (e.g. Frederickson, 1970; Swanston and Swanson, 1976; Beschta, 1978). Most studies have demonstrated increased sediment yields shortly after logging, followed by a gradual return to background levels over several years (e.g. Swanson and Dyrness, 1975). Logging roads have been frequently identified as problematic because of the increased potential for slope failure from cut banks, unconsolidated fills and drainage diversions. Explicit or implicit in these studies, however, is the recognition that sediment production from logged-over drainage basins returns to pre-harvest levels as second-growth forest becomes established. There is far less understanding of the long-term record of debris production from forested basins, or how this compares with the output from basins characterized by a patchwork of second-growth forest of variable age and composition.

Using the geomorphological response to the January 1983 storm as a control, this study investigates various evidence for previous storms and related debris production so as to place the 1983 event in temporal perspective and commence construction of a long-term record of erosion and sedimentation for northwest Washington. The approach involves six lines of evidence, namely historical records, climatic data, tree-ring analysis, old channel deposits, alluvial fan stratigraphy and lake sediments, the last three elements supported by radiometric dating of associated organic remains. The emphasis is on major debris-producing events likely to be more clearly recorded in the erosional and depositional record, not on minor events. It is the infrequent major events that appear to accomplish most work over the long term.

The significance of this study is broadly twofold. First, it establishes a long-term record of erosion and debris production over a broad area which allows a better temporal perspective on the relative importance of natural and human-induced changes in the geomorphology of forested terrain. Second, it demonstrates a multi-faceted approach to the problem of debris production from forest environments that has often been limited previously by narrow approaches (e.g. measurement of sediment yields in streams) or restricted areas (e.g. small contiguous drainage basins). This is not to belittle such studies but to emphasize the need for broader, in this case temporal, perspectives.

STUDY AREA

The study area covers 1500 km^2 of mostly rugged forested foothills lying between the Strait of Georgia and the dormant volcano of Mt Baker (3285 m) just west of the main Cascade Range, and between the Nooksack and Skagit rivers (Figure 6.1). The foothills, whose crests rise inland from 500 to 1500 m, comprise several distinct uplands separated by former glacial troughs now floored by streams, swamps and lakes. The 14 drainage basins examined in this study all have mainstream gradients in excess of 100 m km^{-1}, headwater and tributary gradients of 200–300 m km^{-1}, and sideslopes steepening to 30–50°. All 14 basins lie within the area that experienced more than 15 cm of rainfall on January 9–10, 1983.

The northwest part of the area is underlain by the folded and fractured Chuckanut Formation of Eocene age, some 6000 m of fluvial and deltaic arkosic sandstones interbedded with conglomerates, mudstones and minor coal seams (Johnson, 1984). On weathering, these rocks yield sandstone clasts in a non-cohesive granular matrix. To the southeast, the Chuckanut Formation is in fault

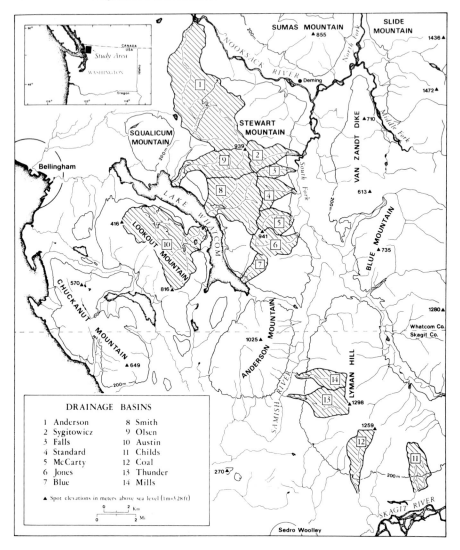

Figure 6.1   Study area, northwest Washington, showing location of principal drainage basins examined

contact with pre-Tertiary phyllites which, on weathering, yield slaty chips and pasty clay-rich residues. In the valleys, bedrock is masked by Pleistocene glacial and glacio-fluvial deposits and by Holocene alluvium.

Average precipitation in the area ranges from 90–120 cm yr^{-1} at the coast, to 160–180 cm yr^{-1} in the Nooksack and Skagit valleys, and probably exceeds 250 cm yr^{-1} in the higher foothills. These averages conceal the effects of infrequent high-intensity rainstorms that approach from the west or southwest and are commonly associated with warming trends in the lower troposphere. The storm of January 9–10, 1983, was one such event.

Coniferous forest in the foothills is dominated by Douglas fir (*Pseudotsuga menziesii*) and subsidiary western hemlock (*Tsuga heterophylla*), with western red cedar (*Thuja plicata*) on lower slopes and other conifers at higher elevations. Periodic lightning-caused fires and windstorms have swept through these forests in prehistoric and historic times affecting the age and vigour of the virgin forest. Timber harvesting and land clearance began along the coast in the mid-nineteenth century and, with the introduction of steam donkeys and logging railroads, spread into nearby valleys and lower foothill slopes after 1880. Around 1900, high-lead cable yarding with steam donkeys made rougher terrain more accessible while the less mobile equipment favoured clearcutting. Since the 1920s, the advent of logging trucks, caterpillar tractors and chainsaws has allowed timber harvesting to penetrate deeper and higher into the Cascade foothills. The most intensive period of logging in the study area was between 1930 and 1950. Today some virgin forest remains but most of the area carries a patchwork of second-growth forest of variable age and composition, with the deciduous red alder (*Alnus rubra*) and bigleaf maple (*Acer macrophyllum*) prominent in recently logged areas pending the emergence of second-growth conifers.

DEBRIS AVALANCHES AND DEBRIS TORRENTS

Debris avalanches are rapid shallow mass movements triggered by several factors that may be grouped in terms of precipitation variables, soil properties, bedrock attitude, slope and sub-colluvial relief. They most commonly occur where, as a result of intense and prolonged rainfall on melting snow, sub-surface drainage converges on bedrock hollows filled with poorly cohesive granular soils and colluvium, on slopes steeper than 28° (Figure 6.2). They are thus common to areas underlain by the Chuckanut Formation, rarer in phyllite areas. Failure most often occurs at the contact between near-cohesionless colluvium and underlying relatively competent bedrock, but may also occur along shear planes within colluvium, notably in the alteration zone along old forest-fire surfaces (Orme, 1987). Debris avalanches accelerate downslope to velocities exceeding 8–10 m s^{-1} and eventually enter stream channels (Figure 6.3).

Debris torrents develop from rapid streamflows, initially low in suspended sediment, which entrain material stored within stream channels or plough into and remobilize debris introduced to channels by debris avalanches and other mass movements. Thereafter a mixture of water, soil, rock and organic debris is supported and further energized by internal fluid turbulence, such that a debris torrent triggered by only 10 m^3 of debris may eventually move more than 10 000 m^3 of material downstream, often in surges reflecting entrainment of

Figure 6.2 Headscarp of a debris avalanche on a 30° slope beneath a mature forest of 120–140-year-old Douglas fir and western hemlock, Smith Creek, Whatcom County. The headscarp (right of figure) revealed much piping, typical of the convergent sub-surface drainage found in the colluvium-mantled swales of the area. Although the headscarp was only 4 m wide and 1.5 m high, the resulting debris quickly gained mass and momentum downslope. (Photo: G. W. Thorsen, September 1984.)

new materials or the breaching of debris jams. As stream gradients lessen and momentum lessens, debris torrents deposit their load on alluvial fans or in nearby lakes and rivers.

In uplands, debris avalanches and debris torrents leave an erosional legacy of avalanache chutes and scoured channels, commonly exposing bedrock. Owing to their episodic development, linked to infrequent major rainstorms, these bare chutes and channels are gradually filled with colluvium and alluvium respectively, and scarred hillslopes are clothed in regenerating forest, at least until the next major storm event again evacuates the accumulated debris. It is the recurrence of this debris production by avalanches and torrents that is now addressed.

RECURRENCE OF DEBRIS PRODUCTION

Historical Records

The best historical measure of debris production is a long-term record of suspended sediment from a gauging station. Such stations do not exist for

Figure 6.3 This old bedrock chute was scoured by a debris avalanche that penetrated advanced second-growth forest in January 1983, removing all colluvial and organic debris en route to join debris torrents pulsing down Smith Creek, Whatcom County. (Photo: A. R. Orme, October 1983.)

drainage basins within the study area, nor is it likely that either automatic or manual samplers could survive debris-torrent surges. In the circumstances, recourse is made to written and oral accounts of uncertain value and to old maps and aerial photographs whose record is usually temporally imprecise.

Local newspapers describe noteworthy floods and debris torrents in January 1855, November 1892, November–December 1893, December 1899, December 1917, November 1932, November 1949 and January 1971. Of these, only the 1917 event appears comparable to the January 1983 event in the study area. For example, the *Bellingham Herald* for December 29, 1917, records that 'the house in which the Nash family lived, on Smith Creek, was washed into the lake by a flood that had been held for hours behind a log jam in the creek.' In a later oral statement, a survivor of that event recalled that the weather had been warm and rainy with some snow in the foothills, and that boulders and large organic debris swept across the Smith Creek alluvial fan in a debris torrent 12 m high to destroy the local railroad bridge. Accounts of the other noteworthy events commonly refer to local landslides, log jams, overbank flooding and deposition, and property damage, but not apparently on a scale comparable to 1917 and 1983.

The above description of the 1917 event is interesting because it states or implies warm, rain-on-snow conditions and the surging of a typical debris torrent once it broke free of a log jam upstream. In recent years it has become common to attribute such log jams to timber harvesting, a belief sometimes supported by the presence of sawn logs in the jams. Long before logging, however, settlers in northwest Washington in the 1850s found most rivers and creeks jammed with large organic debris. The 'Big Jam' on the Nooksack River, noted by government surveyors in 1859 and not removed until 1877, 'extended in one huge conglomeration of logs, stumps and brush for a distance of nearly a mile . . . so long had it been there [that] brush and trees had taken root' (diary quoted in Jeffcott, 1949). In 1881, Morse (1887) noted in the South Fork of the Nooksack River that 'for the next twenty miles there is a jam across the river nearly every mile.' Log jams are important to the development of debris-torrent surges and, because they are recorded before logging began, it must be assumed that debris torrents have long been active in this environment, albeit episodically.

Systematic aerial photography is available for the region from 1943 but, whereas it may reveal avalanche scars and scoured channels in various stages of recovery, the intervals of several years between coverage render this information imprecise. Comparison of old maps may provide a measure of the tracks taken by debris torrents during different surveys, although the timing of change is again uncertain. For example, Olsen Creek redirected its outlet to Lake Whatcom some 460 m to the west between the time of the US Surveyor General's survey of Washington Territory in 1884 and the USGS map of 1905. This change may have been induced or assisted by human activity. To the south, Smith Creek has changed its outlet over a range of 250 m several times between 1884 and today, commonly because debris torrents choke pre-existing outlets.

Climatic Data

With surface winds of $18 \, \text{m s}^{-1}$ and winds of $52 \, \text{m s}^{-1}$ at 5500 m (National Weather Service), the storm of January 9–10, 1983, approached from the southwest and stalled over northwest Washington, yielding heavy rainfall over a wide area. The geomorphological response suggests that this storm was unusual, but the precipitation data alone are less compelling. For example, the maximum 24-hour precipitation at Bellingham had a return interval of only one year (*NOAA Precipitation Frequency Atlas*).

The geomorphological response must therefore be viewed against other variables not immediately apparent. First, a horizontal gradient exists causing precipitation to increase inland without an increase in elevation. Calibration with short-term records indicates that Bellingham rainfall values should be increased by a factor of 1.44 at Smith Creek and 1.63 at Lake Whatcom's southern end. Furthermore, at Nooksack Salmon Hatchery, a station not included in the *NOAA Atlas* but which lies only 122 m above sea level 30 km inland, 18.29 cm of rain fell over the two-day period and the 16 cm that fell over the 24-hour period beginning at 1700 hours on January 9 had a return interval of 80 years. Second, orographic enhancement increases precipitation more or less linearly and conventional equations indicate 24-hour totals of 18–25 cm at higher elevations within the zone of most intense rainfall. Third, the storm in question was the third significant rain producer in January, after events on January 3–5 and 7–8. Nooksack received over 2.5 cm every day except January 6. Thus antecedent moisture in the hillslopes was high prior to January 9. Fourth, on January 5 the snowline lay at 400 m on Stewart Mountain, with 40 cm of snow at 580 m and 75 cm of snow at 900 m (D. Wooldridge, personal communication). At Quillayute on the coast, radiosonde observations indicated that the freezing level dipped to 640 m above sea level on January 8 (it would have been much lower inland) but rose rapidly to over 2000 m on January 9–10. Assuming a snow density of about 0.40 implies that as much as 16 cm of rainfall equivalent may have been available at 580 m and 30 cm at 900 m prior to the storm, although the relatively small areas of upland above 900 m would limit total volume.

Rain-on-snow effects have long been regarded as a major hydrological process responsible for upland erosion and lowland flooding in the Pacific Northwest (Harr, 1981). The snowmelt process is complex but during rainstorms is a result mostly of convection and condensation, heat transfer from warm rain, and long-wave radiation exchange between forest vegetation or low clouds and the snowpack. In the absence of significant wind velocities beneath the forest canopy, these factors are controlled primarily by the vapour pressure gradient and thus air temperature which, as noted, was well above the freezing point of water on January 9–10.

Given the intensity and magnitude of warm rain on melting snow, it appears that considerable water was available on upper slopes and that widespread slope failure and flooding were inevitable. Using Lake Whatcom water levels as a surrogate and conventional peak discharge equations, streamflow from the storm probably far exceeded 100-year recurrence intervals (R. L. Beschta, personal communication). However, it is equally evident that reconstruction of recurrence intervals of both floods and debris production is likely to be very misleading without detailed data on orographic enhancement, temperature gradients, snowpack depth and density, antecedent moisture, as well as debris availability. A comprehensive model could be developed but in view of the large number of assumptions that would need to be made, the model would likely be misleading. In the circumstances, other approaches to the question of debris recurrence should be sought.

Tree-ring Analysis

Colluvial hollows characterized by convergent flows are common sources of debris avalanches. Tree-ring analysis of living trees rooted in the colluvium offers a minimum age for when such hollows were last evacuated, a minimum because such basins may survive beyond one cycle of tree growth. Where hollows contain more or less even-age timber, it is probable that the former tree cover was removed at a specific point in time. Because fire and clearcutting also create even-age second growth, it is important to link an understanding of geomorphology, dendrochronology, fire history and logging practices in the selection of basins to be studied.

Tree-ring increments were obtained from several score colluvial hollows in the study area. In basins evacuated by debris avalanches but in which some colluvium remains for rooting needs, red alders are re-established within months of failure. In basins evacuated to bedrock, establishment of new trees is slower and tree age rarely reflects precisely the year of failure. In both cases, however, the increased light found in exposed avalanche scars promotes rapid initial growth and broad tree rings, but the latter become more compact as competition and a closing forest canopy slow growth. On steep sites prone to frequent failure and in wet areas, alder and associated marsh plants are not normally succeeded by coniferous trees.

Tree-ring analysis of colluvial hollows reveals two peaks in the data. First, a peak in the 1935–45 period occurs in the basins but is also found on nearby hillslopes and old logging roads, and is thus reflective of widespread clearcutting during this period. A second peak in the 1917–22 period is found in numerous basins but not on adjacent hillslopes, many of which carry old timber dating from just after a major fire that selectively swept the area in 1868. This peak indicates widespread slope failure during the December 1917 storm, with subsequent re-establishment of alder, now over-mature, bigleaf maple and conifers. No evidence survives for slope failure between 1868 and 1917. Many failures in January 1983 occurred within 1917 failures, such that recent avalanche scars are rimmed with 65–70-year-old forest contained in turn within 115–120-year-old forest.

Tree-ring data may also indicate a minimum age for when stream channels last scoured. For example, several tributary channels contain Douglas fir dated to the 1850s which, owing to lower valley locations, escaped the 1868 fire. In two Smith Creek tributaries, massive Douglas fir and western red cedar

260–280 years old, growing in mid-channel, effectively checked scour in January 1983, creating debris jams 6–10 m high. In an area where forest growth, decay and regeneration are relatively rapid, it has not been possible to extend the tree-ring record back beyond AD 1700.

Debris-torrent Deposits in Channels

By nature, debris torrents entrain and remove most pre-existing debris in their paths, leaving channels scoured to bedrock. Thus residual accumulations of debris are rare within upland channels and sequences of debris-torrent deposits are exceptional. A search of the area found an unambiguous in-channel sequence in only one locality, in Anderson Creek 3 km NNW of the 939 m summit of Stewart Mountain (Figure 6.1). Over a 100 m stretch of channel, 1–2 m of 1983 debris rests upon an older structureless deposit, 2–4 m thick, typical of a debris torrent. Two tree fragments at similar levels within the lower deposit yielded ^{14}C ages of 1150 ± 80 years BP (1950) and 1110 ± 80 years BP (Teledyne Isotopes I-13,737 and I-13,738, courtesy of G. W. Thorsen). Thus a major debris torrent passed through Anderson Creek around AD 820.

Alluvial Fan Stratigraphy: Distributary Deposits

Alluvial fans are natural depositories for debris-torrent materials, at least for that portion of the load not flushed into lakes and major rivers. Despite lateral and vertical changes in deposition, alluvial fans should thus retain some record of erosion and debris-torrent occurrence in their contributing drainage basins. To test this, an investigation was made of the Mills Creek alluvial fan which was the recipient of a devastating debris torrent in January 1983 (Figure 6.1). The contributing basin is less than 1.8 km wide and descends 4 km westward from 1295 m on Lyman Hill to the fan apex at 95 m. The symmetrical fan has a 700 m radius and descends at 3° to the Samish River 80 m above sea level. The upland basin, where clearcutting began in the late 1930s, is underlain by phyllite which weathers readily along cleavage planes to provide slaty chips to the stream load. The alluvial fan, characterized by slaty silt loams, was first logged in 1889–90 to provide right of way for a local railroad, and then progressively cleared for agriculture over the next 70 years.

Using caterpillar tractor and backhoe in October 1985, five shallow trenches were excavated to depths of 5 m in a radial pattern across mid-fan. As expected, there was much stratigraphic variation between trenches. However, cross-correlation established the occurrence of seven discrete debris-torrent or flood deposits and three distinct palaeosols within 5 m of the surface. The lowest palaeosol, 3.8 m below the surface in Trench 4, was a mottled grey-brown silty clay with abundant charcoal that yielded a ^{14}C age of 1720 ± 90 years BP or around AD 230 (Beta-15256). This palaeosol had developed in an eighth flood deposit whose base was not observed. Table 6.1, showing the sequence in Trench 4, exemplifies the type of data observed. Debris-torrent deposits are normally distinguished from flood deposits by the paucity of bedding and the presence of large clasts and organic debris throughout. Soil development represents periods of quiescence between events. Erosion surfaces at frequent intervals indicate that the debris torrent or flood reaching the fan had sufficient residual energy to erode and entrain at least a portion of a pre-existing deposit. The nature and rate of alluvial fan aggradation revealed by these five trenches are discussed in detail elsewhere (Orme, 1989).

Without better age control it is impossible to date individual events or assess the time interval between successive events. Suffice it to say that a reasonably long episode of stability and forest pedogenesis ended some 1720 ± 90 years ago, perhaps by fire, perhaps by *in situ* carbonation of organics, and was followed by a debris torrent and flood deposit. The last-but-one of the events recognized was a major debris torrent followed by weak soil development and subsequent agricultural disturbance. The torrent is probably attributable to the 1917 storm event. It is larger in mass than the uppermost deposit at the site which is clearly attributable to the 1983 event.

Lake Sediments

Lake Whatcom is a natural sump for sediment and organic debris flushed from several drainage basins in the study area (Figure 6.1). The lake occupies a NW–SE trending glaciated trough, 16.5 km long and up to 1.7 km wide. Its bedrock floor lies well below sea level but, owing to late Pleistocene and Holocene sedimentation, its present floor extends only to 10 m below sea level. The lake has a surface elevation around 95 m above sea level, an area of 2036 ha, and a volume of $9.21 \times 10^8 \, m^3$. Its outlet to Bellingham Bay through Whatcom Creek is now controlled by a low dam. The lake bottom comprises four basins: two small shallow basins to the northwest and two large deep basins farther southeast, the latter containing 95 per cent of the water and separated in part from one another by convergent fan deltas of Smith and Austin creeks, possibly resting on a bedrock sill (Figure 6.4). It was to be anticipated that the lake sediments would reveal episodes of greater and lesser sedimentation. In January 1983, turbidity levels rose from around 1 NTU before the storm at all levels to 12.8 NTU at the surface and 212 NTU at 180 m in mid-January (Brakke, 1984). Attempts to assess recent sedimentation rates using ^{137}Cs and ^{210}Pb have been confounded by the mobility of the former and by sampling problems.

To evaluate the longer-term sedimentation record, a coring program was conducted in August 1984 using a 2.5 m gravity piston corer from a barge. Drilling locations were established by triangulation to shore stations and water depths recorded by electronic sounding. The carefully devised sampling plan involved retrieving ten cores off Smith Creek, together with two cores from

Table 6.1 Mills Creek alluvial fan: stratigraphy in Trench 4

Depth from surface (m)	Dominant composition	Interpretation
0–0.15	Grey phyllite sand and gravel	Debris torrent, January 1983
0.15–1.16	Poorly sorted grey phyllite sand and gravel with truncated and disturbed *palaeosol* (plough layer) in upper 0.25 m	Debris torrent, December 1917 (?)
1.16–1.28	Grey phyllite fine sand and silt with oxidation and clay skins, *palaeosol*	Flood deposit
1.28–2.00	Unsorted phyllite gravel with abundant erratic pebbles and root casts	Debris torrent
2.00–3.05	Graded and moderately well bedded phyllite sand and gravel with erratic pebbles	Debris torrent
3.05	Wedge of well bedded to structureless phyllite sand and silt, thickening from 0 to 0.15 m	Flood deposit
3.05–3.66	Poorly sorted, poorly bedded to structureless sand and gravel, with coarse gravel lenses	Debris torrent with some flood deposition
3.66–4.88+	Unsorted phyllite sand and gravel passing upwards into bedded sands with *palaeosol* including charcoal fragments	Debris torrent passing up into flood deposit. Palaeosol dated at 1720 ± 90 B.P.

∼∼∼∼ = erosion surface (unconformity) between sets

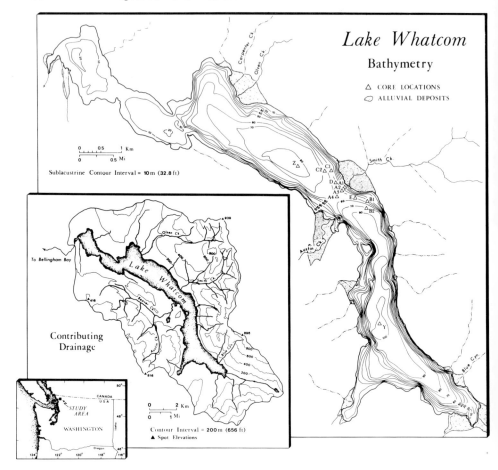

Figure 6.4   Lake Whatcom basin, northwest Washington, showing İake bathymetry, core locations and contributing drainage. Cores Y and Z were retrieved from the deepest southern and northern basins respectively

the flat floors of the deep basins, named Y and Z in Figure 6.4. Cores 5 cm in diameter were retrieved in plexiglass tubes, X-rayed, opened for analysis, and logged. Only cores Y and Z are discussed in this context because they are far enough removed from individual drainage basins to provide integrated data for a broader environment, and because in these locations their stratigraphy is less likely to be disturbed by subaqueous sediment flows. Except in the top few centimetres, there was little variation in compaction throughout the cores.

Core Y retrieved 1.96 m of sediment from the deepest part of the most southerly basin at 101 m below the surface. It comprised a laminated mud with a good mix of organic and inorganic sediment but without significant sand layers, the latter a reflection of distance from coarse sediment sources. X-ray analysis revealed 48 significant events and numerous minor ones, but no reliable datable material was recovered.

In contrast, Core Z retrieved 2.03 m of material from 85 m below the surface in the lake's centre and included wood shards 3–5 cm above the core's base, that is at 1.98 to 2.00 m into the lake's bottom. The wood was dated by ^{14}C to 3370 ± 100 years BP, or about 1420 BC (Radiocarbon Dating Laboratory, Washington State University). The sediments again comprised well-laminated muds that revealed 43 distinctly important events and many lesser ones. In addition, six very large events were represented by organic-rich sands or sandy silts and muds, some with small wood shards. An accumulation of 200 cm of sediment in 3370 years gives a mean sedimentation rate of 1 cm every 16.85 years. Although fully cognizant of the dangers of this assumption, this value would place the 1 cm thick silty sand with wood found 67 cm down the core around 1129 BP, or within the range of the debris torrent recognized in nearby Anderson Creek. A very fine sand layer found 98 cm down the core would place deposition around 1651 BP, or within reasonable distance of the burial of the Mills Creek palaeosol at 1720 ± 90 BP. Such correlations are suggestive, not conclusive.

Including the basal wood shard occurrence, seven very large events in 3370 years represent a recurrence interval of 481 years. Adding the 43 strong laminations to these seven events gives a major event every 67 years. Coincidentally, the 1917 and 1983 events, certainly the largest this century based on other evidence, are 66 years apart.

INTERPRETATIONS AND CONCLUSIONS

Using several lines of evidence, it has been possible to assemble a record of debris production dating back nearly 3400 years (Figure 6.5). During this time, debris avalanches and debris torrents have been recurrent features of the landscape. Of the major events, apart from January 1983, that of December 1917 is emphasized by historical records and tree-ring analysis. A major debris torrent occurred in Anderson Creek around AD 820 and may be reflected in Core Z from Lake Whatcom. The burial of a well-developed palaeosol with charcoal occurred on the Mills Creek fan sometime after AD 230 and this event may also be reflected in Core Z. The seven events recorded in the Mills Creek fan over 1720 years differ in frequency from the seven very large events found in Core Z over 3370 years, but this is explainable. The Mills Creek fan is close to the source of sediment production and, apart from the problems of cut and fill, is likely to record most major events in its basin. In contrast, the deep floor of Lake Whatcom will likely receive the product of most events but only the very largest will be strongly emphasized in the stratigraphic record. Nor can it be assumed that the two subareas had identical erosion histories.

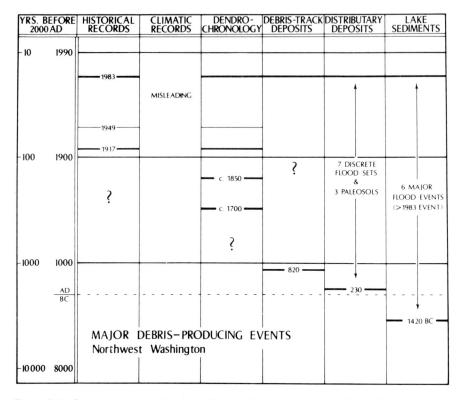

YRS. BEFORE 2000 AD		HISTORICAL RECORDS	CLIMATIC RECORDS	DENDRO-CHRONOLOGY	DEBRIS-TRACK DEPOSITS	DISTRIBUTARY DEPOSITS	LAKE SEDIMENTS
10	1990						
		1983	MISLEADING				
		1949					
100	1900	1917			?	7 DISCRETE FLOOD SETS & 3 PALEOSOLS	6 MAJOR FLOOD EVENTS (>1983 EVENT)
		?		c. 1850			
				c. 1700			
				?			
1000	1000				820		
	AD / BC					230	
							1420 BC
		MAJOR DEBRIS–PRODUCING EVENTS Northwest Washington					
10000	8000						

Figure 6.5 Six approaches to the recognition and timing of major debris-production events, northwest Washington

   The first conclusion that may be drawn from this study is that debris avalanches and debris torrents have been recurrent features of the forest environment of northwest Washington for at least 3400 years, and probably throughout postglacial times. Steep slopes, poorly cohesive soils and colluvium, periodic intense rainstorms and rain-on-snow events have all combined to promote slope failure and sediment mobility, especially where surface and subsurface drainage converges on colluvial hollows or is directed towards steep channels.
   Secondly, in general terms, debris production has occurred under old-growth forest during prehistory, just as it occurs today under both old growth and a patchwork of second-growth forest produced by timber harvesting. Other studies have demonstrated a short-term relationship between timber harvesting and increased sediment yields, followed by a return to background levels over several years. This study demonstrates that a longer-term correlation between timber harvesting and debris production cannot be substantiated. The environment, though often potentially unstable, has an innate capacity to heal after temporary dislocations.

Thirdly, the event of December 1917 was probably greater in terms of the magnitude of erosion and debris production than the 1983 event. Debris avalanches of January 1983 are commonly smaller than, and nested within the 1917 failure scars, while the sites of several 1917 failure scars remained intact during subsequent events. That the steeper parts of the area were yet to experience clearcutting in 1917 again suggests that timber harvesting is not a primary cause of debris production during major events. Nevertheless, several 1983 slope failures were associated with abandoned logging roads and this source of sediment often added to the volume of debris conveyed downstream.

Fourthly, the surging nature of debris torrents, responsible for much overbank flooding and deposition, cannot be attributed solely to timber harvesting because historical evidence cites the occurrence of log jams prior to logging. The extent to which timber harvesting may have modified the calibre of materials contributing to log jams has been the focus of some research but needs more investigation, as indeed does the issue of whether or not log jams should be allowed to remain in stream channels.

Fifthly, the issue of recurrence intervals remains unclear on the evidence presented but some tentative conclusions may be drawn. Specifically, the lake sediment record suggests that very large debris-producing events in the area may be expected every 500 years or so. A broader range of evidence, including historical records, tree-ring analysis and lake sediments, suggests that major events such as those experienced in 1917 and 1983 have a recurrence interval of some 60–70 years. Reconstructions using limited climatic data generally involve too many assumptions to be of value, particularly in view of the rain-on-snow effect that is not readily revealed by simple precipitation data alone.

The foregoing study has used a multifaceted approach to a problem of wide interest to scientists, government agencies concerned with forest management, timber companies involved with forest resources, and property owners beyond the forest margins. In general terms, major debris production is a recurrent feature of the landscape and land-use decisions related to both timber harvesting and lowland development should be made only after careful consideration of the potential dangers of recurrent erosion and deposition.

ACKNOWLEDGEMENTS

The assistance of the Department of Natural Resources (DNR), State of Washington, is acknowledged. G. W. Thorsen and D. Thompson (DNR), R. L. Beschta (Oregon State University) and C. F. Mass (University of Washington) made valuable contributions to discussion in the field and beyond.

REFERENCES

Beschta, R. L. (1978). Long-term patterns of sediment production following road construction and logging in the Oregon Coast Range. *Water Resources Research*, **14**, 1011–16.

Brakke, D. F. (1984). Sediment inflows and water quality in an urbanizing watershed. *Lake and Reservoir Mgmt.*, US Env. Prot. Agency, pp. 239–42.

Frederickson, R. L. (1970). Erosion and sedimentation following road construction and timber harvest on unstable soils in three small western Oregon watersheds. *US Dept. Agric., Forest Serv. Res. Paper*, PNW-104.

Harr, R. D. (1981). Some characteristics and consequences of snowmelt during rainfall in western Oregon. *J. Hydrology*, **53**, 277–304.

Jeffcott, P. R. (1949). *Nooksack tails and trails*, Sedro-Woolley Courier Times.

Johnson, S. Y. (1984). Stratigraphy, age, and paleogeography of the Eocene Chuckanut Formation, northwest Washington. *Can. J. Earth Sci.*, **21**, 92–106.

Morse, E. (1887). Down the Nooksack. *The Overland Monthly*, 2nd Ser., **10**(58), 590–1.

Orme, A. R. (1987). Initiation and mechanics of debris avalanches on steep forest slopes. In R. L. Beschta, T. Blinn, G. E. Grant, G. G. Ice, and F. J. Swanson (eds), *Erosion and Sedimentation in the Pacific Rim*. Int. Assoc. Hydrol. Sci., vol. 165, pp. 139–40.

Orme, A. R. (1989). The nature and rate of alluvial fan aggradation in a humid temperate environment, northwest Washington. *Physical Geography*, **10**, 131–46.

Swanson, F. J. and C. T. Dyrness (1975). Impact of clearcutting and road construction on soil erosion by landslides in the western Cascade Range, Oregon. *Geology*, **3**, 393–6.

Swanston, D. N. and F. J. Swanson (1976). Timber harvesting, mass erosion and steepland forest geomorphology in the Pacific Northwest. In D. R. Coates (ed.), *Geomorphology and Engineering*, Dowden Hutchinson and Ross, pp. 199–221.

7 Modelling the Effect of Vegetation on Air Flow for Application to Wind Erosion Control

ROY P. C. MORGAN
Silsoe College

SUMMARY

A simple model is proposed for simulating the effects of vegetation on soil detachment rate and transport capacity of wind. Wind erosion is determined by whichever is the limiting factor. The most sensitive plant parameters are the projected foliage area facing the wind, leaf area density and leaf alignment. Although the model has not been validated, it gives results which are in accord with expectations from recent research. As examples, wind erosion simulations are presented for fields of bare soil, sugar beet, and sugar beet with barley strips as in-field shelter.

INTRODUCTION

Despite considerable research on the mechanisms of wind erosion and on the transfer of momentum between the atmosphere and the earth's surface, work on the modelling of wind erosion processes has lagged behind that on the physically based modelling of water erosion (Nickling, 1987). Soil loss due to wind can be predicted using the USDA Wind Erosion Equation but this is largely empirical. For example, the effect of vegetation is modelled through relationships between biomass and an equivalent residue of flattened wheat straw (Lyles and Allison, 1981) and between soil loss and the flattened wheat straw equivalents (Woodruff and Siddoway, 1965). With this approach it is not possible to evaluate

Vegetation and Erosion
Edited by J. B. Thornes
©1990 John Wiley & Sons Ltd

the properties of a vegetation cover which most control its ability to protect the soil from erosion.

In fact, very little seems to be known about the salient vegetation properties. Work on shelterbelt design has emphasized the importance of the height and porosity of the plant cover (Caborn, 1965; Seginer and Sagi, 1972; Skidmore and Hagen, 1977). Marshall (1971) has demonstrated the role played by the diameter-to-height ratio of the plant elements in a vegetation stand. Other vegetation properties have not been studied in detail (Heisler and De Walle, 1988). None of the methods described by Nickling (1988), as being used to develop physically based models, adequately deals with the role of vegetation.

This chapter suggests an approach to rectify this situation by incorporating the results of recent research on drag exerted by single plant rows or barriers (Morgan, Finney and Williams, 1988) in a simple mathematical simulation of the wind erosion process. The simulation is used to evaluate the importance of various vegetation properties, namely biomass, plant height, area of foliage and alignment of the leaves, in the early stages of plant growth. Such evaluation is particularly necessary in view of the ability of vegetation sometimes to enhance the risk of erosion instead of protecting the soil (Morgan and Finney, 1987). The data on which the simulation is based are valid for plants up to about 0.4 m in height.

MODEL FORMULATION

The mathematical simulation adopted here for wind erosion follows that proposed by Meyer and Wischmeier (1969) for water erosion. Erosion is considered as a two-phase process of detachment of soil particles from the soil mass and their subsequent transport. The rate of soil loss is controlled either by the rate of soil detachment or by the transport capacity of the wind, whichever is the limiting factor. Both are expressed per unit width. The simulation operates for an individual crop or plant row aligned at right angles to the wind. Transported sediment is routed downwind from one crop row to the next where it is added to the soil detached in that row to determine the amount of soil available for further transport. Figure 7.1 shows a flow chart of the proposed model.

Soil Detachment

The rate of soil detachment (D) is expressed as a function of the shear or drag velocity of the wind (u_*) in excess of a threshold velocity (u_{*_t}) for soil particle movement, raised to the power of 2.8 (Sørensen, 1985). Thus

$$D = k(u_* - u_{*_t})^{2.8} \tag{1}$$

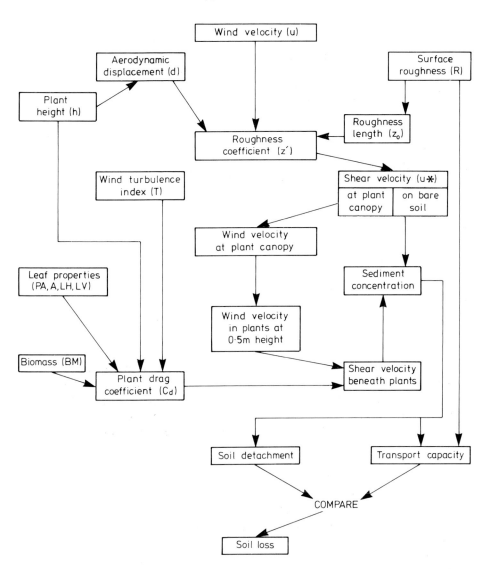

Figure 7.1 Approach used to simulate wind erosion

where k is an experimental proportioning constant. Using the data in Sørensen (1985), $k = 3.3$ if u_* is in $m\,s^{-1}$ and D is in $g\,cm^{-2}\,s^{-1}$. Total detachment in a unit centimetre width of single crop row can then be calculated by multiplying D by the length of the row (L) in the downwind direction.

Transport Capacity

The transport capacity of the wind is expressed by the equation developed by Hsu (1973) to describe aeolian sediment discharge on dunes in the USA, Ecuador and Libya. This equation describes transport capacity (T: $g\,cm^{-1}\,s^{-1}$) as a function of the shear velocity of the wind and the average grain diameter of the material (d_m). Thus

$$T = \exp(4.97d_m - 0.47)\ (u_*/(gd_m)^{0.5})^3 \qquad (2)$$

where g is the acceleration due to gravity. For the simulation developed here, the term d_m is extended to cover the average size of the roughness elements (R) on the soil surface.

Shear Velocity

Without Vegetation

Shear velocity over a bare soil surface is obtained from Bagnold's (1941) equation:

$$u_* = (k/2.3)\ (u(z)/\log\ (z/z_0)) \qquad (3)$$

where k is the von Karman universal constant for turbulent flow, assumed equal to 0.4, u is the measured mean wind velocity at height z, and z_0, often termed the roughness length, is the height above the mean aerodynamic surface at which wind velocity is zero. The value of z_0 can be approximated as an order of magnitude lower than the height (R) of the roughness elements (Monteith, 1973), i.e.

$$z_0 = 0.1R \qquad (4)$$

Equation (3) can be applied to clear air but does not allow for the shear stress exerted by saltating grains. Once sediment is in motion, the grain-borne shear stress should be added. This is equal to the product of the total soil dislodgement rate and the average amount of momentum extracted from the air by the saltating grains (Sørensen, 1985). Close to the bed the grain-borne shear stress amounts to about 14 per cent of the air-borne shear stress. Thus shear velocity in the presence of moving sediment (u_{*s}) can be approximated (Sørensen, 1985) by:

$$u_{*s} = u_* + (0.14\ \varrho_a\ u_*^2) \qquad (5)$$

where ϱ_a is the density of the air.

With Vegetation

A vegetation cover increases the roughness length (z_0) and displaces the height of the mean aerodynamic surface above the ground by a distance (d), which is termed the zero plane displacement. The plane of zero wind velocity is therefore displaced upwards by an amount z', where

$$z' = z_0 + d \qquad (6)$$

Both z_0 and d can be approximated as functions of the vegetation height (h) such that:

$$z_0 = 0.1h \qquad (7)$$
$$d = 0.7h \qquad (8)$$

The term z_0 may be considered a measure of the bulk effectiveness of the vegetation cover in absorbing momentum and the term d as a measure of the mean height at which the absorption takes place (Thom, 1975). As such the terms indicate how vegetation affects air movement close to and above the plant canopy but do not describe the conditions within the vegetation or close to the ground surface.

The effect of vegetation on wind within the plant layer is expressed in the simulation model by the drag coefficient (C_d). This is obtained by balancing the drag force (τ) exerted by the vegetation at height (h) with the extraction of momentum due to the frictional surface area of the individual foliage elements. This gives (Wright and Brown, 1967):

$$\tau(h) = 0.5 \; \varrho_a \int_0^h (C_d \, A(z) \, u(z)^2 \, dz) \qquad (9)$$

where A is the leaf area per unit volume.
Since $\tau = \varrho_a u_*^2$,

$$u_* = 0.71 \int_0^h (C_d \, A(z) \, u(z) \, dz)^{0.5} \qquad (10)$$

This equation was then modified using equation (5) if sediment was present in the wind.

Drag Coefficients for Vegetation

Morgan and Finney (1987) used equation (10) to calculate the drag coefficients in the lowest 0.05 m of single crop rows from field measurements of wind velocity. Morgan, Finney and Williams (1988), using field and wind tunnel data, developed empirical relationships between the values of the drag coefficient and properties of the vegetation. The two relationships which were obtained are used here. One applies to conditions when the drag coefficient decreases in value with increasing wind speed, presumably due to the streamlining of the foliage downwind. The other applies to conditions when the drag coefficient increases its value with increasing wind speed. The reason for this increase is not fully understood. It has been observed before in rice (Uchijima, 1976) and in maize (Wright and Brown, 1967), but with respective crop heights of 0.9 and 1.4 m, much taller than the crops considered here. It may be related to the continuous movement of the foliage which disturbs the surrounding atmosphere and creates a 'wall effect'. This explanation was invoked by Uchijima (1976) for the rice fields. Very little information is available on the behaviour of shorter plants or crops in their early stages of growth but there seems no logical reason why the same explanation should not apply.

The wind tunnel studies described by Morgan, Finney and Williams (1988) show that simulated small-bladed leaves, about 4 cm long, will move continuously through the range of 12 to 38 degrees either side of their angle at rest in a wind speed of $5.9 \, \text{m s}^{-1}$. General field observations certainly confirm that crops such as onions and sugar beet, even when only 2 to 5 cm tall, oscillate continuously in steady winds.

Which of the two conditions described above prevails depends upon the level of wind turbulence as expressed by the index (T), defined as the ratio of the standard deviation to the mean of a series of consecutive wind velocity recordings. The respective relationships are:

$$\log C_d = -1.648 - 1.4060 \log u - 378.4 \, \text{PA} + 0.00466 \, H + 0.01045 \, V \qquad T > 0.2 \qquad (11)$$

$$\log C_d = -0.139 + 0.316 \log u - 369.1 \, \text{PA} + 0.1167 \, \text{BM} - 1.757 \, T \qquad T \leqslant 0.2 \qquad (12)$$

where u = wind speed at 0.05 m height (m s^{-1})

PA = projected area of the vegetation facing the wind in the lowest 0.05 m of a representative 0.1 m wide section of crop row (m^2)

H = downwind alignment of the leaves in relation to a line full-face to the wind (degrees)

V = across-wind alignment of the leaves in relation to the vertical (degrees)

BM = biomass of the lowest 0.05 m of a representative 0.1 m wide section of crop row (kg DM m^{-3})

T = index of wind turbulence (defined above).

The terms H and V are illustrated in Figure 7.2. It should be noted that since an increase in the drag coefficient with increasing wind speed is associated with

Downwind alignment (H)

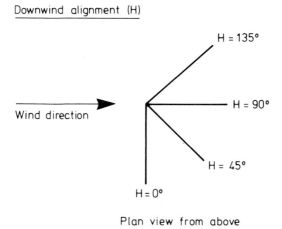

Plan view from above

Cross-wind alignment (V)

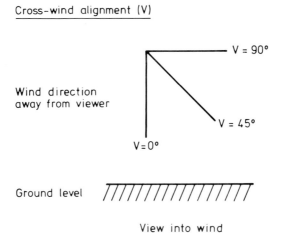

View into wind

Figure 7.2 Diagrammatic representation of downwind (*H*) and cross-wind (*V*) alignments of leaves

continuous winds, when the value of the turbulence index is low, and also that, under these conditions, the drag coefficient then increases in value as the turbulence index decreases, air turbulence cannot be invoked as an explanation of the high drag.

Since the above relationships depend on knowing the wind velocity at 0.05 m height, a procedure was developed to predict this from measured wind velocity data. Equation (3) is used to estimate the wind velocity at the top of the canopy of the vegetation after substituting z' for z_0. To avoid double-counting the effect of momentum absorption by the vegetation through C_d and z_0, z_0 is expressed only in terms of the ground roughness elements. Thus z' becomes the summation of equations (4) and (8).

Wind tunnel studies with simulated plants showed that for single crop rows up to 0.15 m height, the wind velocity at 0.05 m height could be estimated from the velocity at canopy height by:

$$u = 0.9 + 0.1 \ln (0.05/h)u_c \tag{13}$$

where u = wind speed at 0.05 m height
$\quad u_c$ = wind speed at canopy height
$\quad h$ = canopy height.

This relationship is adopted here in preference to other expressions of the wind profile within a plant layer (Cionco, 1965; Thom, 1971) which fitted the data for short plant rows less well.

MODEL OPERATION

The model can be operated over space for a succession of crop rows or segments of bare soil and over time for a sequence of recorded wind speeds. Operation of the model requires input data over time on wind velocity and wind turbulence index, and over space, for each crop row, on ground surface roughness, vegetation height, biomass, projected area of the foliage, leaf area density and downwind and across-wind leaf alignment.

SENSITIVITY ANALYSIS

A sensitivity analysis of the simulation model was carried out to determine which parameters need to be assessed most accurately in order to operate the model and which parameters could be changed most effectively in order to control wind erosion. The analysis was based on varying the value of each input parameter in turn by 10 per cent either side of a pre-selected base value (Table 7.1). The results are expressed in terms of the percentage change in predicted soil loss as a result of a 1 per cent change in the value of the input parameter. Although this type of analysis does not allow for possible interactions between the parameters or for non-linear relationships between input and output over an extended range of parameter values, it does provide an indicator of relative sensitivity within a likely range of operational values. Separate analyses were undertaken for bare soil and for vegetation under turbulent ($T>0.2$) and non-turbulent ($T\leq0.2$) wind conditions.

Table 7.1 Sensitivity of soil loss predictions to a 1 per cent change in the value of the input parameters around a base value

Input parameter	Base value	Percentage change in	
		Soil detachment	Transport capacity
Bare soil			
Wind velocity (u)	$5.7 \, \text{m s}^{-1}$	$+5.5$	$+5.2$
Soil roughness (R)	$0.00015 \, \text{m}$	$+0.3$	-1.4
Tillage roughness (R)	$0.15 \, \text{m}$	$+0.7$	-0.2
Vegetation cover—low turbulence index ($T \leq 0.2$)			
Projected foliage area (PA)	$0.00135 \, \text{m}^2$	-5.2	-2.1
Across-wind leaf alignment (LV)	$45°$	$+4.7$	$+1.9$
Leaf area density (A)	$2.7 \, \text{m}^2 \, \text{m}^{-3}$	$+4.2$	$+1.8$
Downwind leaf alignment (LH)	$40°$	$+1.5$	$+0.7$
Soil roughness (R)	$0.00015 \, \text{m}$	0	-1.7
Wind velocity (u)	$5.7 \, \text{m s}^{-1}$	$+0.5$	$+0.5$
Plant height (h)	$0.005 \, \text{m}$	-0.05	-0.02
Vegetation cover—high turbulence index ($T > 0.2$)			
Turbulence index (T)	0.33	-4.3	-2.2
Projected foliage area (PA)	$0.00135 \, \text{m}^2$	-3.3	-2.1
Leaf area density (A)	$2.7 \, \text{m}^2 \, \text{m}^{-3}$	$+2.8$	$+1.8$
Soil roughness (R)	$0.00015 \, \text{m}$	0	-1.8
Wind velocity (u)	$5.7 \, \text{m s}^{-1}$	$+0.6$	$+0.6$
Biomass (BM)	$0.399 \, \text{kg DM m}^{-3}$	$+0.2$	$+0.2$
Plant height (h)	$0.005 \, \text{m}$	-0.1	-0.1

Wind velocity is the most sensitive parameter for bare soil conditions but is only moderately sensitive when vegetation is present. Projected foliage area, across-wind leaf alignment and leaf area density are the most sensitive of the vegetation parameters, the sensitivity being greater when erosion is detachment-limited than when it is transport-limited. In contrast, soil surface roughness is moderately sensitive when erosion is transport-limited but insensitive when erosion is detachment-limited. Small changes in biomass and plant height have little effect on predicted soil loss. The wind turbulence index is highly sensitive when $T < 0.2$, i.e. when winds are steady, but, by virtue of its exclusion from equation (11), is not sensitive when $T \geq 0.2$, i.e. when winds are gusty. Crossing the critical turbulence value of 0.2, however, from 0.21 to 0.19, increases erosion by 106 per cent when it is detachment-limited and by 30 per cent when it is transport-limited.

These results imply that good quality data on wind velocity, foliage area and leaf alignment are required for wind erosion prediction. When considering the results from the perspective of erosion control, the effect of the vegetation can be seen to depend on the extent to which the adverse effects of increasing leaf area density and biomass are offset by increases in projected foliage area and height. These relationships emerge from previous work (Morgan, Finney and Williams, 1988), which shows that greater leaf area density, resulting in greater contact length between the leaf surface and the air in a downwind direction, brings about higher drag coefficients and an increase in the risk of wind erosion. Plants with high foliage area facing the wind are therefore preferred for erosion control to plants where the foliage area is aligned downwind (Morgan, 1989). Account should also be taken of the feasibility of changing parameter values. For example, biomass generally increases more rapidly than height and both will change more rapidly than leaf alignment as the vegetation grows. Also, orders of magnitude changes in surface roughness can be effected by tillage.

MODEL APPLICATION

As a theoretical example, the model is applied to a 100 m long field of young (5 cm high) sugar beet on a sandy loam soil (critical shear velocity for particle movement $= 0.017$ m s^{-1}) tilled to a very fine smooth seed bed ($R = 0.00015$ m). The wind is assumed to be non-turbulent ($T = 0.17$) so as to give the greatest risk of erosion. Two conditions are simulated for a wind velocity of 11 m s^{-1} at 1 m height: one for sugar beet alone and the other for an in-field shelter system using 0.15 m high live barley strips. The typical spacings, as used in the in-field shelter experiments of the UK Agricultural Development and Advisory Service and illustrated in Figure 7.3, are simplified by assuming successive downwind lengths of 0.30 m for each crop row. For comparison, the results of a simulation for bare soil are also presented. Since conditions stabilize downwind, only the results for the first 9 m of the field need to be considered.

For the bare soil (Table 7.2), soil loss is simulated as detachment-limited initially and increases with distance downwind to reach 197.2 g cm^{-1} after 9 m. Extrapolation of the results reveals that transport capacity (691.14 g cm^{-1} s^{-1}) is reached after 31.5 m, beyond which soil loss will remain constant.

The simulation for young sugar beet shows that soil loss stabilizes after 1.2 m to conditions that are transport-limited in the rows of sugar beet and detachment-limited on the bare soil. Soil particles detached in the bare soil rows are therefore

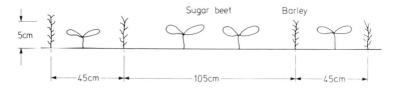

Figure 7.3 Typical in-field shelter system used with sugar beet

Table 7.2 Results of model simulation

Distance downwind (cm)	Soil detachment	Transport capacity $(\mathrm{g\,cm^{-1}\,s^{-1}})$	Soil loss
Bare soil (smooth seed bed)			
30	6.57	691.10	6.57
60	6.57	691.14	13.14
90	6.57	691.14	19.72
120	6.57	691.14	26.29
810	6.57	691.14	177.44
840	6.57	691.14	184.01
870	6.57	691.14	190.59
900	6.57	691.14	197.16
Young sugar beet (0.05 m tall, in rows perpendicular to wind)			
30	6.57	691.10	6.57
60	0.05	6.32	6.32
90	0.05	6.32	6.32
120	6.57	691.25	12.89
150	0.05	6.32	6.32
810	0.05	6.32	6.32
840	0.05	6.32	6.32
870	6.57	691.25	12.89
900	0.05	6.32	6.32
Young sugar beet with barley strips (sugar beet as above, barley strips 0.15 m tall grown in intervening rows)			
30	0.00	0.82	0.00
60	0.05	6.32	0.05
90	0.05	6.32	0.10
120	0.00	0.82	0.10
150	0.05	6.32	0.15
810	0.05	6.32	0.82
840	0.05	6.32	0.87
870	0.00	0.82	0.82
900	0.05	6.32	0.87

Parameter values used in the simulations: downwind length of crop row or bare soil segment $(L) = 30$ m; wind velocity $(u) = 11$ m s^{-1} at 1 m height; soil surface roughness $(R) = 0.00015$ m; wind turbulence index $(T) = 0.17$; plant height $(h) = 0.05$ m for sugar beet and 0.15 m for barley strips; projected foliage area $(PA) = 0.00135$ m^2 for sugar beet and 0.0044 m^2 for barley; leaf area density $(A) = 2.7$ m^2 m^{-3} for sugar beet and 8.8 m^2 m^{-3} for barley; across-wind leaf alignment $(LV) = 45°$ for sugar beet and 50° for barley; downwind leaf alignment $(LH) = 40°$ for sugar beet and 15° for barley; biomass $(BM) = 0.399$ kg DM m^{-3} for sugar beet and 0.590 kg DM m^{-3} for barley.

effectively deposited within the sugar beet so that soil loss is 12.89 g cm^{-1} s^{-1} if the last row is bare and 6.32 g cm^{-1} s^{-1} if the last row is sugar beet.

With the combined sugar beet and barley strip system, soil loss is detachment-limited throughout the length of the field. Detachment is reduced to zero in the barley strips but there is sufficient transport capacity to carry the soil detached within the sugar beet rows through successive barley strips. The soil loss at the end of the field is reduced to $0.87 \, \text{g cm}^{-1} \text{s}^{-1}$. This is the highest value that soil loss will attain because if the field ends in a barley strip, the soil loss will be limited to $0.82 \, \text{g cm}^{-1} \text{s}^{-1}$ by the transport capacity through the strip.

The effectiveness of the barley strips in this simulation is due firstly to the increase in projected area of 325 per cent compared with the young sugar beet. Although this is partly offset by the effect of a 325 per cent increase in leaf area density, the greater sensitivity of soil loss to area makes the change in area more important. Secondly, the increase of 300 per cent in plant height produces a dramatic effect through an otherwise relatively insensitive parameter.

If the soil loss estimates are multiplied by 10 000 to convert values per centimetre-width to a hectare for the 100 m long field, erosion is predicted at $6.9 \, \text{t ha}^{-1} \text{s}^{-1}$ for the bare soil, $63.2 \, \text{kg ha}^{-1} \text{s}^{-1}$ with sugar beet and $8.2 \, \text{kg ha}^{-1} \text{s}^{-1}$ for the combined sugar beet and barley strips. Although there are no measured data against which to compare these figures, they do not seem unreasonable for such a strong wind over a fine seed-bed in terms of either absolute magnitude or relative values. Zachar (1982), for example, predicts an erosion rate of $187.2 \, \text{t ha}^{-1} \text{s}^{-1}$ (assuming a bulk density of $1.0 \, \text{Mg}^{-3}$ for a $12 \, \text{m s}^{-1}$ wind over highly erodible sand dunes.

Extrapolation of the results to longer time periods can only be a guide to erosion rates because continuous wind speeds of $11 \, \text{m s}^{-1}$ at 1 m height will be maintained only for very short durations. Extending the duration to 30 minutes would therefore produce a very extreme event with predicted erosion rates of $12 \, 420 \, \text{t ha}^{-1}$ for the bare sandy loam soil, $113.7 \, \text{t ha}^{-1}$ for the soil with young sugar beet and $14.8 \, \text{t ha}^{-1}$ for the in-field shelter system with barley strips. Since the soil loss cannot increase with distance because it is transport-limited, these rates reduce to 6 410, 56.9 and $7.4 \, \text{t ha}^{-1}$ for a 2 ha field of the same width. According to Hayes (1965) the mean annual rate of wind erosion for a sandy loam soil in the Great Plains is $190 \, \text{t ha}^{-1}$. If this figure is taken as an acceptable indicator, the predicted bare soil erosion rate for the 2 ha field is thus equivalent to maintaining the annual rate for only 34 years. Using barley strips to control the erosion under the young sugar beet reduces the soil loss by 99 per cent.

LIMITATIONS

The operating equations for the model were chosen for their ease of use and convenience in data requirements, as well as their ability to describe the processes involved. It is recognized that individual researchers have their own favourite equations, for example for expressing transport capacity. The merits of one equation over another can only be decided after model validation.

The main limitation of the model presented here is that it has not been validated. In practice, validation would be quite difficult because of the lack of suitable field data. Such data are also required for a realistic assessment of acceptable time and space steps for model operation. The model cannot be tested by comparing its predictions with those of the USDA Wind Erosion Equation or any other model because there is no way of knowing whether the latter are correct.

Certain sections of the model are simplifications of reality and their scientific validity is open to question. Thus, more work is needed on the effect of saltating grains on the wind profile and therefore on shear velocity, particularly under varying conditions of turbulence. The method proposed for estimating the drag coefficient of vegetation close to the ground assumes that the coefficient value is independent of wind velocity. In reality, the value both increases and decreases with wind speed, depending upon the turbulence index. This is because the effective area controlling drag changes either through streamlining or through the setting-up of the 'wall' effect. The determination of shear velocity within the plant layer depends upon a relationship obtained from wind tunnel studies of young simulated crops and may not apply when the vegetation grows taller. The model assumes steady-state conditions with turbulence generated by frictional forces and form drag at the surface and not from air movements induced by vertical temperature gradients. The model applies to the lowest 0.05 m of the atmosphere which is treated, almost certainly unrealistically, as a single isolated layer.

CONCLUSIONS

A simulation model is presented to show the effect of vegetation on wind erosion. The most sensitive parameters of a vegetation cover are the projected foliage area facing the wind, leaf area density and across-wind leaf alignment. Although relatively insensitive, plant height is also important because this may change dramatically as a crop or vegetation cover grows. Despite lack of validation and a number of simplifying assumptions in the operating equations, the model gives results, in both trends and magnitudes, which are in accord with recent research on how vegetation can sometimes protect the soil and at other times enhance erosion. The model applies to the lowest 0.05 m of the atmosphere and to vegetation up to 0.4 m in height. The results indicate that research on the spatial distribution of detachment-limited and transport-limited conditions could provide a better understanding of the wind erosion process as well as confirming whether the patterns simulated by the model are reasonably realistic.

ACKNOWLEDGEMENT

The enthusiastic assistance of Richard Morgan with some of the programming is acknowledged.

REFERENCES

Bagnold, R. A. (1941). *The Physics of Blown Sand*, Methuen, London.
Caborn, J. M. (1965). *Shelterbelts and Windbreaks*, Faber & Faber, London.
Cionco, R. M. (1965). A mathematical model for air flow in a vegetative canopy. *J. Appl. Met.*, **6**, 185–93.
Hayes, W. A. (1965). Wind erosion equation useful in designing northeastern crop protection. *J. Soil & Wat. Conserv.*, **20**, 153–255.
Heisler, G. M. and D. R. De Walle (1988). Effects of wind break structure on wind flow. *Agriculture, Ecosystems and Environment*, **22/23**, 41–69.
Hsu, S. A. (1973). Computing eolian sand transport from shear velocity measurements. *J. Geol.*, **81**, 739–43.
Lyles, L. and B. E. Allison (1981). Equivalent wind erosion protection from selected crop residues. *Trans. Am. Soc. Agric. Engnrs.*, **24**, 405–9.
Marshall, J. K. (1971). Drag measurements in roughness arrays of varying density and distribution. *Agric. Met.*, **8**, 269–92.
Meyer, L. D. and W. H. Wischmeier (1969). Mathematical simulation of the process of soil erosion by water. *Trans. Am. Soc. Agric. Engnrs.*, **12**, 754–8, 762.
Monteith, J. L. (1973). *Principles of Environmental Physics*, Edward Arnold, London.
Morgan, R. P. C. (1989). Design of in-field shelter systems for wind erosion control. In Schwertmann, V., Rickson, R. J. and Auerswald, K (eds), *Soil Erosion Protection Measures in Europe*, Soil Technology Series 1, 15–23.
Morgan, R. P. C. and H. J. Finney (1987). Drag coefficients of single crop rows and their implications for wind erosion control. In V. Gardiner (ed.), *International Geomorphology 1986, Part II*, Wiley, Chichester, pp. 449–58.
Morgan, R. P. C., H. J. Finney and J. S. Williams (1988). Leaf properties affecting crop drag coefficients: Implications for wind erosion control. Paper presented to 5th International Soil Conservation Conference, Bangkok, Thailand.
Nickling, W. G. (1987). Recent advances in the prediction of soil loss by wind. In I. Pla (ed.), *Soil Conservation and Productivity*, Sociedad Venezolana de la Cienca del Suelo, Maracay, pp. 1163–86.
Nickling, W. G. (1988). Prediction of soil loss by wind. Paper presented to 5th International Soil Conservation Conference, Bangkok, Thailand.
Seginer, I. and R. Sagi (1972). Drag on a windbreak in two-dimensional flow. *Agric. Met.*, **9**, 323–33.
Skidmore, E. L. and L. J. Hagen (1977). Reducing wind erosion with barriers. *Trans. Am. Soc. Agric. Engnrs.*, **20**, 911–15.
Sørensen, M. (1985). Estimation of some aeolian saltation transport parameters from transport rate profiles. In O. E. Barndorff-Nielsen, J. T. Møller, K. Rømer Rasmussen and B. B. Willetts (eds.), *Proceedings of International Workshop on the Physics of Blown Sand*, University of Aarhus, 141–90.
Thom, A. S. (1971). Momentum absorption by vegetation. *Quart. J. Roy. Met. Soc.*, **97**, 414–28.
Thom, A. S. (1975). Momentum, mass and heat exchange of plant communities. In J. L. Monteith (ed.), *Vegetation and the Atmosphere*, Volume 1, Academic Press, London, pp. 57–109.
Uchijima, Z. (1976). Maize and rice. In J. L. Monteith (ed.), *Vegetation and the Atmosphere*, Volume 2, Academic Press, London, pp. 33–64.
Woodruff, N. P. and F. H. Siddoway (1965). A wind erosion equation. *Soil Sci. Soc. Am. Proc.*, **29**, 602–8.
Wright, J. L. and K. W. Brown (1967). Comparison of momentum and energy balance methods of computing vertical transfer within a crop. *Agron. J.*, **59**, 427–32.
Zachar, D. (1982). *Soil Erosion*, Elsevier, Amsterdam.

8 The Role of Simulated Vegetation in Soil Erosion Control

R. JANE RICKSON
Department of Rural Land Use,
Silsoe College

SUMMARY

The application of geotextiles for soil erosion control has overtaken the research into their effectiveness in controlling the processes of erosion.

Five different types of geotextile and a control (with no geotextile protection) were tested on a laboratory based runoff rig, which simulated overland flow. Their performance in controlling runoff volume and soil loss was evaluated.

The results were analysed to see whether the selected geotextiles control runoff and soil loss by imitating the salient properties of vegetation, which are known to be effective in controlling soil loss. It appears that the geotextiles do not control erosion by reducing runoff volumes, but by modifying runoff velocities with a 'geotextile-induced roughness'.

INTRODUCTION

A great deal of research has shown how effective vegetation is in controlling the processes of soil erosion by water. However, little regard is given to the important period when vegetation is emerging and establishing on man-made slopes. During these weeks or even months severe erosion may occur. This critical period of low vegetation protection rarely occurs in natural vegetation successions, or on agricultural land where mulching of previous crop residues or cover crops are used. However, on road construction and urban development sites little emphasis is given to these critical times, and yet these environments are extremely susceptible to high rates of erosion in excess of $480\,t\,ha^{-1}\,yr^{-1}$ (Wolman and Schick, 1967). This is because of the disturbed and compacted soils, exposed erodible subsoils, steep slopes and lack of vegetation cover.

Vegetation and Erosion
Edited by J. B. Thornes
©1990 John Wiley & Sons Ltd

Simulated vegetation in the form of geotextiles can provide immediate erosion protection on such sites and thus prevent both the on-site and off-site consequences of erosion.

THE USE OF GEOTEXTILES IN SOIL EROSION CONTROL

Geotextiles are defined as 'permeable textiles used in conjunction with soil or rocks as an integral part of a man-made project' (John, 1987). Increasing use is being made of such products as construction safety standards and specifications are becoming more stringent throughout the world. In this chapter, only geotextiles used in soil erosion control are studied. These types of geotextile are usually in the form of erosion mats and can be classified according to their composition (natural or synthetic—which determines their durability on site) and mode of installation (surface or buried).

By controlling erosion, the geotextiles perform two functions. They reduce off-site problems such as sedimentation, non-point pollution and contaminated water supplies, and provide a stable, non-eroding environment in which vegetation can establish and ultimately provide long-term biological erosion control (sometimes in combination with a permanent geotextile). The geotextiles control erosion by mimicking the salient properties of live vegetation that have been shown to reduce soil loss rates.

Surface-laid geotextiles often have high percentage cover simulating canopy interception and storage effects, and rough fibres simulating stem effects. Buried geotextiles simulate the root effect of vegetation. Some geotextiles are even made of vegetative matter which will biodegrade as the permanent vegetation establishes, thus contributing to the succession that occurs in natural vegetation.

The ways in which the different geotextiles control erosion processes can be studied by taking a geomorphological approach. Indeed the applied geomorphologist has an increasingly important role to play in the assessment and control of soil erosion on man-made slopes as conventionally trained civil engineers often have little background knowledge of the detailed erosion processes operating there (Morgan and Rickson, 1988).

This chapter outlines a study into the effectiveness of selected geotextiles in modifying surface hydrology and thereby controlling sediment detachment and transport rates of runoff. The study is the second stage of an evaluation of selected geotextiles in controlling erosion. The first stage of experiments looked at how the different geotextiles reduce soil detachment by raindrop impact (splash erosion). The results of that study can be found elsewhere (Rickson, 1988).

EXPERIMENTAL DESIGN

A runoff rig was used in this experiment to simulate field conditions in the laboratory (Figure 8.1). Larger-scale field trials are proposed for the future. At present the laboratory results are used to show the *relative* effectiveness of the different geotextile treatments in controlling erosion by runoff.

A highly erodible sandy loam soil (Table 8.1) was placed in the soil box and the slope was set at 10° (17.63 per cent). Road embankments and cuttings are often steeper than this, and it would be erroneous to extrapolate the present results to steeper slopes, given the non-linear relationship between slope steepness and erosion rate. It follows that the effectiveness of the erosion control treatments might also be dependent on slope steepness. Further trials need to quantify the effect of slope on geotextile performance. However, these trials were aimed at producing simple comparisons of geotextile performance under given conditions. Runoff discharge was set by means of a gapmeter at a flow rate of 2.358 l/min (39.3 ml/sec).

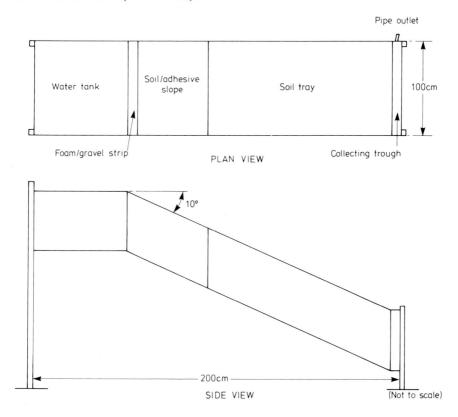

Figure 8.1 The runoff rig

Table 8.1 Properties of the test soil

Coarse sand	$> 600 \, \mu m$	8.15%
Sand	$> 212 \, \mu m$	41.94%
Fine sand	$> 63 \, \mu m$	18.09%
Total sand		68.18%
Clay	$< 2 \, \mu m$	9.77%
Silt	$< 63 \, \mu m$	22.10%

Uniform sheet flow was generated by passing the overflow from the discharge tank (Figure 8.1) over a coarse gravel 'dam' placed on top of a saturated foam sponge. Flow was spread out by the rough soil/adhesive slope located downslope of the foam/gravel section. The time taken for runoff to reach the bottom of the slope was recorded, although this was difficult to assess for some of the geotextiles, as their thickness obscured the runoff observations. Sediment and runoff samples were collected from an outlet pipe running from the trough at the end of the soil slope. Infiltration during each experimental run was collected, although there was a considerable time lapse between the actual experimental run and corresponding infiltration of that particular run. This delay is due to the low infiltration rates of the compacted soil layers (needed to generate runoff over short slope lengths) and the distance between the soil box and the infiltration outlet pipe.

The geotextiles used in this experiment are shown in Table 8.2.

The different treatments were tested in a random sequence, along with a control treatment (with no geotextile). The geotextiles were laid on the soil slope, which had been compacted, levelled and wetted up to ensure initial moisture conditions were the same at the beginning of every experimental run. These experimental conditions made it difficult to incorporate buried geotextiles into this particular set of trials. Over the given slope lengths, runoff will not occur unless the soil is compacted. Buried geotextiles are filled with relatively uncompacted soil and therefore another technique would be required for testing

Table 8.2 Geotextiles and their properties

Geotextile	Materials	Natural/synthetic	Surface mat/buried mat
Geojute Fine mesh	Jute	Natural	Surface
Geojute	Jute	Natural	Surface
Enviromat	Wood chips, light-sensitive plastic mesh	Natural/synthetic	Surface
Enkamat	Polymer mesh	Synthetic	Surface or buried
Bachbettgewebe	Coir	Natural	Surface

their effect on runoff. Enkamat can be used as a surface or buried geotextile. The American specification of laying the geotextile on the surface was applied in these experiments.

The selected geotextiles were pinned down with wire staples at the sides and end of the slope. The upslope edge of the geotextile is crucial in that it must not be undermined by the surface runoff. This problem was resolved by burying the top edge of the geotextiles in the gravel 'dam' at the top of the soil adhesive slope (Figure 8.1).

Runoff volume and sediment weight were recorded and sediment concentration was calculated for all treatments throughout the experiments at 15 seconds, 30 seconds, and at 1, 2, 4, 6, 8 and 10 minutes from start of runoff. The experiments were replicated three times.

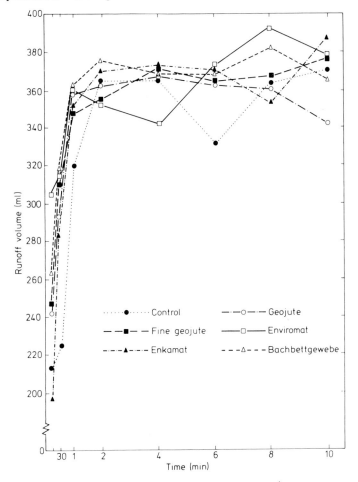

Figure 8.2 Mean runoff volumes over time

RESULTS

The mean runoff and sediment for each treatment at each time interval was
calculated from the three replications (Figures 8.2 and 8.3). Mean total runoff
and sediment for each treatment was also calculated. Variability between the
replicates for any one treatment was quite high. This was because a new geotextile
sample had to be used in each replication, to ensure independent sampling. It
is unlikely that each length of geotextile cut from the roll will be identical in
terms of physical properties (percentage cover, fibre thickness and density, and
weave pattern). Also, often it was difficult to ensure spontaneous uniform sheet

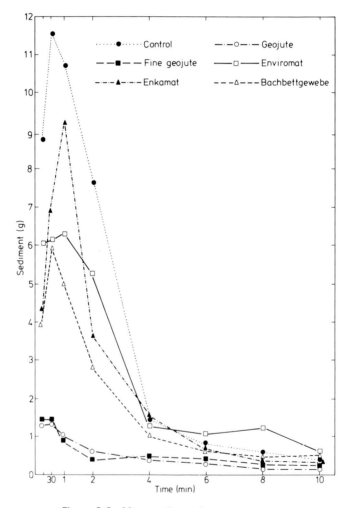

Figure 8.3   Mean sediment losses over time

flow over the soil adhesive slope (Figure 8.1) and uniform soil moisture content for each replication. However, it is assumed that the variability is cancelled out when averaging the results.

The mean total runoff and soil loss results were then analysed to test for significant differences between the six treatments.

Runoff

There was no significant difference in the runoff volumes when comparing the different geotextile treatments with the control ($p < 0.05$). This is surprising as earlier research had shown that the natural geotextiles could hold more water in their fibres, which would result in lower runoff amounts (Rickson, 1987). This did not appear to be evident from these tests, although the Geojute (with the highest water-holding capacity of all the geotextiles tested) did have the lowest mean total runoff of all.

The differences in the runoff observed in the first 15 seconds are more marked than at any other time throughout the experimental tests (Figure 8.2). The Enkamat and control treatments have much lower volumes of runoff at the beginning of the tests. This was because single flowlines formed rather than a uniform sheet of flow as was seen for the other geotextile plots. These flowlines moved faster downslope, so that flow off the soil plot occurred quickly (at 46 and 26 seconds respectively), but not with any great volume. These initial single flowlines may have been obscured by the density of geotextile in the other treatments, so that runoff was collected much later (and therefore in greater volume) than when first generated. It is also thought that the thicker geotextiles helped to spread the flow uniformly as a sheet rather than in single flowlines.

The highest mean total runoff volume was observed for the Enviromat and the Bachbettgewebe. In the first few minutes these did not wet up very effectively, nor impede any downslope movement of runoff as the contact between geotextile and soil surface was not good. This allowed a large total of runoff volume to flow along the geotextile/soil interface. Time to runoff was very quick, at 42 and 28 seconds respectively. Only the control plot produced runoff after a shorter time period than this.

Enviromat has the most variable runoff amounts over time. The high initial runoff volume after one minute fell drastically at four minutes. It may be that as the wood chips were wetted up and became heavier, they impeded flow and runoff was held behind the mulch elements. As time went on, and runoff built up behind the elements, the wood chips became aligned downslope, and so no longer formed a barrier or obstruction to the flow, and runoff increased dramatically, especially at eight minutes. The increase runoff may then also be due to the release of water behind the mulch elements.

The Geojute and Fine Geojute did show signs of reducing runoff with time. This is probably due to the wetting up of the highly absorbent jute fibres. After six minutes, runoff volume began to increase for the Fine Geojute, suggesting that the product had reached saturation. Even after ten minutes the thicker and coarser Geojute continued to decrease runoff—presumably this product was still to reach saturation. Despite these observations, the total runoff 'absorbed' seems to be insignificant.

Overall, it was surprising that there was no statistical difference in the runoff observed for the different geotextile treatments when compared with the control. Many workers have noted the water-holding capacities and ability to dam runoff as being important in reducing runoff volumes on slopes protected by geotextiles, but these current experiments do not wholly support these theories. More importantly, in terms of erosion control, it appears that geotextiles do not control soil loss (which was significantly different for some of the treatments) by controlling runoff volumes.

Sediment Losses

Highest mean total sediment losses (Figure 8.3) of 41.90 g were observed for the control plot, with no geotextile protection. Some microrills were observed on the control plot, but these were of insufficient depth to concentrate flow. Greatest sediment loss occurred after 30 seconds, by which time the runoff had increased sufficiently to detach and transport the easily available soil particles. Indeed, all treatments exhibited this increase in sediment over the first minute of the experimental run. This trend was particularly noted for the treatments with higher sediment losses. Enkamat in particular yielded over twice as much sediment after one minute as compared to that at 15 seconds of runoff. Sediment loss then fell dramatically within the second minute. It is thought that at first the random roughness of the Enkamat fibres caused retardance in the flow, which reduced the flow's transporting capacity. As flow increased and submerged the fibres this effect was less pronounced and sediment transport and detachment increased dramatically. Also protection of the soil from the runoff was poor because of the rigidity of the geotextile. However, where contact was made, deposition of the coarser particles occurred.

Enviromat produced high rates of sediment loss at the beginning of the experiments, perhaps reflecting the high volume of runoff observed at this time. Unlike the Enkamat, the increase in sediment loss was not as marked after one minute but sediment loss remained high throughout the experiment, especially in the final four minutes; at eight minutes, sediment loss exceeded that observed for the control plot. Sediment loss was more evenly distributed throughout the course of the experiment, being relatively low in the beginning and relatively high at the end. This may reflect the point raised earlier, that the orientation and anchorage of mulch components changed in time; as the wood chips became oriented parallel with the direction of flow, more soil loss occurred. On one run, microrills were formed where wood chips had oriented downslope and were concentrating flow at this point.

Despite the difference in the pattern of sediment yield over time for the Enkamat and Enviromat, their total sediment losses were not significantly different.

The Bachbettgewebe also reduced overall sediment loss to half that observed for the control plot, but this difference was insignificant at the 5 per cent level. Contact with the soil was vital in trapping transported particles. As the coir wetted up in time, so sediment losses declined as a better contact with the soil was made. Again, sediment loss increased after one minute and then fell as easily available loose soil particles were detached and transported by the flow.

Whilst sediment loss was greatest for the control plot, there was no significant difference in the mean total soil loss observed for the control and the Enviromat, Enkamat and Bachbettgewebe treatments. This is surprising and disappointing, and may reflect the problems of experimental variability mentioned earlier. Soil losses declined dramatically for the control, Enkamat, Enviromat and Bachbettgewebe between the first and fourth minute.

There was a significant reduction in sediment loss observed for both the Geojute and Fine Geojute when compared with the control plot. With a total sediment loss of just 5.55 g overall, Geojute yielded the least sediment throughout the experiment, closely followed by the Fine Geojute, with an overall sediment loss of 5.62 g. There was no significant difference between these treatments in controlling total soil losses. As Figure 8.3 shows, these two treatments yielded less soil loss than the other treatments, especially in the first four minutes.

The dramatic increase in sediment loss after one minute was less pronounced for the Geojute and Fine Geojute, whose sediment losses tended to decline gradually over time.

There appears to be little difference between all the treatments after the first four minutes of the experiment. This suggests that the experiments are mainly concerned with the transporting and not detaching power of runoff, or that easily removed material was depleted. Either can be justified because most soil detachment is rainsplash dominated (not simulated here); and large amounts of easily transported, detached material will be found on constructed slopes where geotextiles may be considered as an erosion control technique. Future tests will combine the effect of raindrop splash (detachment) and runoff (transport) processes on the geotextiles, using a rainfall simulator and runoff rig—thus being more realistic to the actual processes operating on slopes.

DISCUSSION

It is thought that geotextiles control runoff and sediment loss by mimicking the salient properties of vegetation that affect erosion processes. The present set of experiments is only concerned with canopy and stem effects, rather than the effect of a root system, which may be simulated by buried geotextiles, which were not tested on this occasion. The experiments were concerned with runoff alone, and it is assumed that the erosion process operating here is transport rather than detachment of soil particles. Therefore, the soil loss results may be explained in terms of the effects of the geotextiles on the transporting capacity of the runoff.

The transport capacity of overland flow is dependent on:

—slope (assumed constant here)
—runoff volume;
—runoff velocity (as affected by 'plant-induced roughness').

Runoff Volume

A simple approach to evaluating the effect of vegetation (or simulated vegetation in the case of geotextiles) on runoff volume is to use runoff related to a percentage cover for each geotextile and express the runoff volume as a percentage of the control or bare plot (Rickson and Morgan, 1987). For most data for live vegetation, this approach gives an exponential decrease in runoff volume with increasing percentage cover. This obviously does not apply in the present study, where (Figure 8.4) a high percentage cover (Enviromat) has a lower runoff ratio than a geotextile with poorer percentage cover (Enkamat).

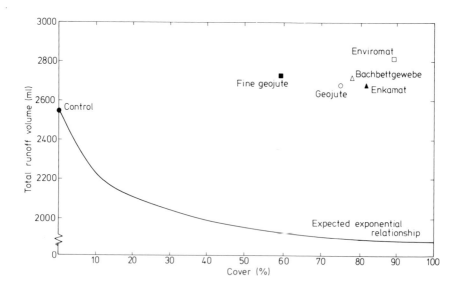

Figure 8.4 Relationship between runoff volume and percentage

Therefore, using real vegetation/runoff volume relationships may not be useful when evaluating the effectiveness of different geotextiles. Indeed, despite the obvious visual differences in the geotextiles (most apparent was percentage cover), there was no significant difference in the runoff volumes observed for the different geotextile treatments.

The jute geotextiles also transferred water within their fibres, as evidenced by the dripping fibres at the end of the slope. Runoff was thus transferred downslope on the fibres rather than on the soil; this may explain why runoff volume was high, yet sediment loss was low.

The results imply that geotextiles are not effective in controlling runoff volumes. This may be of importance when designing drainage ditches and soakaway areas at the bottom of slopes protected by geotextiles. It also appears that geotextiles do not encourage infiltration (as at first suspected) on slopes of this degree. This may have implications for the available soil water during the critical periods of vegetation establishment on protected slopes. Also, depression storage is minimized if the geotextile is not in good contact with the soil.

Runoff Velocities

Another factor affecting the transport capacity of overland flow is the effect of plant-induced roughness which reduces runoff velocities, and therefore erosive power. Morgan (1986) states that the transporting capacity of runoff varies with the fifth power of runoff velocity. Therefore, if the geotextile treatments can reduce runoff velocities at all, the effect on the transport capacity will be significant.

Ideally, quantified roughness parameters such as Manning's n should be calculated for each of the geotextiles, and used to predict a 'geotextile-induced roughness to flow' as has been done with some mulch materials (Foster, Johnson and Moldenhauer, 1982). This was difficult, however, because runoff velocities were not easily measured beneath the geotextiles. However, the factors affecting Manning's n—hydraulic radius (assumed equal to flow depth), slope and velocity—were qualitatively assessed to give a measure of geotextile-induced roughness (GIR) (Table 8.3).

The Manning equation is particularly good in explaining the effectiveness of the geotextiles in controlling erosion. First, observations revealed that ponding (= depth of flow) was important for the Geojute and the Fine Geojute. Runoff was held within the mini-cells created by the weave of the jute fibres. Surface tension retained the runoff and a meniscus formed within these mini-catchments. Runoff was observed to flow in small circular motions as it was retained between the fibres. Bachbettgewebe also has a woven appearance, but contact with the soil surface was not as good as observed for the more 'drapable' jute products, so that ponding and increased depth of flow were less apparent. Also, the more

Table 8.3 Qualitative assessment of factors affecting geotextile-induced roughness (GIR) to flow

Treatment	Depth of flow[1] (mm)	Slope	Velocity[2] (m/sec)	GIR[3]	Rank	Soil loss rank
Control	0.68	10	0.0577	5.525	6	6
Geojute	3.00	10	0.034	25.784	2	1
Fine geojute	2.00	10	0.024	27.838	1	2
Enviromat	2.00	10	0.036	18.559	3	5
Enkamat	0.68	10	0.033	9.827	5	4
Bachbettgewebe	2.00	10	0.0535	12.372	4	3

[1]Depth of flow was calculated for the control plot as 0.68 mm. Relative values (based on visual observations) were then assigned to the geotextile treatments to the nearest millimetre.
[2]Velocity was calculated by the time it took for discharge to reach the bottom of the slope.
[3]GIR = Geotextile-induced roughness coefficient, calculated by substituting the estimated flow depths and measured velocities into the Manning equation.

regular weave of the Bachbettgewebe tended to concentrate runoff down the warp fibres—shown by the lines of saturation of the geotextile. Enviromat also has poor contact with the soil, and despite the fact that flow velocity was observed to be slow for this product, lack of ponding meant that the GIR was lower than with the jute products.

Time to runoff gave an indication as to the effect of the geotextiles on velocity. Again, the jute products showed slow velocities as their fibres retarded the movement of runoff downslope by ponding and absorption to some extent, as reflected in the GIR.

These qualitative results show good agreement with the observed soil losses, although Enviromat ranked third according to the GIR and yet was fifth in soil erosion control. The high runoff volume observed for this treatment may help to explain this discrepancy.

The strong agreement between actual soil losses and GIR supports the argument that roughness is important in controlling the transport capacity of overland flow by reducing velocity. Runoff volume reduction is not the key to erosion control when using geotextiles. Factors such as moisture retention and absorption by the geotextile fibres do not appear to be directly related to erosion by runoff, although indirectly these are important as they increase geotextile weight, and therefore surface contact is good and the geotextiles will be able to pond and retain water.

CONCLUSION

Experimental conditions were limited for this set of experiments. However, the results do give insights into how different types of surface geotextile perform in controlling soil erosion associated with overland flow. The jute products performed best under these specific conditions, but not through their ability to 'soak' up the runoff as expected but by retarding runoff within the mini-catchments created by the woven fibres. Runoff velocity is consequently reduced, so reducing transport capacity greatly. The detachment process by runoff was not simulated in these experiments, but given the results of the rainsplash trials (Rickson, 1988), the jute products do seem to be highly effective in controlling the processes of both detachment and transport of soil particles.

REFERENCES

Foster, G. R., C. B. Johnson and W. C. Moldenhauer (1982). Hydraulics of failure of unanchored cornstalk and wheat straw mulches for erosion control. *Trans. Am. Soc. Agric. Eng.*, **25**, 940–7.

John, N. W. M. (1987). *Geotextiles*, Blackie and Son, Glasgow.

Morgan, R. P. C. (1986). *Soil Erosion and Conservation*, Longman.

Morgan, R. P. C. and R. J. Rickson (1988). Soil erosion control: Importance of geomorphological information. In J. M. Hooke (ed.), *Geomorphology in Environmental Planning*, Wiley, Chichester, pp. 51–60.

Rickson, R. J. (1987). Geotextile applications in steepland agriculture. In *Proceedings of the International Conference on Steepland Agriculture in the Humid Tropics*, Kuala Lumpur, August 1987 (in press).

Rickson, R. J. (1988). The use of geotextiles in soil erosion control: Comparison of performance on two soils. In Rimwanich, S. (ed.) *Land Conservation for Future Generations*. Proceedings of the Fifth International Soil Conservation Conference, Department of Land Development, Ministry of Agriculture and Cooperatives, Bangkok.

Rickson, R. J. and R. P. C. Morgan (1987). Approaches to modelling the effects of vegetation on soil erosion by water. In R. P. C. Morgan and R. J. Rickson (eds), *Erosion Assessment and Modelling*, Proceedings of a workshop, held in Brussels, December 1986. Commission of the European Communities, EUR 10860, pp. 237–54.

Wolman, M. G. and A. P. Schick (1967). Effects of construction on fluvial sediment, urban and suburban areas of Maryland. *Water Resources Research*, **3**, 451–64.

9 Boundary Layers under Salt Marsh Vegetation Developed in Tidal Currents

JOHN PETHICK, D. LEGGETT
Institute of Estuarine and Coastal Studies, University of Hull

and L. HUSAIN
University of Pertanian, Malaysia

SUMMARY

Salt marsh accretion has been studied for almost a century now without any attempt to measure or even observe the actual processes acting. Observations have merely been made of the long-term *results* of accretion and inferences drawn as to the processes which cause such changes in surface elevation. In part this is due to the extremely slow rates involved and the difficulty of direct observation in the field.

This chapter reports on some work carried out on salt marsh vegetated surfaces within a laboratory flume. The measurement of boundary shear stress under simulated tidal currents shows the importance of the vegetation to the development of a deep Z_0 layer and this in turn is shown to relate to the deposition of sediments from suspension. A further finding suggests that the development of such a boundary layer depends on the turbulent mixing length within the vegetation.

INTRODUCTION

Deposition on tidal salt marshes has been the subject of field study for most of this century. The early work of Steers (1935, 1948), Richards (1934) and later work by Ranwell (1964) in Britain and Pestrong (1965) in the USA made

Vegetation and Erosion
Edited by J. B. Thornes
©1990 John Wiley & Sons Ltd

fundamental additions to our knowledge of the processes of salt marsh accretion. Field methods have become, of course, more sophisticated and are essential to our understanding of the complex processes of sedimentation on salt marshes. Workers such as Letzsch and Frey (1980) and Stumpf (1983) have shown, from field observations, that marsh accretion is often an episodic, storm-driven, process. Carr (1983) and Carr and Blackley (1986) have described more accurate methods of field determination of sedimentation rates. Stevenson, Ward and Kearney (1988) have collated field data on the flux of sediment within salt marsh creek systems which provide an alternative approach to the problems of accretion within salt marshes. Yet, despite this period of protracted and intensive study, little progress has been made in understanding the physical processes which result in deposition on a vegetated surface. In part this has been due to the impracticality of working on an extremely slow-acting process in a difficult field environment. Precise measurement of the flow patterns of tidal waters around salt marsh vegetation has not been possible and the measurement of the morphological response to accretionary processes, that is the rise in height of a marsh surface, has traditionally been undertaken over periods of months or years and thus constitutes only the net response of the surface.

The movement of suspended sedimentary particles through a vegetation stand has, however, been studied on sand dunes. The pioneering work of Bagnold (1941) showed that the vegetation acted as macro-roughness elements and that these caused the velocity profile to steepen and to become raised off the bed. Bagnold's work showed that the increase in shear stress associated with the vegetated surface tended to steepen the velocity profile through a point of focus, so that, as well as an increase in the shear velocity (U^*), an increase in the depth of the roughness length (Z_0) was observed.

The application of such basic laws of flow and associated sediment transport to suspended sediment within salt marsh vegetation has not been extensive. As noted above, the main impediment has been the difficulties of precise measurement in the field and, in order to overcome these difficulties, it seems appropriate to attempt to isolate the problem within a laboratory model. Only one attempt has been made to do this: Fonesca, Fisher and Zeimen (1977) produced a laboratory flume model of a *Zostera* bed and measured velocity profiles within the vegetation. Their work confirmed the general conclusions concerning the influence of vegetation on fluid dynamics outlined above from sand dune work. They showed that both the shear velocities and the intercept or Z_0 layer increased as the leaf area index of the vegetation was increased, thus substantiating Bagnold's proposal that modifications of the profile took place around a pivotal point.

This chapter uses a similar laboratory approach to that of Fonesca *et al.* (1977) to the problem of salt marsh sediment dynamics. The work was initiated as a response to a specific problem encountered in the field. Field observation of *Spartina anglica* showed that its vigour in many sites is not as pronounced now as it appeared to be in previous years. In the Humber estuary, for example, despite repeated planting attempts during the period 1940–62, *Spartina* failed to colonize the mudflats successfully and remained as isolated clumps approximately 2 m in diameter. The interesting point about these clumps was that, almost without exception, they were located in a hollow in the surrounding mudflat. There are three possible explanations for this observation. First, *Spartina* may preferentially colonize hollows in the initial unvegetated mudflat; second, *Spartina* could induce erosion once it is established; or third, accretion rates under these small *Spartina* clumps may be lower than on the surrounding mudflats. The first of the possibilities may be rejected on the grounds that the initial mudflat is not characterized by surface irregularities of the scale noted under the *Spartina* clumps. The other hypotheses need more careful examination. If *Spartina* fails to colonize in a continuous sward and forms isolated clumps which either create eroding conditions or decrease the rate of deposition, then any coastal protection policy which uses *Spartina* planting may actually create the reverse of those conditions it intends to accelerate.

METHODS

The experiments described here were performed in a 7.5 m long laboratory flume with a width of 0.3 m and a depth of 0.5 m. A bed of estuarine mud, 5 cm thick, was introduced into the flume and allowed to consolidate over 1 week. A stand of *Spartina anglica* was collected from an estuarine salt marsh by clipping the individual stems off at ground level. The stems were collected in late summer when they were in full leaf with the height of the stems averaging 0.45 m. These stems were then inserted into the mudflat bed on a regular grid intersection of 0.05 m starting at 2 m from the flume input; the mudflat for this first 2 m was left bare of vegetation. Once the *Spartina* had been introduced into the bed, a 0.4 m column of estuarine water containing 150 ppm of suspended sediments was allowed to stand over the bed for 2 days so that sediment was deposited over the surface.

Estuarine water containing 150 ppm suspended sediments having a grain diameter range of between 10 μm and 100 μm was then allowed to flow through the flume to a depth of 0.4 m and at an average velocity of 0.05 m/sec. Suspended sediment concentration was monitored continually during the experiment using a turbidity meter to ensure that concentration remained equal throughout.

Velocity profiles were then taken at 0.5 m intervals on the bare mudflat and at 0.3 m intervals within the *Spartina* zone, using an ultra-sonic current meter with an accuracy of ± 1 mm/s. Measurements were taken at 2 mm increments from the bed to height of 0.3 m.

Accretion rates on the bed were monitored at each alternate profile position using an ultra-sonic bed profiler accurate to $\pm 100 \mu$m. Measurements of accretion were carried out for 1 hour at each point.

RESULTS

Field Morphology

Field observations show that small clumps of *Spartina*, less than 2 m in diameter, stand in hollows of approximately 10 cm depth below the level of the surrounding mudflats. These hollows have a slight lip or ridge around their edge standing 1 cm above the mudflat level. Most of the hollows have standing water within them at low tide. More extensive stands of *Spartina*, however, show a gradual, but steady rise in height from the level of the fronting mudflat, levelling off at 5 m to 10 m from the edge of the vegetation to give a convex edge to the front of the marsh.

Laboratory Accretion Rates

Measurements taken in the flume with the ultra-sonic bed profiler (Table 9.1) show a similar pattern to the field observations. Figure 9.1 shows that a peak in the accretion rate occurs immediately inside the *Spartina* zone, but that this is followed by a drop in accretion for the next 1.5 m before increasing once more. This pattern of accretion would result in a surface morphology almost identical to that observed in the field. The results are reasonable, too, when compared with measured salt marsh accretion rates in the field. For example, Ranwell (1964) measured accretion rates of 0.1 m/year on a low *Spartina* marsh in Bridgewater Bay, Somerset. The rates of accretion measured in the flume experiment attains values of 0.08 μm/s at distances of 2 to 3 m within the marsh vegetation. Using a figure of 0.5 hours inundation per tide, the measured rate extrapolates to give an accretion rate of 0.105 m/year.

Table 9.1 Laboratory results

Distance from vegetation edge (m)	Upper U^* (m/s)	Lower U^* (m/s)	Z_0 (m)	Accretion rate (μm/s)
-1.50	0.0056	Un-vegetated zone:	0.0018	0.080
-0.50	0.0053	no lower profile	0.0013	0.081
0.20	0.0078	0.000033	0.0398	0.085
0.50	0.0026	0.0033	0.0347	0.083
0.80	0.0073	0.0025	0.0285	0.070
1.10	0.013	0.0011	0.0363	0.075
1.40	0.012	0.0031	0.0363	0.073
1.70	0.014	0.0037	0.0389	0.079
2.00	0.011	0.0011	0.0468	0.080
2.30	0.022	0.0023	0.0611	0.083
2.60	0.029	0.0036	0.1026	0.085

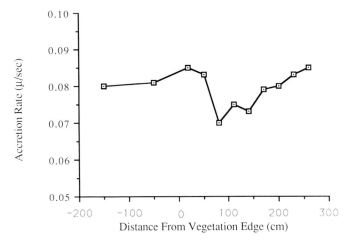

Figure 9.1   Accretion rates measured in the flume

Flow Characteristics

The results of the flow measurements in the flume are shown in a series of diagrams (Figure 9.2). The most obvious characteristic is the abrupt change from a simple logarithmic profile over the fronting mudflat to a series of complex profiles within the vegetation. The velocity profile over the bare mudflat under a mean flow velocity of 0.05 m/s is characterized by a $U*$ of 0.0054 m/s and has $Z_0 = 150 \mu$m. The characteristics of the velocity profiles within the vegetation, however, cannot be so easily described; the most obvious features are:

1. A general convex upward profile.
2. A marked discontinuity which occurs at between 0.03 m and 0.1/m from the bed.
3. An upper profile above this discontinuity which approximates a log form.
4. A lower profile which approximates a linear form.
5. An increase in shear velocity in the upper part of the profile from 0.0053 m/s on the bare mudflat to an average value of 0.013 m/s within the vegetated zone (Table 9.1).
6. A secondary velocity profile, beneath the discontinuity, which is characterized by a much lower shear velocity than the upper section. The average $U*$ here is 0.0022 m/s (Table 9.1).

These characteristics may be interpreted as showing the effects of the increase in bed roughness due to the presence of vegetation. The increase in shear stress caused by this increased drag has increased the Z_0, which is now marked by an abrupt discontinuity at heights ≥ 0.03 m. The term Z_0 is no longer strictly

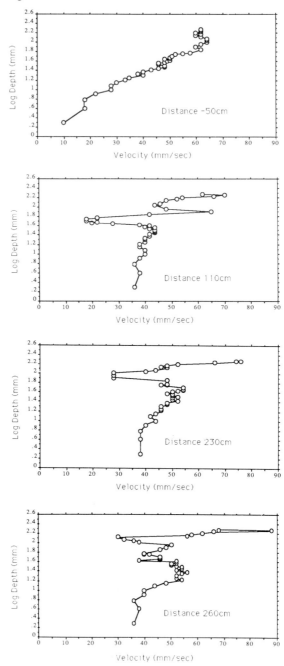

Figure 9.2 Velocity profiles at given distances from vegetation edge

applicable here, of course, since the profile does not actually cut the Y-axis; nevertheless it will continue to be used as a convenient shorthand. Perhaps the most interesting and important feature of the profiles within the marsh vegetation is the considerable depth of this Z_0 layer. If a semi-log regression is fitted to the velocity profile above the discontinuity then the intercept on the depth axis may be regarded as the direct analogy to Z_0 in a non-vegetated flow. Such a procedure shows that the Z_0 layer is 0.039 m deep at the leading edge of the vegetation, rising to 0.102 m at 2.6 m distance inside the vegetated zone (Table 9.1).

The lower part of the profile, that is beneath the discontinuity, may be regarded as a secondary velocity profile, perhaps equivalent to the viscous sub-layer described by McCave (1970) and Dyer (1986). This section of the flow is character-ized by low shear velocities and an apparently linear form, although this may be due merely to the absence of velocity measurements below 0.0015 m from the bed.

DISCUSSION

The intention of this chapter has been to offer some physical explanation for the pattern of accretion observed both in the laboratory results and in the field morphological data. It has always been assumed that the influence of vegetation on accretion rates on tidal salt marshes has been to reduce the velocity of flow, thereby allowing sediments to drop out of suspension. This is clearly altogether too simplistic a view and a more refined hypothesis might be that shear stress within the vegetation is reduced to some value below the critical shear for suspension, thus increasing accretion rates. Such a model would result in the type of equation for deposition rates suggested by Dyer (1979):

$$R = -C\,W_s\,2(1 - t/t_{\text{crit}})$$

where R = deposition rate, C = sediment concentration, W_s = sediment fall velocity, t = observed shear stress, t_{crit} = critical shear stress.

The results shown here, however, do not support such a generalized hypothesis. Instead of a simple modification of the slope of the velocity profile the presence of vegetation results in a two-stage profile. In the upper part of the profile shear velocities are in excess of the critical values needed to maintain a suspended sediment transport (Figure 9.3) and it is clear, therefore, that deposition cannot take place from this upper profile, which will maintain sediment transport across the upper surface of the tidal flow.

The lower velocity layer, corresponding to the viscous sub-layer of McCave (1970), on the other hand, does exhibit much lower shear velocities than either the upper profile or the profile outside the vegetation and it is here that the potential for increased accretion may occur. Table 9.1 and Figure 9.3 show that

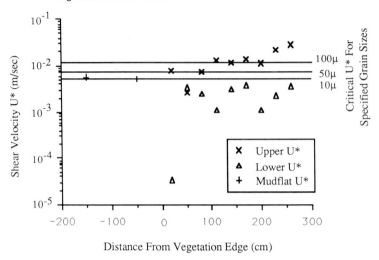

Figure 9.3   Observed and critical shear velocities for suspended sediment transport

the shear velocities here are always lower than the critical values needed for suspension so that deposition may be expected to take place from within this section of the flow. On the basis of the predictive equation for deposition rates quoted above it may be expected that the rate of deposition would show a direct correlation with the shear velocity in this lower zone but, as Figure 9.4 indicates, the spatial variation in this shear velocity is complex and cannot be directly related to the observed deposition rates

If the deposition rate at the bed is not directly related to the shear velocity despite the fact that these are always sub-critical in this zone, then an alternative

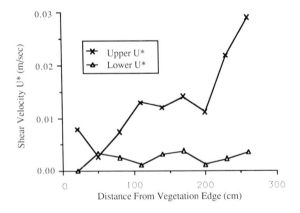

Figure 9.4   Variation in U^* for upper and lower profiles with distance from vegetation edge

hypothesis must be forwarded. It may be argued that the total amount of deposition which takes place on a marsh during a tidal inundation will be dependent on the height of the water column above the surface—in other words, the total amount of sediment held in suspension above a unit area of bed. In the case of the profiles described in this chapter, it is not the height of the water column which is relevant here but the height of the sub-layer to the velocity profile within which the shear velocities are sub-critical for suspended sediment. As reported above, the Z_0 position may be fixed by extrapolating the best fit regression line to the upper profile until it cuts the Y-axis (Table 9.1) and this provides a convenient method of defining the upper limit of the lower section of the profile. Figure 9.5 shows the spatial variation in this depth and Figure 9.6

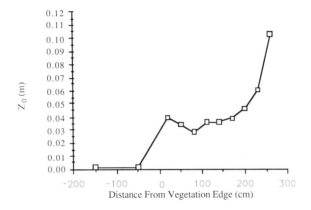

Figure 9.5   Variation in Z_0 with distance from vegetation edge

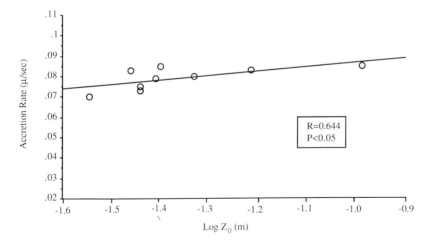

Figure 9.6   Semi-log relationship between accretion rate and Z_0

indicates that a relationship does exist between this depth and the observed accretion rates ($R = 0.644; P < 0.05$).

The implication of this finding is that the deposition rate is dependent on the height of the sub-layer, but this can only be true if the sub-layer is isolated from the suspended sediment in the upper velocity profile. If this were not so then the supply of sediment accross the interface between the two sections of the profile would mean that the height of the lower layer is independent of the volume of sediment available for deposition during a tidal inundation. If exchange of sediment across the interface was taking place then the deposition rate would be influenced by the height of the entire water column which is constant at all points over the marsh surface. If the lower layer is effectively isolated from the sediment supply in the upper layer then only those sediments contained within the lower layer can be deposited and the height of the layer becomes a critical factor. It is clear that the absence of exchange across the interface would soon result in very much lower suspended sediment concentrations in the lower layer—and work on the vertical distribution of sediments within this lower 3 cm is proceeding. The present chapter merely offers this as one explanation of the observed relationships.

Finally some discussion may be made here of the reasons for the observed spatial variation in the depth of the lower velocity profile (Figure 9.5). Following Bagnold (1941), many authors (see, for example, Mulligan, 1988) have described the existence of a 'focus' for the velocity profile about which the profile rotates as its gradient changes. The existence of such a focus means that as shear stress increases, due, for example, to an increase in surface roughness, so the profile slope flattens and the section of the profile to the left of the focus rises. Thus Z_0 must increase as shear stress increases so that the presence of a roughness element such as vegetation must result in a deeper Z_0 or, as we show in this chapter, a deeper sub-layer to the profile. The measurements taken from the flume experiment show that the development of the complex velocity profile within the vegetated zone takes place over a distance of approximately 2 m. In the leading edge of the vegetation, therefore, the profile is changing rapidly and an initial sudden rise in the Z_0 layer is followed by a fall with a parallel rise and fall in the accretion rates here. This means that clumps of vegetation which are smaller than this 2 m mixing length will experience lower accretion rates overall than either on the fronting mudflat or on a more expansively vegetated marsh surface and hence such clumps will produce the depressions on the marsh surface observed in the field.

CONCLUSION

The work reported here indicates that the effect of vegetation on salt marsh deposition is more complex than was previously assumed. The deposition rate is shown to depend on the depth of a secondary velocity profile which occupies the lower 0.03–0.10 m of the water column. The shear velocities calculated for the upper and lower layers lie either side of the critical values for suspended sediments so that deposition cannot take place from the upper layer but will do so from within the lower zone. The implication of the relationship between the depth of the lower layer and the deposition rate is that there is little input of sediment particles from the upper into the lower layer—an important hypothesis which clearly requires further research.

The mixing length for the development of the velocity profile in the laboratory flume is approximately 2 m and vegetated zones smaller than this length can be expected to receive lower accretion rates than more extensively vegetated surfaces or even on bare mudflats. This has important implications for the use of *Spartina* for coastal protection works especially in view of the poor development of *Spartina* swards now described in many localities. In such cases a patchy *Spartina* sward may actually depress accretion rates to levels below those which would have taken place without the presence of vegetation.

REFERENCES

Bagnold, R. A. (1941). *The Physics of Blown Sand and Desert Dunes*, Morrow & Co., New York.

Carr, A. P. (1983). A new technique for determining the surface level of salt marshes and other compacted sedimentary environments. *Earth Surface Processes and Landforms*, **8**, 293-6.

Carr, A. P. and M. W. L. Blackley (1986). Seasonal changes in surface level of a salt marsh creek. *Earth Surface Processes and Landforms*, **11**, 427-39.

Dyer, K. R. (ed.) (1979). *Estuarine Hydrography and Sedimentation*. Cambridge University Press.

Dyer, K. R. (1986). *Coastal and Estuarine Sediment Dynamics*, Wiley.

Fonesca, M. S., J. S. Fisher and J. C. Zeiman (1977). Influence of the seagrass, *Zostera marina L.*, on current flow. *Estuarine, Coastal and Shelf Science* (1982), **15**, 351-64.

Letzsch, S. W. and R. W. Frey (1980). Deposition and erosion in a Holocene salt marsh, Sapelo Island, Georgia. *J. Sediment. Petrol.*, **50**, 529-42.

McCave, I. N. (1970). Deposition of fine-grained suspended sediment from tidal currents. *J. Geophys. Res.* **75**, 4151-9.

McCave, I. N. (1979). Suspended sediment. In K. R. Dyer (ed.), *Estuarine Hydrography and Sedimentation*.

Mulligan, K. R. (1988). Velocity profiles measured on the windward slope of a transverse dune. *Earth Surface Processes and Landforms*, **13**(7), 573-82.

Pestrong, R. (1965). The development of drainage patterns on tidal marshes. *Stanford Stud. Geol.*, **10**, No. 2.

Ranwell, D. S. (1964). Spartina marshes in southern England, II.Rate and seasonal patterns of sediment accretion. *J.Ecol.*, **52**, 79-94.

Richards, F. S. (1934). The salt marshes of the Dovey Estuary, IV. The rates of vertical accretion, horizontal extension and scarp erosion. *Ann. Bots.*, **18**, 225-59.

Steers, J. A. (1935). A note on the rate of sedimentation on a salt marsh at Scolt Head Island, Norfolk. *Geol. Mag.*, **72**, 443–5.

Steers, J. A. (1948). Twelve years' measurements of accretion on Norfolk salt marshes. *Geol. Mag.* **85**, 163–6.

Stevenson, J. C., L. G. Ward and M. S. Kearney. Sediment transport and trapping in marsh systems: Implications of tidal flux studies. *Mar. Geol.*, **80**, 37–59.

Stumpf, R. P. (1983). The process of sedimentation on the surface of a salt marsh. *Estuarine Coastal Shelf Science*, **17**, 495–508.

10 Effects of Vegetation on Riverbank Erosion and Stability

COLIN R. THORNE
Department of Geography,
University of Nottingham

SUMMARY

Recent studies of the hydraulic geometry of natural channels have highlighted the importance of bank vegetation in affecting bank processes and width adjustment. Physically, this may be explained by considering the effects of vegetation on the processes of erosion and deposition responsible for width adjustment and the establishment of a stable cross-section.

Serious bank retreat usually occurs by a combination of fluvial erosion of intact bank material and mass failure under gravity, followed by basal clean-out of disturbed material. Vegetation can significantly affect both flow erosion and mass stability. Bank accretion occurs by the accumulation of sediment as a basal wedge or berm. Berm building is encouraged by the ingress of vegetation, which tends to retard near-bank velocities and induce deposition.

However, vegetation effects are complex, and vegetation cannot be classed simply as benefit or liability to bank stability without detailed consideration of other factors including the processes responsible for retreat or advance, bank material properties and bank geometry, and the type, age, density and health of the vegetation.

INTRODUCTION

The influence of bank properties on channel geometry has been clearly demonstrated both in empirical studies (for some recent examples see Charlton, 1982; Andrews, 1984; Bray, 1984; Hey and Thorne, 1986), and on theoretical

Vegetation and Erosion
Edited by J. B. Thornes
©1990 John Wiley & Sons Ltd

grounds (Thorne and Osman, 1988). Despite this, it is still argued by some researchers that the relative success of regime equations that ignore bank characteristics demonstrates that bank effects must be negligible (Bettess, White and Reeve, 1988). Although this is a minority view, and most geomorphologists and engineers accept that the bank material properties do influence hydraulic geometry, there is still no consensus of opinion on the linkage between bank characteristics, bank processes and the stable channel geometry.

Within this area of uncertainty there exists further controversy concerning the role of bank vegetation in affecting bank properties, processes and stability. Much of this controversy arises because the role of vegetation is often subtle and complex, while statements describing its role tend to be sweeping and generalized. Progress depends on recognizing that it is not sufficient to consider banks simply as vegetated or unvegetated. On vegetated banks factors such as the type and density of vegetation, its age and health are also important because they directly control the vegetation's influence, which may be either to enhance, or reduce bank stability.

In this chapter an attempt is made to establish a framework within which to consider the effects of vegetation on bank processes and hydraulic geometry. To achieve this, it is necessary first to review the processes and mechanisms responsible for bank retreat or advance. An exhaustive account of these topics may be found in Thorne (1982), but a short summary is included here to allow this chapter to stand alone. The remainder of the chapter concentrates on an examination of the role of bank vegetation in affecting each of these processes and mechanisms. On this basis, a framework is presented that could form the basis for future research aimed at quantifying vegetative effects.

EROSION PROCESSES

Bank erosion occurs when grains or assemblages of grains are removed from the bank face by the flow. Erosion consists of two distinct events: detachment and entrainment. Sufficiently strong forces of lift and drag exerted on the bank by the flow may detach and entrain grains directly from the intact soil, but more commonly grains are loosened and even detached prior to entrainment by weakening and weathering under sub-aqueous or sub-aerial conditions. The nature of the processes responsible, and the form of grain or grain assemblage entrained, depend on the geotechnical properties of the bank material. Of particular importance is the presence or absence of cohesion.

Non-cohesive Banks

Non-cohesive bank material is usually detached and entrained grain by grain. Stability depends on the balance of forces acting on surficial grains. Motivating forces are the downslope component of submerged weight and the applied fluid forces of lift and drag. Resisting forces are the slope-normal component of submerged weight and inter-granular forces due to friction and inter-locking. Inter-locking can be a major source of erosion resistance in imbricated alluvial deposits. When analysing the stability of a non-cohesive bank with respect to flow erosion, the lift and drag forces can be represented by the boundary shear stress, because this scales on the same parameters of flow intensity (Wiberg and Smith, 1987). This approach yields reliable estimates of bank stability for non-cohesive materials. However, for natural rivers, the practical applications of analyses for non-cohesive banks are limited by the fact that most alluvial bank materials exhibit some cohesion. This may be real cohesion due to the presence of silt and clay fractions, or apparent cohesion due to either capilliary suction in the unsaturated zone, or the binding effect of vegetation roots and rhizomes.

Cohesive Banks

Cohesive bank material is usually eroded by the detachment and entrainment of aggregates or crumbs of soil. The motivating forces are the same as those for non-cohesive banks, but the resisting forces are primarily the result of cohesive bonds between particles and aggregates. The bonding strength, and hence the bank's erosion resistance, depends on the physico-chemical properties of the soil and the chemistry of the pore and eroding fluids (Arulanandan, Gillogley and Tully, 1980). Field and laboratory experiments show that intact, undisturbed cohesive banks are much less susceptible to flow erosion than are non-cohesive banks (Thorne, 1982).

Usually, serious erosion of cohesive banks takes place through the loosening or detachment of aggregates by sub-aerial or sub-aqueous processes, followed some time later by entrainment of the disturbed material by the flow (Wolman, 1959). The processes responsible for loosening aggregates are mostly driven by the dynamics and physical state of soil moisture close to the bank face.

If the bank is poorly drained, positive pore water pressures act to reduce the effective cohesion and weaken the soil. In extreme cases, loss of strength may be complete, leading to bank failure by liquefaction. The most favourable conditions for high pore pressures occur in saturated banks following heavy and prolonged precipitation, snowmelt, and/or rapid drawdown in the channel. Where the soil surface is exposed, surface processes of raindrop impact and overland flow may also be important.

Most alluvial soils are expansive, that is they swell and shrink significantly during cycles of wetting and drying. This generates a ped fabric with desiccation cracks between peds and a crumb structure to the soil. Cohesion between peds and crumbs is much weaker than within them, so that a heavily desiccated soil may have little erosion resistance.

Freezing of soil moisture can seriously reduce erosion resistance. This has been demonstrated by Lawler (1986) for the case of needle ice formation. At larger scales, ice lenses and wedges heave apart soil peds and blocks destroying inter-ped cohesion, and ice cantilevers left on the banks during the spring thaw can do serious structural damage (Church and Miles, 1982).

FAILURE MECHANICS

Flow erosion generates bank retreat directly, but it also promotes mass instability leading to more rapid and more spectacular retreat. Failure occurs when flow erosion of the bank and/or the bed adjacent to the bank reduces the factor of safety with respect to the most critical mode of failure to unity. The type of failure depends on the geometry of the bank, the geotechnical properties of the bank material and the bank stratigraphy. Again, the presence or absence of cohesion is particularly important.

Non-cohesive Banks

Mass failure of non-cohesive banks occurs by shearing along shallow, planar or slightly curved surfaces. The motivating force is shear stress on the potential failure plane due to the downslope component of weight. Resisting force is the shear strength of the potential failure plane, due to the slope-normal component of weight, friction and granular inter-locking. Deep-seated failures are rare because in a non-cohesive material the shear strength increases more quickly with depth than does the shear stress (Terzaghi and Peck, 1948). In well-drained banks, failure occurs when the bank slope angle exceeds the friction angle. This can result from bank weakening due to the loss of imbrication or from over-steepening of the bank angle by basal scour. In poorly drained banks, failure can occur owing to a reduction in the effective friction angle due to positive pore pressures. Consequently, mass failure may be triggered by rapid drawdown or heavy precipitation. Generally though, the coarse texture of non-cohesive banks makes them well drained. The stability of non-cohesive banks with respect to mass failure may be analysed quite reliably using Taylor's infinite slope approach, which may be found in standard geotechnical engineering texts (Terzaghi and Peck, 1948).

Cohesive Banks

For cohesive banks the motivating force is the downslope component of weight of the potential failure block. Shear strength depends not only on frictional forces, but also on cohesion. Deep-seated failures are common on unstable cohesive riverbanks. This is the case because in a cohesive bank the shear strength increases less quickly with depth than

Figure 10.1   (a) Slab-type failure of a steep, eroding river bank. (b) Rotational slip of a less steep, eroding river bank. Red River near Plain Dealing, Louisiana

does the shear stress (Terzaghi and Peck, 1948). The shape of the critical failure surface depends mostly on the geometry of the bank. Low, steep banks fail by sliding downwards and outwards along an almost planar surface (Figure 10.1a). Usually, the upper half of the potential failure block is separated from the intact bank by a near-vertical tension crack. This crack is the result of tensile stress that exists in the upper part of the bank adjacent to a steep slope (Terzaghi and Peck, 1948). During failure, a slab or block of soil topples forward into the channel. This is called a slab-type or toppling failure. High, less steep banks fail by rotational slip along a curved surface passing close to, or just above, the toe of the bank. The failure block is back-tilted away from the channel (Figure 10.1b). Generally, slab failures abound on unstable banks steeper than about 60° and rotational slips on banks with slopes less than 60° (Lohnes and Handy, 1968; Thorne, 1988).

Most mass failures of cohesive banks occur following rather than during high flows in the channel. This is because the switch from submerged to saturated conditions that accompanies drawdown in the channel approximately doubles the bulk unit weight of the bank material, increasing the motivating force on the potential failure surface in about the same proportion. This is the case even in the absence of significant excess pore water pressures. In undrained banks the probability of failure being triggered by rapid drawdown is even greater.

The stability of cohesive banks with respect to rotational slip may be analysed using well-established techniques such as the Bishop simplification of the method of slices (Bishop, 1955). This has been found to give results which are very close to those of more sophisticated methods, and is much simpler to use. Slab-type failures have received relatively less attention and their analysis is less well established, but a method developed by Osman and Thorne (1988) seems to give reliable estimates of bank stability with respect to this failure mode (Thorne, Biedenharn and Combs, 1988).

Stratified Banks

Alluvial banks often consist of layers of non-cohesive and cohesive materials. Erosion processes on stratified banks proceed by a combination of those found on single material banks. Generally, non-cohesive layers are eroded more quickly than cohesive ones. This leads to the generation of berms and terraces where cohesive material underlies non-cohesive material, and an overhanging or cantilevered bank where cohesive material overlies non-cohesive material (Thorne, 1978). Bank failures can be complex. The failure surface may lie entirely within one layer, or may cut across several layers. Each case must be analysed individually, depending on the sequence, thickness and geotechnical properties of each layer in the stratified bank.

BANK ACCRETION

The opposite of bank erosion and retreat is accretion and advance. This occurs when the bank is stable and sediment input to the basal area from longstream and cross-stream transport is greater than sediment output to downstream. The deposited sediment forms a basal accumulation that has been variously termed a basal wedge (Thorne, 1978) or berm (Harvey and Watson, 1986). The wedge or berm further stabilizes the bank by protecting intact bank material from flow attack and loading the toe against mass failure.

Basal accumulation may be driven by general bed aggradation due to sediment overloading downstream of a rapidly eroding reach (Harvey and Watson, 1986). In this case berms develop along both banks through a reach (Figure 10.2). Often though, a berm appears behind an attached bar, such as an alternate bar in a straight channel. In this case berms develop on alternate sides of the channel, while flow attack and retreat of the bank opposite continue and are intensified by flow deflection by the bar–berm couplet (Thorne, 1978). This process leads to the development of a sinuous channel from an initially straight one.

Figure 10.2   Bank accretion by the formation of a wedge or berm at the base. *Note:* (i) former cut bank lines behind the berms, now stabilized naturally because of the protection afforded by the berms; and (ii) volunteer species invading berms and accelerating berm growth by trapping wash load. James Wolf Creek near Senatobia, Mississippi

BASAL ENDPOINT CONTROL

The concept of basal endpoint control explains the linkage between sedimentary processes operating exclusively on the banks and those operating in the channel as a whole. The concept characterizes the balance in sediment coming into and going out of the basal area. Bank processes and failures input sediment to the basal area, in a more or less disturbed state. The removal of this material depends entirely on its entrainment by the flow. Consequently, the amount and residence time of sediment stored in the basal area depend on the balance between the rates of supply from the bank and removal by the flow. There are three possible states for this balance: input greater than output (impeded removal); input equal to output (unimpeded removal); and input less than output (excess basal capacity). The rate of bank retreat adjusts to the state of basal endpoint control as follows.

Impeded Removal

Bank processes, plus any sediment inputs from upstream and laterally across the channel, supply material to the basal area at a higher rate than it is removed by the flow downstream. Basal accumulation (berm building) results, decreasing the bank angle and height, and buttressing the bank. Bank stability increases and the rates of sediment input and bank retreat decrease, tending towards the second state.

Unimpeded Removal

Processes delivering and removing sediment are in balance. No temporal change in basal elevation or storage takes place. The bank retreats by parallel retreat at a rate determined by the degree of fluvial activity at the base. If the sediment load at the base is zero, then the state of basal endpoint control is *static* and the bank retreat rate is zero.

Excess Basal Capacity

Basal scour has a greater capacity to transport sediment than that supplied by bank processes, failures and fluvial inputs. Basal lowering and under-cutting result, increasing bank height and angle, and decreasing bank stability. The rates of sediment input and bank retreat increase, tending towards the second state.

The concept demonstrates that the long-term rate of bank retreat is fluvially controlled, regardless of the nature of the bank and the processes and mechanisms actually involved in bank retreat. That does not mean that these factors are irrelevant, however, because they are important in controlling the bank geometry and setting limiting values for the stable bank height and angle.

The implications of this concept for channel evolution towards a stable, or regime, geometry are discussed at length in a recent paper (Thorne and Osman, 1988).

VEGETATION EFFECTS

Bank erosion processes and failure mechanisms act in different ways to produce bank retreat. Vegetation can significantly affect either, or both, facets of retreat in any particular riverbank. It is convenient to consider these affects separately, while bearing in mind that in the real world they operate interactively to produce a net impact on bank stability that may be positive or negative. Vegetation also plays a central role in bank accretion and berm building. This is considered after the discussion of flow erosion and bank stability. These effects have important implications for the use of vegetation in schemes to protect and stabilize riverbanks which are discussed briefly at the end of this section.

Flow Erosion

Retardance of Near Bank Flow

Bank vegetation increases the effective roughness height of the boundary, increasing flow resistance and displacing the zero plane of velocity upwards, away from the bank. This has the effect of reducing the forces of drag and lift (and their surrogate, boundary shear stress) acting on the bank surface. As the boundary shear stress is proportional to the square of near bank velocity, a reduction in this velocity produces a much greater reduction in the forces responsible for erosion. Quantification of this effect is difficult, however.

A great deal of sound theoretical and empirical work on vegetation and flow resistance has been undertaken by Nicholas Kouwen at the University of Waterloo, Canada (for example see Kouwen, 1970; Kouwen and Unny, 1973; Kouwen and Li, 1979; Kouwen, Li and Simons, 1980). This has shown that the extremely complicated nature of flow over a vegetated surface precludes a complete mathematical description based on the physics of flow. Much of the complication arises because when natural vegetation is subjected to a streamwise drag force it tends to bend. The amount of bending depends on the flexural stiffness of the stem and the magnitude of the fluid drag. But the drag depends on the flow velocity, which in turn depends to some extent on degree of bending. The velocity distribution remains approximately logarithmic throughout, the major effect of bending being to move the virtual origin of the distribution (the height above the bed at which the velocity goes to zero) progressively nearer the boundary as bending increases. Using these facts it has been possible to develop a model which is process based and incorporates the basic flow and boundary parameters, and which on this basis is an improvement over purely empirical methods (Kouwen, 1987). Kouwen's model is the basis for a procedure to calculate the effective roughness height and velocity distribution for vegetated channels used by the US Department of Agriculture (1980). The input parameters are: vegetation height; vegetation stiffness (a function of type, height and health of plant (US Department of Agriculture, 1954)); and channel geometry and slope. This approach could form the basis for development of a physically based method to calculate and predict the response of near bank velocity distributions to the presence of a particular type of bank vegetation.

A further effect of bank vegetation on the near bank flow field can be to damp turbulence. It is now recognized that detachment and entrainment usually occur during turbulent sweeps, when velocities and stresses may, for short durations, attain values triple the time-averaged mean (Jackson, 1976; Leeder, 1983). If vegetation reduces the magnitude of instantaneous velocity and shear stress peaks by suppressing meso- and macro-scale eddies, then this would further reduce the erosive attack on the bank.

Clearly, the type of vegetation is very important. While grasses and shrubs are effective at low velocities, their impact decreases as velocities increase, and is all but eliminated once the stems are prone. Conversely, the stems of woody species continue retarding the flow up to very high velocities, but may generate serious bank scour through the local acceleration of flow around their trunks. In this regard, the density of vegetation is probably an important factor too.

Also of importance is the spacing of trees or shrubs along the channel. Single trees or small groups of trees are impediments to the flow that generate large-scale turbulence and severe bank attack in their wakes. Hence, the flow is usually able to isolate and flank hard points in the bank resulting from the effects of widely spaced trees or groups of trees. For trees to be effective in reducing flow attack on the bank they must be spaced sufficiently closely that the wake zone for one tree extends to the next tree downstream, preventing re-attachment of the flow boundary to the bank in between. In this regard the effects of trees continue even after the death of the plant. An isolated, downed tree may generate local scour and, unless removed, can become a locus of serious channel instability. But a dense accumulation of downed timber on a bank can be quite effective in protecting the bank from flow scour (Figure 10.3).

Scale must also be considered. For example, on a small river fallen trees divert the flow and generate significant scour of the bed and opposite bank, effects obviously negligible in large rivers.

The extra flow resistance associated with a dense stand of vegetation on the banks of a river is seen as a disbenefit by many river engineers. This view is based on the assumption that by increasing roughness, bank vegetation

Figure 10.3 Downed trees slowing the rate of bank retreat by forming a natural crib-wall at the base of an eroding bank. Red River near Fulton, Arkansas

significantly reduces channel capacity, thereby promoting flooding. On these grounds, bank vegetation is often removed in a 'clearing and snagging' operation, the stated aim of which is to increase the channel's conveyance. In fact, in most natural channels at high in-bank flows the contribution of bank roughness to total channel resistance is small. This is the case because in channels of high width-to-depth ratio flow ($w/d > 30$) resistance depends mostly on bed roughness and channel shape, not bank roughness. For such channels any increase in conveyance achieved through clearing bank vegetation is more than lost when bank erosion due to reduced erosion resistance leads to widening, an increase in width-to-depth ratio, and a reduction in the hydraulic efficiency of the channel cross-section.

Reduction of Soil Erodibility

Compared to an unvegetated or fallow state, slopes covered by a good stand of close-growing vegetation experience an increase in erosion resistance of between one and two orders of magnitude (Carson and Kirkby, 1972; Kirkby and Morgan, 1980). Vegetation not only protects the soil surface directly, but the roots and rhizomes of plants bind the soil and introduce extra cohesion over and above any intrinsic cohesion that the bank material may have. The presence of vegetation does not render a bank immune from flow erosion, but the critical condition for erosion of a vegetated bank is the threshold of failure of the plant stems by snapping, stem scour, or uprooting rather than that for detachment and entrainment of the bank material itself. Vegetation failure is usually associated with much higher levels of flow intensity than soil erosion *per se.*

Vegetation also reduces the effectiveness of weakening and loosening processes which are often the precursors to flow entrainment. All else being equal, vegetated banks are better drained and drier than unvegetated ones, so that the impact of moisture-related processes is immediately reduced. Further, reinforcement by roots and rhizomes means that grains that would have been detached by, say, needle-ice formation on an unvegetated bank, remain bonded to the bank when vegetation is present. Again the type, age and health of vegetation are important.

Infiltration capacity at the surface of a vegetated soil is generally much higher than for the unvegetated state. Consequently, the volume of surface runoff for a given precipitation event is reduced, decreasing its effectiveness in generating surface erosion.

Generally, a species with a dense network of fibrous roots is of more benefit than one with a sparse network of woody roots. Woody roots may disturb the structure and any imbrication of the bank material, and weaken it through root wedging, though research by Gray (1978) suggests that this is at most a second-order effect.

To be effective, vegetation must extend down the bank at least to the average low water plane, otherwise the flow will undercut the root zone during significant flow events. In this respect, plants which are tolerant of inundation are more effective than terrestrial species. For overall bank protection, a mixture of riparian and terrestrial species provides the best solution.

Quick-growing species are better able to respond to erosion events and to recolonize eroded areas of bank. Plants that are slow to mature are likely to be attacked and washed out if the time needed for them to become established is longer than the return period for flows that significantly eroded the bank. Species that die back in winter provide little or no protection during that season, which may well be the period of most significant bank erosion.

Dead vegetation leaves relic roots in the bank. The effectiveness of relic roots in binding the soil decreases with time and eventually the voids and holes left by dead roots may reduce cohesion to a level below that for fallow soil. The rapidity with which relic roots decay depends on conditions within the bank and cannot easily be predicted.

Bank Stability

The impact of vegetation on slope stability has been the subject of careful and sustained research by Donald Gray at the University of Michigan (see for example, Gray, 1978; Gray and Leiser, 1983; Gray and MacDonald, 1989). Riverbanks constitute a particular class of slopes in general, and in this context many of the findings from slope studies can be applied to riverbank stability. However, care must be exercised to take full account of the characteristics of bank erosion and failure when applying results and conclusions based on studies of other types of slope.

Bank Drainage

Compared with unvegetated banks, vegetated banks are drier and much better drained. Both facts enhance the stability of the vegetated banks with respect to mass failure. In general terms, vegetated banks are drier for three major reasons. First, the canopy prevents between about 15 and 30 per cent of the precipitation from ever reaching the soil surface, by intercepting it and re-evaporating it to the atmosphere. Second, plants draw water from the soil and transpire it to the atmosphere, reducing soil moisture levels between precipitation events and lowering the water table in the bank. Third, suction pressures in the soil are increased by water abstraction at the roots, so that the height of the capilliary fringe is increased and water is drawn towards the surface from greater depths than in an unvegetated bank. This has the effect of increasing evaporative loss from the soil surface. Drier banks are more stable because the bulk unit weight of the soil is reduced while effective and apparent cohesions are both increased. Also, lower antecedent moisture levels reduce the frequency of saturated conditions in vegetated banks.

The better drainage of vegetated banks occurs because of the more open structure of a soil penetrated by plant roots produces order of magnitude increases in bulk hydraulic conductivity. By reducing the magnitude of positive pore pressures that often trigger failure following rapid drawdown, vegetation can make a major contribution to bank stability.

Dead vegetation can be detrimental to drainage conditions within the bank. Relic roots and root holes present ready pathways for rapid seepage that can lead to piping. In this context piping is the erosion of a pipe-shaped cavity, usually followed by catastrophic failure of the overburden. The term is also applied to a failure resulting from loss of strength due to high excess pore pressures, here termed liquefaction.

Soil Reinforcement

Soil is strong in compression, but weak in tension. Plant roots are weak in compression, but strong in tension. When combined, the soil–root matrix produces a type of reinforced earth which is much stronger than the soil or the roots separately. The theory of reinforced earth was first established by Vidal (1969). Roots are effective both in adding tensile strength to the soil and, through their elasticity, distributing stresses through the soil, so avoiding local stress build-ups and progressive failures. Careful work over many years by Larry Waldron at the University of California at Berkeley has shown that compared with the fallow state, increases of 100 per cent and more in the shear strength of soils are associated with the development of a mature stand of willows (Waldron, 1977). Gray (1978) and Gray and MacDonald (1989) report similar results for other species. These increases come about primarily due to increases in the apparent cohesion of the soil. Root reinforcement seems to have little impact on the apparent friction angle.

Increases in tensile strength can be even greater. In studies of alluvial bank materials in Mississippi bluff-line streams, Thorne, Murphey and Little (1981) found the tensile strength of rooted samples to be on average ten times that of unrooted ones. Hence, vegetated banks are better able to resist the development of tension cracks due to desiccation and to tensile stresses behind steep banks that often trigger both slab-type and cantilever failure of unvegetated banks.

An important point that must be borne in mind is that reinforcement extends only down to the rooting depth of the vegetation. For grasses this is just a few centimetres, for shrubs some tens of centimetres and even for trees is seldom more than a metre. Bank height is therefore an important factor. If the bank height is less than or equal to the rooting depth, then roots almost certainly cross the potential shear surface and reinforce it against failure. However, if the bank height exceeds the rooting depth then slip surfaces for toe failures will pass beneath the zone of root reinforcement. Roots will continue to prevent shallow slips and still bind the failure block together during and after collapse— so that failed blocks are more likely to remain at the toe and protect the intact bank from further erosion, but the stabilizing effect of root reinforcement on deep-seated failures is lost (Gray and MacDonald, 1989). This phenomenon is clearly demonstrated on the banks of streams that are subject to severe degradation (Figure 10.4). Trees switch from holding the banks up to dragging them down in response to increasing bank heights and angles caused by basal lowering and toe erosion (Simon and Hupp, 1986; Harvey and Watson, 1986).

Slope Buttressing and Soil Arching

A buttress is a massive structure at the toe of a slope which retains the slope and loads the toe against shear failure. Arching occurs when soil is prevented from sliding between or around piles that are firmly anchored in an underlying, and unyielding layer. Trees can act as gravity buttresses and/or cantilever piles with soil arches between them. The theory of slope stability under these circumstances has been analysed by Wang and Yen (1974). Gray (1978) reports field examples of buttressing and arching in slopes in Idaho. For buttressing and arching to be effective on a riverbank requires a well rooted and closely

Figure 10.4 Channel degradation has increased the bank height so that it exceeds the rooting depth of this tree. The reinforcing effect of the roots has been lost, as the critical failure surface for either slab-type or rotational slip failure would pass below the root zone. Furthermore, undercutting below the root zone means that the surcharge effect of the weight of the overhanging tree is likely to cause a mass failure. Red River near Fulton, Arkansas

spaced stand of trees extending along the bank toe. This should increase bank stability with respect to shallow and deep-seated slips, and soil creep.

Surcharging

The surcharge weight of vegetation, and especially trees, is generally recognized as reducing bank stability. This is certainly the opinion of most river engineers, who have seen sections of an otherwise stable bankline dragged down by the weight of overhanging trees. However, surcharging can also be beneficial to bank stability. This depends on the slope angle and the position of the tree on the bank. On gently sloping banks, the contribution of surcharge weight to the downslope component of weight is small compared with that of the slope-normal component. Consequently, the net effect of surcharging is to increase stability through increasing frictional resistance to shearing. On steep banks the converse is true, and surcharging decreases stability.

Trees at the top of steep banks are usually a major liability with respect to surcharging, because their weight tends to produce both a shear force and a turning moment that are highly effective in promoting toppling failure (Figure 10.4). This effect is especially pronounced when bank-top trees lean over into the channel as a result of asymmetry due to grazing only on the bankward side, or to wind loading. These problems are most strongly associated with bands of trees only one or two trees deep and hence a wide stand of trees is preferable in terms of their impact on bank stability.

Bank Accretion

Bank accretion depends on the deposition of sediment on the bank and in the basal area. Theoretical and empirical results suggest that vegetation plays a central role in promoting sediment deposition on advancing banks. On natural banks volunteer species colonize berms, enhancing their stability and accelerating their growth rate (Figure 10.2). Increased stability comes about because of the effects discussed already in the preceding paragraphs. Accelerated growth occurs for two reasons.

Firstly, as the capacity of flowing water to transport bed material load increases approximately with the sixth power of the velocity, the effect of bank vegetation in retarding near bank velocities by increasing local flow resistance can be startlingly effective in promoting sediment deposition.

Secondly, vegetation is also effective in trapping fine material carried as wash load. Wash load is sediment carried by the river which is finer than that commonly found in the bed. Its transport is almost independent of the flow velocity, and usually wash load travels quickly through the fluvial system to be deposited at the downstream limit, either in an estuary or a lake. But wash-load deposition can be significant on vegetated banks where dense stands of stalks and stems damp turbulence and filter out fine material. For example, Harvey and Watson (1986) noted that bank advance by berm building in streams in Mississippi was greatly enhanced by the establishment of willows that produced wash-load deposition manifest as clay drapes in the deposited sediment (Figure 10.2).

Bank Protection and Stabilization

Clearly, the potential exists for the use of vegetation in protecting riverbanks against erosion and stabilizing them with respect to mass failure. However, no consensus exists as to the practical usefulness of vegetation in this regard. For example, experience gained in the Section 32 Streambank Protection Demonstration Program (US Army Corps of Engineers, 1981) showed that vegetation can be an important and integral part of a scheme. On the other hand, Corps' design and maintenance manuals for bank protection schemes that use riprap specifically exclude the use of vegetation at the construction stage, and require that any volunteer species that subsequently colonize the stabilized bank be removed. The reasons for the wholesale exclusion and removal of vegetation from riprap banks are fourfold. First, the stems of woody vegetation might cause local acceleration and scour of the riprap in the wake zone. Second, growth of vegetation might disrupt the riprap blanket. Third, the roots of dead vegetation leave voids in the bank that weaken it and may cause local subsidence. Fourth, and most important, vegetation obscures the bank surface making annual inspection to check the integrity of the riprap difficult. These are all valid arguments against using certain types and species of vegetation, but the point that has been missed is that they do not apply equally to *all* types and species.

Experience in stabilizing riverbanks in a wide range of fluvial environments in the USA has established the salient and universal rule that vegetation protection alone is usually insufficient and cannot be relied upon. In practically all cases, it is essential to fix the toe of the bank using an engineering structure. This may be achieved using a revetment, longitudinal dyke, hardpoints, or a series of transverse dykes. However, it is seldom necessary to carry this structural protection up to the top of the bank. In fact, experience shows that 'minimum toe protection' allied to some bank grading is all that is required structurally, and that upper bank protection using a suitable combination of planted and volunteer species provides sufficient protection. This approach has been found to be compatible with sound engineering, while comparing favourably with more traditional alternatives economically (US Army Corps of Engineers, 1984). Perhaps its main benefit is the great potential for improving the environmental impact of bank protection schemes on in-stream and riparian ecosystems.

CONCLUSIONS

This chapter has demonstrated that the role of vegetation in affecting bank erosion and stability is complex. At this stage it is not possible to quantify the effects of vegetation in any general fashion, although theoretical models and practical experience that might be used to this end do exist. It is possible, though, to identify the parameters defining the bank, channel and vegetation that must be considered in developing a viable approach. Table 10.1 sets out these parameters on the basis of the content of this chapter. Future research should be aimed at improving the definition of these variables, resolving areas of

Table 10.1 Framework for the assessment of vegetation effects on bank erosion and stability

A. BANK PARAMETERS	B. CHANNEL PARAMETERS
Bank geometry	*Flow intensity*
Height	Near bank velocity
Slope	Bank shear stress
Presence of berms	Level of macro-turbulence
Bank materials	*Channel geometry*
Non-cohesive	Cross-sectional size and shape
Cohesive	Planform
Stratified	Sediment load
Level of fertility	Degradational or aggradational?

C. VEGETATION PARAMETERS

Type	*Density*	*Position*
Grasses	Sparse	Bank toe
Shrubs	Open	Mid-bank
Trees	Dense	Top bank
Diversity	*Age*	*Spacing*
Mono-stand	Immature	Continuous
Mixed	Mature	Close
Climax-vegetation	Old	Wide
Health	*Height*	*Extent*
Healthy	Short	Wide
Fair	Medium	Medium
Poor	Tall	Narrow

uncertainty concerning their effects on bank processes, and developing quantitative methods of representing and measuring them.

REFERENCES

Andrews, E. D. (1984). Bed material entrainment and hydraulic geometry of gravel-bed rivers in Colorado. *Geological Society of America Bulletin*, **95**, 371–8.

Arulanandan, K., E. Gillogley and R. Tully (1980). Development of a quantitative method to predict critical shear stress and rate of erosion of natural undisturbed cohesive soils. *Report GL-80-5*, US Army Engineer Waterways Experiment Station, Vicksburg, MS 39180, USA.

Bettess, R., W. R. White and C. E. Reeve (1988). On the width of regime channels. In W. R. White (ed.), *River Regime*, Wiley, Chichester, pp. 135–47.

Bishop, A. W. (1955). The use of the slip circle in the stability analysis of slopes. *Geotechnique*, **19**(3), 7–17.

Bray, D. I. (1984). Study of channel changes in a reach of the North Nashwaaksis stream, New Brunswick. *Proceedings Sixth Canadian Hydrotechnical Conference, Ottawa*, Canadian Society of Civil Engineers, pp. 107–27.

Carson, M. A. and M. J. Kirkby (1972). *Hillslope Form and Process*, Cambridge University Press, Cambridge, UK.

Charlton, F. G. (1982). River stabilization and training in gravel-bed rivers. In R. D. Hey, J. C. Bathurst and C. R. Thorne (eds), *Gravel-Bed Rivers*, Wiley, Chichester, pp. 635–57.

Church, M. A. and M. J. Miles (1982). Riverbank stability in arctic and subarctic rivers. In R. D. Hey, J. C. Bathurst and C. R. Thorne (eds), *Gravel-Bed Rivers*, Wiley, Chichester, pp. 259–68.

Gray, D. H. (1978). The role of woody vegetation in reinforcing soils and stabilizing slopes. *Symposium on Soil Reinforcing and Stabilizing Techniques in Engineering Practice*, Sydney, Australia.

Gray, D. H. and A. T. Leiser (1983) *Biotechnical Slope Protection and Erosion Control*, Van Nostrand Reinhold Co., New York.

Gray, D. H. and A. MacDonald (1989). The role of vegetation in river bank erosion. In M. A. Ports (ed.), *Hydraulic Engineering*, Proceedings of the ASCE Conference, ASCE, New York (in press).

Harvey, M. D. and C. C. Watson (1986). Fluvial processes and morphological thresholds in incised channel restoration. *Water Resources Bulletin*, **22**(3), 359–68.

Hey, R. D. and C. R. Thorne (1986). Stable channels with mobile gravel beds. *Journal of Hydraulic Engineering, American Society of Civil Engineers*, **112**(8), 671–89.

Jackson, R. G. (1976). Sedimentological and fluid dynamic implications of the turbulent bursting phenomenon in geophysical flows. *Journal of Fluid Mechanics*, **77**, 531–60.

Kirkby, M. J. and R. P. C. Morgan (1980). *Soil Erosion*, Wiley-Interscience, London and New York.

Kouwen, N. (1970). *Flow retardance in vegetated open channels*. Unpublished PhD thesis, University of Waterloo, Ontario, Canada.

Kouwen, N. (1987). Velocity distribution coefficients for grass-lined channels. Discussion of paper 20435 by D. M. Temple, *Journal of Hydraulic Engineering, ASCE*, **113**(9), 1221–4.

Kouwen, N. and R. M. Li (1979). Biomechanics of vegetated channel linings. *Journal of the Hydraulics Division, American Society of Civil Engineers*, **106(HY6)**, 1085–1203.

Kouwen, N., R. M. Li and D. B. Simons (1980). Velocity measurements in a channel lined with flexible plastic roughness elements. *Technical Report CER79-80NK-RML-DBS11*, Colorado State University, Fort Collins, CO 80523, USA.

Kouwen, N. and T. E. Unny (1973). Flexible roughness in open channels. *Journal of the Hydraulics Division, American Society of Civil Engineers*, **99(HY-5)**, 713–28.

Lawler, D. M. (1986). River bank erosion and the influence of frost: A statistical examination. *Transactions of the Institute of British Geographers*, **NS11**, 227–42.

Leeder, M. R. (1983). On the interactions between turbulent flow, sediment transport and bedform mechanics in channelized flows. In J. D. Collinson and J. Lewin (eds), *Modern and Ancient Fluvial Systems*, Special publication, International Association of Sedimentologists, vol. 6, pp. 5–18.

Lohnes, R. A. and R. L. Handy (1968). Slope angles in friable loess. *Journal of Geology*, **76**(3), 247–58.

Osman, A. M. and C. R. Thorne (1988). Riverbank stability analysis. I: Theory. *Journal of Hydraulic Engineering, American Society of Civil Engineers*, **114**(2), 134–50.

Simon, A. and C. R. Hupp (1986). Channel evolution in modified Tennessee channels. *Proceedings of the Fourth Interagency Sedimentation Conference*, Las Vegas, Nevada, **2**, 5-71–5-82.

Terzaghi, K. and R. B. Peck (1948). *Soil Mechanics and Engineering Practice*, Wiley, New York.

Thorne, C. R. (1978). Processes of bank erosion in river channels. Unpublished PhD thesis, University of East Anglia, Norwich, NR4 7TJ, UK.

Thorne, C. R. (1982). Processes and mechanisms of river bank erosion. In R. D. Hey, J. C. Bathurst and C. R. Thorne (eds), *Gravel-Bed Rivers*, Wiley, Chichester, pp. 227–71.

Thorne, C. R. (1988). Analysis of bank stability in the DEC watersheds, Mississippi. *Report to the European Research Office, US Army, contract number DAJA45-87-C-0021.*

Thorne, C. R., D. S. Biedenharn and P. G. Combs (1988). Riverbank instability due to bed degradation. In S. R. Abt and J. Gessler (eds), *Hydraulic Engineering, Proceedings of the 1988 Conference on Hydraulic Engineering, Colorado Springs*, American Society of Civil Engineers, pp. 132–8.

Thorne, C. R., J. B. Murphey and W. C. Little (1981). Bank stability and bank material properties in the bluff line streams of northwest Mississippi. *Report to the Vicksburg District, US Army Corps of Engineers*, USDA Sedimentation Lab., Oxford, MS, USA.

Thorne, C. R. and A. M. Osman (1988). The influence of bank stability on regime geometry of natural channels. In W. R. White (ed.), *River Regime*, Wiley, Chichester, pp. 135–47.

US Army Corps of Engineers (1981). The streambank erosion control evaluation and demonstration act of 1974, section 32, public law 93-251. *Final Report to Congress*, Office of the Chief of Engineers, Washington DC.

US Army Corps of Engineers (1984). Streambank protection. *Short course reference notes*, US Army Engineer Waterways Experiment Station, Vicksburg, MS 39180, USA.

US Department of Agriculture (1954). Handbook of channel design for soil and water conservation. *SCS-TP-61*, Washington DC, pp. 1–34.

US Department of Agriculture (1980). *Watershed and Stream Mechanics*, short course lecture notes prepared by D. B. Simons *et al.*, Colorado State University for the USDA Soil Conservation Service.

Vidal, H. (1969). The principle of reinforced earth. *Highway Research Record*, **282**, 1–16.

Waldron, L. J. (1977). Shear resistance of root permeated homogeneous and stratified soil. *Soil Science Society of America Journal*, **41**, 843–9.

Wang, W. L. and B. C. Yen (1974). Soil arching in slopes. *Journal of the Geotechnical Engineering Division, American Society of Civil Engineers*, **100**(GT1), 61–78.

Wiberg, P. L. and J. D. Smith (1987). Calculations of the critical shear stress for motion of uniform and heterogeneous sediments. *Water Resources Research*, **23**(8), 1471–80.

Wolman, M. G. (1959). Factors influencing the erosion of cohesive river banks. *American Journal of Science*, **257**, 204–16.

11 The Recovery of Alluvial Systems in Response to Imposed Channel Modifications, West Tennessee, USA

ANDREW SIMON and C. R. HUPP
US Geological Survey, Nashville

SUMMARY

Dredging and straightening of alluvial channels between 1959 and 1978 in West Tennessee caused a series of morphologic and vegetative responses along modified reaches and tributary streams. Degradation occurred for 10 to 15 years at sites upstream of the area of maximum disturbance and lowered bed levels by as much as 6.1 meters. Following degradation, reaches upstream of the area of maximum disturbance experienced a secondary aggradation phase in response to excessive incision and gradient reduction. Downstream of the area of maximum disturbance, aggradation began after the channel work and continued for at least 20 years.

The adjustment of channel geometry and phases of channel evolution are characterized by six process-oriented stages of morphologic development—premodified, constructed, degradation, threshold, aggradation, and restabilization. Downcutting and toe removal during the degradation stage cause bank failure by mass wasting when the critical height and angle of the bank material are exceeded (threshold stage). Channel widening continues through the aggradation stage as the 'slough line' develops as an initial site of lower-bank stability. Pioneering riparian species such as black willow, river birch, and silver maple, begin to re-establish and proliferate

Prepared in cooperation with the Tennessee Department of Transportation

Vegetation and Erosion
Edited by J. B. Thornes
©1990 A. Simon
Published by John Wiley & Sons Ltd

upslope with time. Alternate channel bars form during the restabilization stage and represent incipient meandering of the channel.

A threshold bank condition at a 1.55 factor of safety ($p = 0.06$),
and a minimum, stable factor of safety of 2.0 for bank slopes and
designs are calculated. In the context of channel adjustment, quantitative data indicating factors of safety, probabilities of failure,
patterns of riparian species distribution, and bank configuration fully
support the six-stage conceptual models of bank-slope development
and channel evolution.

INTRODUCTION

Natural or man-induced changes imposed on a fluvial system tend to be absorbed
by the system through a series of channel adjustments (Gilbert, 1880; Mackin,
1948; Lane, 1955; Hack, 1960; Schumm, 1973; Bull, 1979). Because of the drastic
changes in energy conditions imposed by channel dredging, enlarging, and
straightening, a rejuvenated condition is established within the fluvial system
much like that imposed by uplift or a natural lowering of base level. The
predominant difference is one of time scale. Channel adjustments caused by
climatic changes or uplift may be exceedingly slow and progressive, practically
imperceptible during man's period of observation. Conversely, large-scale
channel modifications by man result in a sudden and significant shock to the
fluvial system that causes migrating knickpoints and observable morphologic
changes. This shortened time scale presents an opportunity to document
successive process-response mechanisms through the course of fluvial adjustment
over time and space. Channel modifications from 1959 to 1978 through much
of three major river systems in West Tennessee created a natural laboratory
for the study of channel adjustments in rejuvenated fluvial networks. This
chapter describes rate and trends of adjustment processes during channel
evolution. The shifting significance of fluvial and mass-wasting processes is
addressed in the context of channel response towards a more stable morphology.

BACKGROUND

West Tennessee is an area of 27 500 km² bounded by the Mississippi River on
the west and the Tennessee River divide on the east. The region is characterized
by unconsolidated, highly erosive formations, predominantly of Quaternary age
(US Department of Agriculture, 1980). The major rivers of the region (Obion,
Forked Deer, and Hatchie) flow in channels formed of medium-sand beds and
silt-clay banks (Figure 11.1). In contrast, many small tributary streams flow
through extensive silt deposits of reworked Wisconsin loess and have strikingly
similar bed- and bank-material properties (Simon, 1990). Those tributaries that

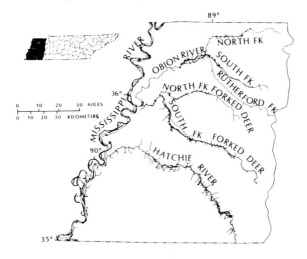

Figure 11.1   Location of study streams in West Tennessee

head in the eastern more sandy Tertiary and Cretaceous formations flow on medium-sand beds. A lack of bedrock control of base level throughout these three major river systems assures unrestricted bed-level adjustment.

Accounts of river conditions following land clearing throughout the region (mid to late 1800s) describe extremely sinuous, sluggish sediment-choked streams that frequently overflowed their banks for 3 to 10 days at a time (Hidinger and Morgan, 1912). This channel infilling, attributed to major deforestation and severe upland erosion in the late 1800s, prompted channel dredging and straightening in West Tennessee near the turn of the century. The most recent channel modification program in West Tennessee occurred between 1959 and 1978 and is the subject of this study. Channel lengths were shortened as much as 44 per cent, gradients were increased as much as 600 per cent, and beds were lowered as much as 5.2 m (Simon, 1990).

DATA COLLECTION AND METHODS

Channel morphology data were collected and compiled from previous surveys to determine channel change with time and to interpret stages of channel evolution and bank-slope development. These data consisted of bed elevations and gradients, channel top-widths, and channel lengths before, during, and after modification. Where available, gaging station records were used to record annual changes in the water-surface elevation at a given discharge (specific gage; Robbins and Simon, 1983). Changes in the water-surface elevation at a given

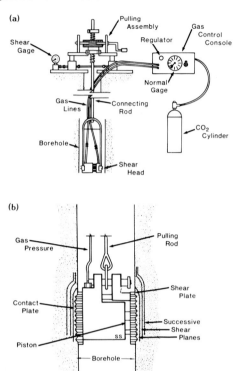

Figure 11.2   (a) Schematic drawing of borehole shear test assembly. (b) Detail of shear head in borehole. (Modified from Thorne *et al.*, 1981.)

discharge imply similar changes on the channel bed and can be used to document bed-level trends (Blench, 1973; Robbins and Simon, 1983).

Soil mechanics properties, cohesion (*c*) and friction angle (ϕ) were determined from 169 field tests by using a borehole shear test device (BST). This instrument is used to conduct drained, *in situ*, direct-shear tests on the walls of a 7.6 cm-diameter borehole (Figure 11.2; Handy and Fox, 1967). The BST has a number of advantages over the more conventional, laboratory triaxial shear tests:

1. A number of separate trials (5 to 12) on the same material are run for each test, to produce a single cohesion and friction angle value.
2. Data obtained from the instrument are plotted on site, allowing for repetition if the results are unreasonable.
3. Tests can be carried out at various depths in the bank to locate weak areas (Thorne, Murphey and Little, 1981).

Trials at successively higher normal pressures produce a series of points on the Mohr–Coulomb line which, by linear regression, are used to calculate c and ϕ. For a good review of BST techniques and applications, the reader is directed to Luttenegger and Hallberg (1981), and Thorne *et al.* (1981).

Bank stability was modeled assuming low streamflow conditions by using software developed by Huang (1983). A factor of safety (FS; ratio of resisting to driving forces) of 1.0 indicates that driving and resisting forces are equal and that bank failure is imminent. Even natural slopes with forested conditions often exhibit factors of safety between 1.0 and 1.5 (Sidle, Pearce and O'Louthlin, 1985). Under conditions of such marginal stability, a probabilistic approach is useful in estimating the chance of failure from factor-of-safety data. On the basis of a normal probability distribution, failure probabilities were calculated for each site (Pender, 1976; Huang, 1983).

All site-specific slope-stability analyses were associated with the stage of bank-slope development observed in the field. Changes in bank stability over the course of fluvial adjustment were then established on a time-independent basis to reflect variable force and resistance relations.

The identification and dating of various geomorphic surfaces can be diagnostic in determining the relative stability of a reach and the status of bank-slope development (Simon and Hupp, 1986; Hupp and Simon, 1986). Dendro-chronologic and dendrogeomorphic analyses of riparian vegetation yielded information about channel widening, bank accretion, and timing of initial bank stability. Estimates of channel widening are made by determining ages of stem deformation associated with bank failure, then measuring the width of the slump block, or the distance between the affected stems and the present top-bank edge. Species presence and cover were estimated at each site by tallying the number of species present and noting the areal canopy coverage over a bank, respectively.

Stems that remained buried for at least a growing season form root tissue instead of steam tissue, with adventitious roots sprouting in what was once stem wood. Exhuming buried stems to the depth of the original germination point, measuring the depth of burial, and determining the age of the stem through cores or cross-section, provide an estimate of sediment-accretion rate.

System-wide data on channel widening, bank accretion, and vegetal cover combined with species presence data were analyzed with multivariate statistical methods including binary discriminant (BDA; Strahler, 1978; Hupp, 1987), and detrended correspondence analyses (DCA; Hill and Gauch, 1980). Standardized residuals (D-values) from the BDA were computed in terms of standard deviations (from contingency tables) that indicate species 'preference' (positive values) or 'avoidance' (negative values) for particular site variables. DCA is an ordination technique that clusters similar sites (based on morphologic characteristics) according to species presence–absence data. Ordination thus allows for the interpretation of relations among geomorphic characteristics and riparian vegetation.

BED-LEVEL ADJUSTMENT

Adjustment of the channel bed at a site immediately follows channel modifications and can be described through time. Channel-bed adjustments through time (irrespective of the temporal and spatial scales applied) are best described mathematically by nonlinear functions, which asymptotically approach a condition of minimum variance.

Bed-level changes at a site through time are depicted by a simple power equation that reflects the initially rapid rate of degradation (or aggradation), as well as the asymptotic nature of the process with increasing time. The exponent to this equation (*b*) indicates the intensity of processes on the channel bed and is directly related to the magnitude of the imposed disturbance and the site's distance from the area of maximum disturbance (equation on Figure 11.3).

Degradation exponents (− *b*) generally range from − 0.005 to − 0.040, with the greatest rates of change occurring near the upstream side of the area of maximum disturbance (AMD), usually the upstream terminus of the channel work (A in Figure 11.3). Increases in downstream channel gradient and cross-sectional area by man result in stream power that is more than sufficient to transport the bed material delivered from upstream. The bed of the channel, therefore, erodes headward to increase bed-material transport and (or) reduce channel gradient (C in Figure 11.3). Typically, degradation occurs for 10 to 15 years at a site and can deepen the channel up to 6.1 meters.

Aggradation rates (+ *b*) are less than their degradation counterparts when absolute values are compared, and generally range from 0.001 to 0.009 with

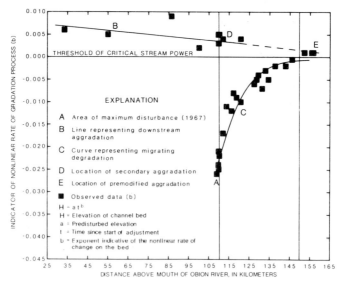

Figure 11.3   Model of bed-level response in the Obion River system

the greatest rates occurring near the stream mouth. This process begins downstream of the AMD immediately following channelization. Aggradation rates may reach 0.12 m/y, and can continue for more than 20 years (B in Figure 11.3). As degradation proceeds upstream from the AMD to reduce channel gradient, aggradation downstream of the AMD similarly flattens gradients as part of an integrated basin response due to the man-induced increases in energy conditions (Simon, 1990).

Aggradation near D in Figure 11.3 occurs at previously degraded sites where gradient has been significantly reduced by incision and knickpoint migration. Flows become incapable of transporting the greater bed-material loads being generated from degradation channel beds upstream. The channel aggrades and thereby increases gradient and transporting capacity.

Effects of the imposed disturbance decrease with distance upstream, resulting in minimal degradation rates at about river kilometer 150 ($b = 0.0$). The absence of net erosion or deposition on the channel bed ($b = 0.0$) is analogous to Bull's (1979) threshold of critical stream power. Further upstream (E in Figure 11.3), channel beds of the Obion River system (including upstream reaches of the North, South, and Rutherford Forks) remain unaffected by the downstream channel work and aggrade at low, 'background' rates.

With the short time scales involved in the application of the disturbance and the channel's consequent response, the evolution of channel form over time and space can be monitored. Numerical results such as those plotted in Figure 11.3 can be derived for most alluvial systems where even limited data are available. Results may be applied over longer time scales with the inclusion of dendrochronologic data (as in Graf, 1977). It has been shown that bed-level adjustment does follow nonlinear trends, both over time at a site, and over time with distance upstream (Figure 11.3).

WIDTH ADJUSTMENT AND BANK-SLOPE DEVELOPMENT

Channel widening by mass-wasting processes is common in adjusting channels of West Tennessee. Bank failure in the loess-derived alluvium is induced when banks are overheightened and oversteepened by degradation, undercutting, and seepage forces at the toe of the bank. Piping and tension cracks in the bank materials enhance bank failures by internally destabilizing the bank (Simon, 1990).

Channel bank material of West Tennessee streams is completely alluvial, composed of loess-derived Quaternary sediments which are predominantly nonplastic, dispersive silts of low cohesion (Simon, 1990). *In situ* shear-strength tests confirm the existence of low cohesive strengths; generally less than 14 kPa. These highly erodable materials tend to maintain high moisture contents, with a mean degree of saturation of 91 per cent (169 tests).

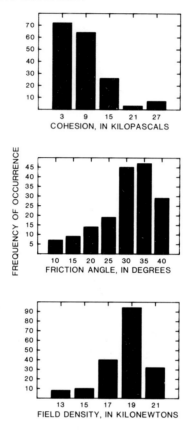

Figure 11.4   Frequency distributions of bank-material properties

Moderate rises in river stage are sufficient to complete saturation of the channel banks and result in mass failures upon recession of river stage. Figure 11.4 shows the frequency distributions of the primary soil mechanics variables: cohesion, friction angle, and field density. Note that the histograms do not display a normal distribution, but are markedly skewed towards lower strength characteristics. Friction angle and cohesion values vary within accepted limits for the types of material tested (Lohnes and Handy, 1968).

The processes and successive forms of bank retreat and bank-slope development reflect the interaction of hillslope and fluvial processes. Interpretations of these processes are based largely on bed-level adjustment trends and botanical evidence. A complete 'cycle' of slope development from the premodified condition through stages of adjustment, to the eventual re-establishment of stable bank conditions, assumes that degradation is of sufficient magnitude to instigate unstable bank conditions, and a 20- to 40-year period

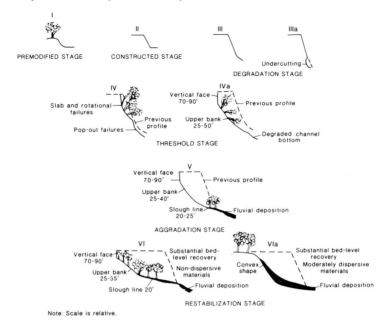

Figure 11.5 Six-stage model of bank-slope development in disturbed channels

of mass wasting. However, in grossly unstable channels composed of silt-clay alluvium, much longer periods of time may be required for the complete restabilization of the channel. A conceptual model of the six stages of bank retreat and bank-slope development (Figure 11.5) represents distinguishable bank morphologies characteristic of the various reach types describing bed-level adjustment. By applying a space for time substitution, longitudinal variation in bank-slope development can be used to denote morphologic change through time. Sand-bed channels will require 20 to 40 years of channel adjustment to pass through the first five stages of the model. In these types of channels, a total of 50 to 100 years is assumed necessary for the restabilization of the channel banks (stage VI) and for the development of incipient meanders.

Reaches that have degraded beyond the critical conditions of the bank material and exhibit frequent failures represent stage IV conditions. Channel widening by slab, rotational, planar, and pop-out failures are common during this stage. Only the relatively small pop-out failures that occur at the base of the bank take place during the previous stage (stage III, degradation). These failures are attributable to an increase in shear stress due to unloading of the bank (degradation) with no corresponding increase in shear strength. Secondary failures generally occur along stage V reaches on low- and mid-bank surfaces in previously failed materials that maintain only residual strengths. These failures are shallow relative to their length and are aided by the additional weight of accreted sediments. Secondary failures are common along the Obion River main stem where bank accretion has occurred for at least 20 years.

Mass wasting of banks is the dominant channel-forming process during stage IV and decreases in prominence during the other stages. Quantitative information on widening rates and factors of safety can be generalized and placed in a logical conceptual framework, from the premodified condition through channel construction and adjustment, to restabilization (Figure 11.6). In this analysis, stages are time independent and represent the assimilation of data from various times (1959–87). Data regarding factors of safety are based on calculated values at saturation.

Factors of safety for both planar and rotational failures are plotted with rates of channel widening to present bank-stability trends over the course of fluvial adjustment. As would be expected, there is a clear, inverse relation between widening rates and FS.

Probabilities of failure (p) were computed for both planar and rotational failures, sorted by stage, and plotted (Figure 11.7). As expected, results reflect the increased mean probabilities of failure during stage IV. Assuming that

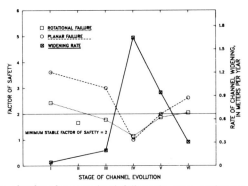

Figure 11.6　Factors of safety for mass bank failures by type, and stage of channel evolution

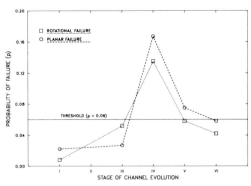

Figure 11.7　Probability of mass bank failures by type, and stage of channel evolution

large-scale rotational and planar failures do not occur during stages I ($p = 0.009$ and $p = 0.024$), III ($p = 0.052$ and $p = 0.026$), and VI ($p = 0.042$ and $p = 0.058$), we can separate figure 11.7 into two sections at an approximate probability of 0.06; an upper section representing generally unstable, failing banks, and a lower section representing generally stable banks. A least square regression analysis using FS as the dependent variable and probabilities of failure as the independent variable was used to calculate the FS for the threshold probability of 0.06. The resulting threshold FS is 1.55, indicating that bank-slope conditions and designs with 1.50 are truly marginal. These results are in general agreement with data reported by Pender (1976) for cutslopes. As with the probability plot (Figure 11.7), Figure 11.6, showing FS trends by stage, can be sectioned. Given that a value of 1.55 represents a threshold FS, some higher value would then correspond to reasonably stable bank conditions. Again by using stages I, III, and VI to represent relatively stable conditions, a horizontal line can be drawn at 2.0 on Figure 11.6 to represent a minimum, stable FS. The results indicate that, for loess-derived alluvium, the use of an FS of 1.5 to designate stability may be dubious.

SLOPE DEVELOPMENT AND CHANNEL EVOLUTION

The six stages of bank-slope development represent a conceptual model of width adjustment. Stages (premodified, constructed, degradation, threshold, aggradation, and restabilization) are induced by a succession of interactions between fluvial and hillslope and processes (Figure 11.5). By associating the six bank-slope development stages with the dominant hillslope and fluvial processes, and with characteristic channel forms, a conceptual model of channel evolution over time and space was developed (Table 11.1). The model does not indicate that each adjusting reach will undergo all six stages but implies that specific trends of bed-level response will result in a series of mass-wasting processes and definable bank and channel forms. However, the conceptual framework of the simultaneous retreat of the vertical face and flattening along surfaces below is supported by the observations of other investigators (Carson and Kirkby, 1972, p. 184).

VEGETATION DISTRIBUTION AND RECOVERY

Patterns of riparian species distribution are strongly associated with the geomorphic stage of adjustment. The most common pioneer species are black willow, river birch, silver maple, sycamore, green ash, and box elder. The older vegetation of the upper reaches (typically stages I and VI) is considerably more diverse and include many species normally associated with flood plains such

Table 11.1 Stages of channel evolution

Stage		Dominant processes		Characteristic forms	Geobotanical evidence
No.	Name	Fluvial	Hillslope		
I	Premodified	Sediment transport—mild aggradation; basal erosion on outside bends; deposition on inside bends	—	Stable, alternate bars, convex top-bank shape; flow line high relative to top bank; channel straight or meandering	Vegetated banks to flow line
II	Constructed	—	—	Trapezoidal cross-section; linear bank surfaces; flow line lower relative to top bank	Removal of vegetation (?)
III	Degradation	Degradation; basal erosion on banks	Pop-out failures	Heightening and steepening of banks; alternate bars eroded; flow line lower relative to top bank	Riparian vegetation high relative to flow line and may lean towards channel
IV	Threshold	Degradation; basal erosion on banks	Slab, rotational and pop-out failures	Large scallops and bank retreat; vertical-face and upper-bank surfaces; failure block on upper bank; some reduction in bank angles; flow line very low relative to top bank	Tilted and fallen riparian vegetation
V	Aggradation	Aggradation; development of meandering thalweg; initial deposition of alternate bars; reworking of failed material on lower banks	Slab, rotational and pop-out failures; low-angle slides of previously failed material	Large scallops and bank retreat; vertical face, upper bank, and slough line; flattening of bank angles; flow line low relative to top bank; development of new flood plain (?)	Tilted and fallen vegetation; re-establishing vegetation on bank; deposition of material above root collars of slough-line vegetation
VI	Restabilization	Aggradation; further development of meandering thalweg; further deposition of alternate bars; re-working of failed material; some basal erosion of outside bends; deposition on inside bends	Low-angle slides; some pop-out failures near flow line	Stable, alternate channel bars; convex–short vertical face, on top bank; flattening of bank angles; development of new flood plain (?); flow line high relative to top bank	Re-establishing vegetation extends up slough-line and upper-bank; deposition of material above root collars of slough line and upper-bank vegetation; some vegetation establishing on bars

as bald cypress and various species of oak, in addition to the pioneer species.

Stages I and VI cannot be clearly separated on a botanical basis; the latter has usually had 40 or more years to recover from the last cycle of channel work. Top banks are low relative to mean water levels and riparian surfaces support a relatively diverse species assemblage (13 species). Banks of stages I and VI tend to be gentle and aggradational as opposed to the higher and steeper banks of stages III, VI, and V.

Stage III is the most vegetatively diverse (21 species), including many riparian species normally associated with natural bank conditions such as ironwood, sweetgum, overcup oak, willow oak, bald cypress, basswood, and various elm species. Banks along stage III reaches usually have been stable for many years because degradation has not yet been sufficient to cause bank failures and channel widening.

Stage IV reaches are characterized by highly unstable banks. Nearly 50 per cent of all stage IV sites have no woody species although a tangle of herbaceous species may be common on slump block surfaces of the upper bank. Only five woody species, all pioneers, occur on stage IV reaches and then only in protected areas such as on relict inside bends where the thalweg is diverted. Stage V reaches are characterized by distinct zones of proliferating pioneer woody species (nine species), on low- to mid-bank surfaces, indicating the initiation of low-bank stability.

D-values for six selected species, silver maple, river birch, and black willow representing pioneer species, and willow oak, bald cypress, and ironweed representing species of stable sites, are shown in Figure 11.8. Note that all species have positive associations for low widening rates and a negative association with high widening rates, suggesting a pervasive influence of widening characteristics in patterns of species distribution. The pioneer species (Figure 11.8a) have positive values for medium widening rates and medium cover as would be expected in species that initially occupy disturbed areas. Conversely, the natural-site species (Figure 11.8b) cannot tolerate even medium widening rates, and as components of the mature forest, are positively associated with only high cover values. Black willow, perhaps the most pioneering of all species, is strongly associated with bank accretion; a testament to this species' ability to survive the rapidly aggrading low-bank areas during stage V. The distinct site preference or avoidance pattern of selected species suggests that they are indicators of specific bank conditions, which allows simple vegetation reconnaisance of an area to be used for at least preliminary evaluation of site conditions.

The results of the multivariate DCA of species presence versus site conditions are shown in Figure 11.9, where the site-variable categories are ordinated by significant principal components: DCA axes I and II. This ordination, based entirely on species presence, accurately reflects the hydrogeomorphic characteristics outlined in the six-stage model of channel evolution (Table 11.1). Note that site conditions naturally cluster in groups that can be identified with

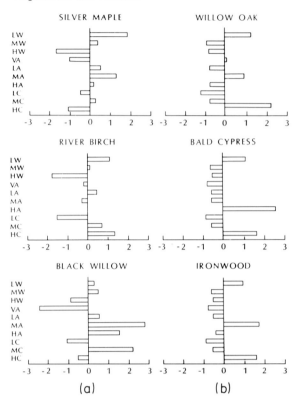

Figure 11.8 *D*-values for selected species showing preference (+) or avoidance (−) for given site characteristics (V = very low, L = low, M = medium, H = high, W = widening, A = accretion, C = cover)

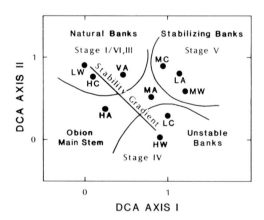

Figure 11.9 DCA ordination of site variables based on species presence data alone (V = very low, L = low, M = medium, H = high, W = widening, A = accretion, C = cover)

specific stages of the model (Figure 11.9). Thus, the independent vegetation ordination fully supports the conceptual framework of the bank-slope model (Figure 11.5) and the model of channel evolution (Table 11.1). These results strongly suggest that patterns of species distribution may be used to infer ambient bank stability and recent historical conditions.

CONCLUSIONS

The models described in this chapter make the following assumptions: (1) there is no local bedrock control of bed level; (2) overadjustment and secondary response are active processes; (3) the bed and banks are free to adjust to imposed changes; and (4) successive stages of evolution are not interrupted by other disturbances. Extrapolation of the six-stage models of bank-slope development and channel evolution should be particularly appropriate for areas of the Mississippi embayment and the central United States. Application over a broader geographical area is conceptually justified on the basis that similar processes can create similar forms. Variations in time scales and channel forms from the idealized model will occur due to local bedrock control and variations in relief, soil properties, and climatic conditions. Still, the models reflect the overadjustments inherent to fluvial response and may be useful in determining expected changes in alluvial channel morphology over the course of a major adjustment cycle.

REFERENCES

Blench, T. (1973). Factors controlling size, form, and slope of stream channels. 9th Hydrology Symposium, Edmonton, Canada, 1973. *Proceedings, Fluvial Processes and Sedimentation*, pp. 421–39.

Bull, W. B. (1979). Threshold of critical power in streams. *Geol. Soc. of Am. Bull.*, part 1, **90**, 453–64.

Carson, M. A. and M. J. Kirkby (1972). *Hillslope Form and Process*, Cambridge University Press, London.

Gilbert, G. K. (1880). *Report on the Geology of the Henry Mountains* (2), Geographical and Geological Survey of the Rocky Mountain Region, United States Government Printing Office, Washington.

Graf, W. L. (1977). The rate law in fluvial geomorphology. *Am. J. of Science*, **277**, 178–91.

Hack, J. T. (1960). Interpretation of erosional topography in humid temperate regions. *Am. J. of Science*, **258-A**, 80–97.

Handy, R. L. and N. W. Fox (1967). A soil borehole direct-shear device. *Highway Research Board News*, **27**, 42–51.

Hidinger, L. L. and A. E. Morgan (1912). Drainage problems of Wolf, Hatchie, and South Fork of Forked Deer Rivers, in West Tennessee, In *The Resources of Tennessee*, Tennessee Geological Survey, **2**(6), 231–49.

Hill, M. O. and H. G. Gauch (1980). Detrended correspondence analysis: An improved ordination technique. *Vegetatio*, **42**, 47–58.

Huang, Y. H. (1983). *Stability Analysis of Earth Slopes*, Van Nostrand Reinhold Co., New York.

Hupp, C. R. (1987). Determination of bank widening and accretion rates and vegetative recovery along modified West Tennessee streams. *Proceedings International Symposium on Ecological Aspects of Tree-Ring Analysis*, Palisades, New York, August 1986, pp. 224–33.

Hupp, C. R. and A. Simon (1986). Vegetation and bank-slope development. *Proceedings of the 4th Federal Interagency Sedimentation Conference*, Las Vegas, Nevada, March 1986, **2**, 5-83–5-92.

Lane, E. W. (1955). The importance of fluvial morphology in hydraulic engineering, *Proceedings of the ASCE*, **81**, 745.

Lohnes, R. A. and R. L. Handy (1968). Slope angles in friable loess, *J. of Geol.*, **76**(3), 247–58.

Luttenegger, J. A. and B. R. Hallberg (1981). Borehole shear test in geotechnical investigations. *ASTM Spec. Tech. Publ.*, **740**, 566–78.

Mackin, J. H. (1948). Concept of a graded river. *Geol. Soc. of Am. Bull.*, **59**, 463–511.

Pender, M. J. (1976). Probabilistic assessment of a cut slope. *New Zealand Eng.*, **15**, 239–46.

Robbins, C. H. and A. Simon (1983). Man-induced channel adjustment in Tennessee streams. *United States Geological Survey Water-Resources Investigations Report 82-4098*, 129.

Schumm, S. A. (1973). Geomorphic thresholds and the complex response of drainage systems. In M. Morisawa (ed.), *Fluvial Geomorphology*, Binghamton, State University of New York, pp. 229–310.

Sidle, R. C., A. J. Pearce, and C. L. O'Louthlin (1985). Hillslope stability and land use. *Am. Geophys. Union Monograph #12 (Water Resources)*.

Simon, A. (1990). Gradation processes and channel evolution in modified West Tennessee streams: Process, response, and form. *US Geological Survey Professional Paper*, **1470**, 93.

Simon, A., and C. R. Hupp (1986). Channel widening characteristics and bank slope development along a reach of Cane Creek, West Tennessee. In S. Subitsky (ed.), *Selected Papers in the Hydrologic Sciences*, US Geological Survey Water-Supply Paper, 2290.

Strahler, A. H. (1978). Binary discriminant analysis: A new method for investigating species–environment relationships. *Ecology*, **59**, 108–16.

Thorne, C. R., J. B. Murphey, and W. C. Little (1981). Stream channel stability, Appendix D, *Bank stability and bank material properties in the bluffline streams of northwest Mississippi, Stream Channel Stability*, US Dept. Agriculture Sedimentation Laboratory, Oxford, Mississippi, 257.

US Department of Agriculture (1980). *Summary Report Final: Obion-Forked Deer River Basin Tennessee*, Soil Conservation Service, Nashville, Tennessee, 43.

12 The Effect of Land Use on Nitrogen, Phosphorus and Suspended Sediment Delivery to Streams in a Small Catchment in Southwest England

A. L. HEATHWAITE
Department of Geography, University of Sheffield

T. P. BURT
School of Geography, University of Oxford

and

S . T. TRUDGILL
Department of Geography, University of Sheffield

SUMMARY

Changing land use practices may be causing the incidence of infiltration-excess overland flow to increase, particularly in intensively grazed areas. Experimentation with a rainfall simulator and runoff plot monitoring in a South Devon catchment has shown that heavy grazing of permanent grassland resulted in an 80 per cent reduction in the infiltration capacity. Surface runoff from overgrazed permanent grassland was double that from lightly grazed areas, and at least twelve times that from ungrazed areas. Additionally, the removal of the vegetation cover through severe poaching led to an increase in the rate of suspended sediment, total nitrogen, and total phosphorus delivery in surface runoff by 30, 9 and 16 times respectively. Over 90 per cent of the total nitrogen delivered was in inorganic NH_4–N form, whereas for phosphorus, over 80 per cent of the total phosphorus was delivered in organic form. Riparian zones are important in regulating the sediment and solute flux from agricultural land to the stream. As 60 per cent of riparian zone land use in the study catchment was permanent grassland, heavy grazing of these areas must be minimized to avoid substantial increases in solute and sediment loads.

Vegetation and Erosion
Edited by J. B. Thornes
© 1990 John Wiley & Sons Ltd

INTRODUCTION

Since the early 1970s in the United Kingdom, two major changes in agricultural land use have been recorded: the conversion of grassland to arable (Oakes, Young and Foster, 1981) and increased livestock (mainly dairy) numbers. Both represent a more intensive land use and are associated with greater fertilizer inputs. This in turn may result in increased loads of sediment and solutes, particularly nitrogen and phosphorus, to streams. It is important to determine how different land uses act as sources of sediment and solutes and the way in which they influence stream water quality. In this respect the presence or absence of grazing animals, the amount and condition of the vegetation cover and the timing of ploughing or fertilizer applications are likely to be important land use controls on stream water quality.

This chapter will examine the effect of land use on surface runoff, sediment, nitrogen and phosphorus delivery from the hillslope to the stream system. The objective is to predict delivery in small headwater basins and relate this to delivery in larger drainage basins, and to examine ways of minimizing non-point inputs to streams due to agricultural activities. There are few studies that provide information on the long-term changes in catchment land use and fertilizer inputs and which then link these changes to stream solute and sediment concentrations on a scale that enables linkages between the land and the stream to be established. The 46 km^2 Slapton catchment in southwest Devon is appropriate for such a study, particularly as monitoring began in 1970. Two sub-catchments, Slapton Wood and Stokeley Barton, have been the focus for most research and have been used to represent changes in the Slapton catchment as a whole. Table 12.1 shows the change in land use, fertilizer nitrogen and stream nitrate loads from 1974 to 1986 for these catchments. It is clear that the proportion of arable land has increased significantly. This is paralleled by a doubling of the nitrogen fertilizer inputs and is associated with a similar rise in the stream nitrate load. Burt *et al.* (1988) were able to show that the nitrate concentration for the Slapton Wood catchment had increased significantly over the period 1970 to 1985 by 2 mg l^{-1}.

Table 12.1 A comparison of land use, nitrogen fertilizer and stream nitrate loads in the Slapton Wood and Stokeley Barton catchments for 1974 and 1986

Land use	Slapton Wood		Stokeley Barton	
Catchment area (ha)	93		153	
Land use (%)	1974	1986	1974	1986
Arable	24	46	52	66
Grass	65	43	41	32
Wood	11	11	7	2
kg ha^{-1} N	58	127	108	173
kg ha^{-1} NO$_3$ stream load	28	69	28	42

Past research in the Slapton catchment has shown the importance of subsurface flow in delivering high concentrations of nitrate to streams (Troake, Troake and Walling, 1976; Burt *et al.*, 1983, 1988). The delivery of sediment and solutes in surface runoff from the catchment hillslopes to the streams has not been examined in detail to date. This delivery mechanism may be particularly important for suspended sediment, particulate, adsorbed and organic phosphorus and organic nitrogen loads. This chapter will deal with surface delivery of both inorganic and organic nitrogen and phosphorus in more detail.

THE SITE

Slapton Ley (SX 825 439), a 0.8 km^2 lake, is the largest natural freshwater body in southwest England. It is the sink for sediment and solute inputs from the surrounding 46 km^2 arable and grassland catchment. Since the 1960s there has been concern that the Ley is becoming increasingly eutrophic.

Catchment topography consists of flat-topped ridges dissected by narrow deep valleys (Mercer, 1966). The soils are freely drained acid brown earths overlying impermeable Devonian slates and shales. They have a clay loam texture with a high content of silt-sized particles (30–40 per cent) and fragments of weathered slate (Trudgill, 1983). Soil depth varies between 60 cm and 1 m. These brown earth soils correspond to the Dartington (now Manod) series. The area has a mean annual rainfall of 1035 mm, and a mean annual temperature of 10.5 °C (Ratsey, 1975).

Merrifield, a small (18 ha) catchment (SX 817 475) near Slapton, has been monitored since August 1987. The area forms part of the 27 km^2 Gara catchment (Figure 12.1), which because of its size contributes most of the total runoff and stream sediment and solute load to Slapton Ley. The slopes of the Merrifield catchment are convexo-concave in form (Figure 12.2): there is a small plateau area with maximum slope angles of 1–2°; below this, slope angles increase to 20° but decrease again close to the stream. The stream discharges directly to the River Gara. In the period of measurement the catchment had three different land uses: permanent grass, recently re-seeded temporary grass (previously barley), and arable (kale—a fodder crop for sheep). This catchment was chosen as this land use is representative of the 46 km^2 Slapton catchment as a whole; it has the advantage of being small enough to provide a relatively homogeneous study environment and yet large enough to provide experimental plots.

The 1987 fertilizer application rates for each land use at Merrifield are given in Table 12.2 together with the average rates of application (in 1986) for the Gara catchment as a whole (Johnes and O'Sullivan, in press). The application rate for barley at Merrifield is low as it does not include applications made in

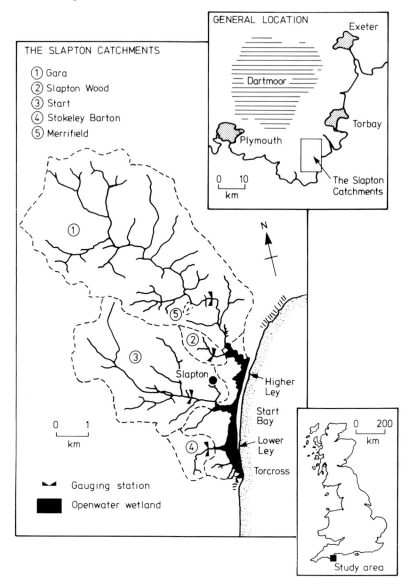

Figure 12.1 Map of the Slapton catchments

1986 at sowing. The barley field was ploughed and re-seeded in September 1987. The nitrogen and phosphorus application to the kale is low in comparison with the average for arable crops in the Gara catchment. This is because the kale was intended for grazing so most of the nitrogen and phosphorus will be returned in excreta, hence a lower fertilizer application is practical.

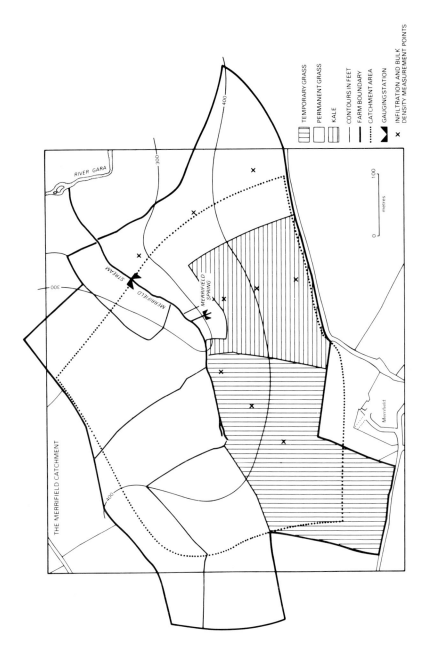

Figure 12.2 Map of the Merrifield catchment

Table 12.2 Fertilizer application rates for each land use at Merrifield

Land use	Month	Application rate (kg ha^{-1})	Total application (kg)	Annual application for Gara catchment (1986)[1] (kg ha^{-1})
Permanent grass	April 1987	74 N 16 P 31 K	325 N 70 P 136 K	
	June 1987	49 N 11 P 21 K	215 N 48 P 92 K	
	August 1987	25 N 5 P 10 K	110 N 22 P 44 K	
	Annual	148 N 32 P 62 K		164 N 37 P
Barley[2]	March 1987	74 N 16 P 31 K	255 N 55 P 107 K	170 N 53 P
Kale	June 1987	124 N 27 P 52 K	450 N 98 P 189 K	206 N 109 P

[1]Johnes and O'Sullivan (in press).
[2]Barley field converted to re-seeded grass in September 1987. No fertilizer was added to the re-seeded grass during the period of measurement.

METHODS

The results of two sets of experiments at Merrifeld are presented in this chapter. The first refers to monthly measurements of infiltration capacity, bulk density and soil moisture from August 1987 to March 1988 for the three different land uses at Merrifield: (1) permanent grass, (2) barley (re-seeded in September 1987) and (3) kale. The second refers to the rainfall simulation experiments at Merrifield in August 1988: here four land-use categories were monitored in order to reflect land use in the Slapton catchment as a whole. Three of the land uses were located in the Merrifield catchment. These were: (1) temporary grass (recently re-seeded); (2) permanent grass subdivided into (a) heavily grazed (extensively poached) and (b) lightly grazed areas; and (3) prepared ground (defined as ground ploughed and rolled within the previous two months) which previously contained the kale crop. For the fourth land use (cereal), a harvested barley crop in an adjacent field was used. This field had the same slope and soil characteristics as the main site. Recently ploughed land was also studied but as no runoff was recorded from these plots due to their high infiltration capacity (averaging $260\,\text{mm}\,\text{hr}^{-1}$), they were not included in the results presented below.

Monthly Infiltration Capacity, Bulk Density and Soil Moisture

For each land use, permanent grassland, kale, and recently re-seeded grass, the dry bulk density, moisture content, and infiltration capacity of the soil were measured at monthly intervals, starting in August 1987. Soil moisture and bulk density were recorded for the upper 5 cm of the soil; samples were collected using Kubiena tins of known volume. Infiltration capacity was measured using a double-ring constant head infiltrometer (Burt, 1978). The results presented are means of nine replicates for each land use taken at top slope, mid slope and base slope positions in the catchment. These sample sites are indicated in Figure 12.2.

Rainfall Simulation

To measure sediment and solute delivery from different land uses a series of rainfall simulation experiments were run at Merrifield in August 1988. A full description of the rainfall simulator used is given in Bowyer-Bower and Burt (1989). This 'drip-type' simulator has over 500 drop formers placed in the lower plate of a sealed Perspex box of dimensions 500 mm width and 1000 mm length. The box is supported in a metal frame, with adjustable legs for levelling and adjusting the box to the required distance of 1.5 m above the ground surface. The principal advantage of this sort of simulator is that it allows great accuracy in replicating rainfall character between experimental sites; this accuracy cannot be achieved using spray-type simulators though these produce droplets which more nearly approach their terminal velocity because of the greater fall height involved. The drip-type simulator used has been extensively tested with regard to replication of drop size distribution and rainfall intensity.

For each land use, the experiments were run in duplicate for a four-hour period at a rainfall intensity of $12.5\,\text{mm}\,\text{hr}^{-1}$. This intensity was used to maximize the amount of runoff from the plots to provide sufficient volumes for chemical analysis whilst still remaining representative of heavy rainfall within the catchment. Surface runoff was collected by means of a trough from a $0.5\,\text{m}^2$ bound plot. The runoff total was measured, and bulked runoff samples at 20 minute intervals retained for chemical analysis.

Inorganic (particularly nitrogen) inputs to the stream system have been covered in detail in previous research at Slapton (Burt et al., 1983, 1988). In the experiments reported here, emphasis was also placed on evaluating the organic nitrogen and particulate, adsorbed and organic phosphorus component. Total nitrogen and phosphorus were determined in plot runoff in addition to inorganic nitrogen, inorganic phosphorus and suspended sediment. This enables some assessment of the organic loading to the stream to be made, particularly through the comparison of grazed and ungrazed plots. Animal excreta and slurry can be highly polluting if they reach the stream system, owing primarily to their high biological oxygen demand but also to their high nitrogen and phosphorus content.

Suspended sediment was measured after filtration under suction through a Whatman GF/C filter paper. The filtrate was analysed for dissolved inorganic NO_3–N, NH_4–N, and PO_4–P using automated versions of the colorimetric techniques of Henriksen and Selmer-Olsen (1970), Crooke and Simpson (1971), and Murphy and Riley (1962) respectively. Additionally surface runoff from the rainfall simulation experiments was analysed for total nitrogen and phosphorus in solution using an alkaline persulphate digestion procedure modified for a microwave digestion technique (Heathwaite, Johnes and Burt, forthcoming).

RESULTS AND DISCUSSION

Infiltration, Bulk Density and Land Use

Figure 12.3 shows the variation in infiltration capacity and bulk density from August 1987 to March 1988 for the three land-use categories (permanent pasture, kale and temporary grass) at Merrifield. Table 12.3 shows the corresponding variation in monthly rainfall and soil moisture.

In general, the kale had a higher infiltration capacity in comparison with the grassland (both temporary and permanent). This is probably due to the

Table 12.3 The variation in soil moisture content (%) through time with land use at Merrifield

Month	Rainfall (mm)	Kale	Barley (temporary grass)	Permanent pasture	Management
Aug 87	13	20	27	23	barley ploughed & re-seeded
Sep	56	25	32	29	
Oct	163	30	37	33	
Nov	106	26	27	26	grazing cattle
Dec	99	38	39	36	in per. past.
Jan 88	240	38	41	37	sheep begin
Feb	104	35	36	35	grazing kale
Mar	108	34	34	30	

[1]Moisture content was measured for the upper 5 cm soil.
[2]Moisture content (%): mean of 9 replicates for each land use.

169

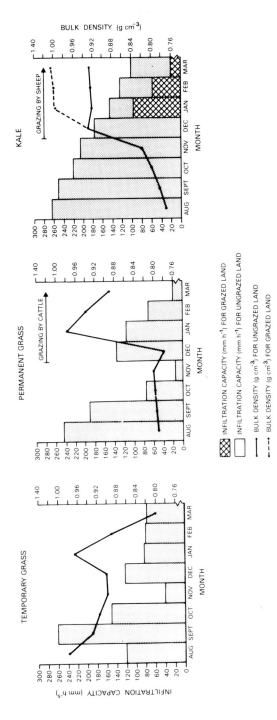

Figure 12.3 The relationship between infiltration capacity and bulk density for different land uses at Merrifield

ploughing and sowing of their field in June 1987. The infiltration capacity of the kale declined steadily through time; once sheep began grazing the kale at the end of November 1987, the infiltration capacity in grazed sections of the kale declined sharply (Figure 12.3). A corresponding increase in the bulk density of the topsoil of the kale field was also recorded as grazing began. Increased rainfall input from October through to March raised the soil moisture content (Table 12.3) which also contributed to the reduction in infiltration capacity measured. The infiltration capacity of the temporary and permanent grassland is more variable and appears to be largely controlled by management practices (Table 12.3). When the barley field was ploughed and re-seeded in September 1987 this, not surprisingly, increased the infiltration capacity and lowered the bulk density. The permanent grassland showed a marked increase in bulk density once cattle began grazing in December 1987. This is matched by a reduction in infiltration capacity which cannot be attributed to increased soil moisture since that actually declined over the period December 1987 to March 1988 (Table 12.3).

Although bulk density and infiltration capacity differ between land uses, these differences are small in comparison with the compaction and subsequent reduction in infiltration capacity caused by grazing animals in the kale and permanent pasture. In the Merrifield catchment, subsurface flow would be expected to dominate flow processes due to the combination of permeable soils over impermeable bedrock. Intensive grazing of the kale throughout the winter and early spring when rainfall inputs were also high (Table 12.3) resulted in the extensive poaching and puddling of the entire hillslope. By the end of the winter, the compacted soil surface and extremely low infiltration capacity (the final infiltration capacity was $0.1 \, mm \, hr^{-1}$) resulted in widespread overland flow, even during relatively low rainfall intensities of less than $3 \, mm \, hr^{-1}$.

In the UK in general, low rainfall intensities mean that the infiltration capacity of the soil is not usually exceeded (Kirkby, 1978), so that the runoff regime should be dominated by saturation excess and subsurface stormflow mechanisms. There is evidence to suggest, however, that infiltration-excess overland flow is becoming more common (Burt, 1987), primarily due to changing land-use practices. Grazing is obviously an important factor in increasing the likelihood of infiltration-excess surface runoff and hence sediment and solute delivery to streams during storms.

Land-use Controls on Solute and Sediment Delivery

Table 12.4 shows the effect of different land uses on surface runoff and infiltration capacity. The final infiltration capacities ranged from $11–12 \, mm \, hr^{-1}$ for cereal and ungrazed temporary grass down to $2 \, mm \, hr^{-1}$ for heavily grazed permanent grass. This meant that the surface runoff from the ungrazed plots was low (less than 10 per cent of the total rainfall input)

Table 12.4   Rainfall simulation of the effect of land use on surface runoff from hillslope plots

Land use	Rainfall intensity (mm hr^{-1})	Total[1] runoff (mm)	Runoff as % total rainfall	Infilt.[2] capacity (mm hr^{-1})	Bulk density (g cm^{-3})	Moisture content (%)
Heavily grazed permanent grass	12.50	26.5	53.0	2.66	1.179	22
Lightly grazed permanent grass	12.50	11.6	23.2	8.38	1.122	21
Temporary grass	12.50	2.3	4.6	11.91	0.958	24
Cereal	12.50	3.7	7.4	11.33	1.076	19
Prepared ground (compacted by rolling)	12.50	10.6	21.2	7.56	0.926	23

[1]Total runoff in the 4-hour experiment period.
[2]Infiltration capacity at 4 hours from the start of the experiment.

whereas the surface runoff from grazed permanent grassland ranged from 23 per cent of the total rainfall input for lightly grazed areas, to over 50 per cent for heavily grazed areas. Even this high percentage surface runoff was lower than might be anticipated because some of the rainfall input was stored in depressions on the soil surface which were created by extensive cattle poaching. Such depression storage has been shown to be less important on steeper slopes in the Merrifield catchment (Heathwaite and Burt, unpublished data). The main control of surface runoff production (Table 12.4) is the amount of surface compaction and the presence or absence of vegetation cover. For prepared ground (recently rolled and so compacted), the surface runoff is more than double that of cereal or temporary grass which have an intact vegetation cover. Grazing compacts the soil surface and so increases the magnitude of surface runoff. For the heavily grazed permanent grassland where the vegetation cover has been completely removed, the surface runoff was 2.5 times that of the prepared ground.

Table 12.5 shows the rate of sediment and solute delivery in surface runoff from the different land uses. The main point is the high rate of suspended sediment, total phosphorus and total nitrogen delivery in surface runoff from the heavily grazed permanent grassland, which is 30, 16 and 9 times respectively the delivery from the lightly grazed areas. For the permanent grassland, only a small fraction (less than 20 per cent) of the total phosphorus was in inorganic form, the remaining 80 per cent being organic, particulate or adsorbed phosphorus; hence the high association with suspended sediment delivery. Most of the nitrogen, however, was in inorganic form, virtually all (98.8 per cent) as NH_4–N. For this grazed permanent grassland much of the nitrogen will be returned to the soil surface as urine or in dung. Although this nitrogen source

Table 12.5 Rainfall simulation of the effect of land use on sediment and solute delivery from hillslope plots

Land use	Total runoff (mm)	Total P (mg/mm)	Inorganic P (mg/mm)	Total N (mg/mm)	Inorganic NH_4-N (mg/mm)	Inorganic NO_3-N (mg/mm)	Suspended sediment (mg/mm)
Heavily grazed permanent grass (12.5 mm hr^{-1})	26.5	4.7	0.8	2.60	2.37	0.03	840
Lightly grazed permanent grass	11.6	0.3	0.2	0.30	0.098	0.002	30
Temporary grass	2.3	0.3	0.1	nd	nd	nd	70
Cereal	3.7	0.2	0.1	nd	nd	nd	80
Prepared ground	10.6	0.2	0.1	0.30	0.043	0.057	480

is spatially variable, it will be in a highly mobile form. Thus any surface runoff crossing such dung or urine patches will obtain a high nitrogen load without having entered the soil.

For cereal and temporary grass, the rate of phosphorus and suspended sediment delivery was low and of a similar magnitude for each land use. No nitrogen could be detected in surface runoff. This could be due to the August timing of the experiments, which is a considerable time since the last nitrogen fertilizer application. However, Bergstrom (1987) found similarly negligible nitrogen delivery from temporary grassland. For the prepared ground, the rate of nitrogen delivery was low and similar to that from the lightly grazed permanent grass. However, suspended sediment delivery was high, and probably related to the lack of vegetation cover. Unlike the heavily grazed grassland (which also had no vegetation cover) this high suspended sediment delivery was not associated with a high total nitrogen or phosphorus input. This implies that it is grazing that is the significant contributor to the high nitrogen and phosphorus load from the heavily grazed grassland.

The results for the permanent grassland suggest that in heavily grazed areas where the surface vegetation is removed and inputs of dung and urine are high, the potential nitrogen load in surface runoff will be extremely high. These extensively poached areas effectively form point sources of nitrogen (and phosphorus and suspended sediment) for the stream. Their importance for stream water quality will depend on their proximity to the stream or the existence of pathways in the field-to-channel flow towards the stream.

Land-use Management to Minimize Sediment and Solute Inputs to the Stream

In order to be of value for land-use management, the results of the rainfall simulation experiments need to be linked to the variation in stream water quality, preferably for storm events when the sources of sediment and solutes in the stream can, to some extent, be identified. It has been shown that the different land-use categories of the Merrifield catchment will contribute to surface runoff, sediment, and solutes at different magnitudes. The next stage would be to extrapolate these results to give predicted loads per hectare, but to do this it is necessary to assume that the rate of surface delivery will be similar for the entire land-use area. This is unlikely, particularly as a very small (0.5 m^2) area was monitored and localized variations in surface delivery within the catchment will occur. Dunne (1983) and Dunne and Black (1970) identify the importance of localized areas adjacent to the stream as contributors of storm runoff. Additionally, partial source areas (Betson, 1964) such as tracks formed by grazing animals are important, particularly when considering the extremely high delivery of nitrogen, phosphorus and suspended sediment shown in these experiments from areas that have been heavily grazed. Although both form only a small proportion at any one time of the total field area, they may be key areas in terms of sediment and solute delivery in surface runoff reaching the stream system. This is especially important if the areas immediately adjacent to the stream or riparian zones are subject to heavy grazing or poaching if, for example, they serve as access points to drinking water for stock. Riparian zones have a high potential for regulating nutrient fluxes between the catchment and the stream. The water table in the riparian zone is usually high, so chemical modification of groundwater entering this zone is likely before it enters the stream (Warwick and Hill, 1988). Soil water displaced from riparian zones may provide up to 90 per cent of storm runoff (Sklash and Farvolden, 1979).

Land use in the riparian zone of the Slapton catchment is given in Table 12.6. Land use in the catchment is given for comparison in Table 12.7. This shows that permanent grassland dominates land use in the riparian zone of the Gara and Start catchments, which together form 93 per cent of the total catchment area. This contrasts with the overall land use (Table 12.7), which has a higher proportion of temporary grassland and arable crops. Arable land use tends to be

Table 12.6   Land use in riparian zones of the Slapton catchment (1986)

Catchment	Total stream length (m)	Per-manent grass	Tem-porary grass	Wood	Arable cereals	Arable roots	Marsh
				Percentage of total stream length			
Gara	20 620	65	19	12	2	2	—
Slapton Wood	1 400	—	29	64	—	7	—
Start	9 860	56	16	16	3	—	9
Stokeley Barton	1 030	8	36	42	—	—	—
Total	32 910	60	19	16	1	1	3

Table 12.7   Land use in the Slapton catchments (1986)

		Grassland Permanent	Grassland Temporary	Arable Cereals	Arable Roots	Woodland
Catchment area (ha)			Percentage of total area			
Gara	2362	39	34	15	5	4
Slapton Wood	93	21	21	33	9	12
Start	1079	25	37	28	6	3
Stokeley Barton	153	9	20	57	4	2

Permanent grass: permanent pasture and long-term leys (7–10 years).
Temporary grass: grass sown within 4 years of survey date and short-term leys (less than 7 years).
Cereals: wheat, barley, oats, rye and maize.
Roots: potatoes, sugar beet, turnips, swede, fodder beet and mangolds.

located in plateau areas. As this means that it is some distance from the stream, its impact on stream water quality through surface sediment and solute delivery may be limited. Both the Slapton Wood and Stokeley Barton catchments have a high percentage of woodland in riparian zones (Table 12.6), which again contrasts with the total catchment land use, where a much greater percentage of cereals is recorded for both catchments. Forested riparian zones appear to be particularly effective in regulating nutrient, especially nitrate, fluxes to the stream from adjacent cultivated land (Lowrance, Todd and Asmussen, 1984; Peterjohn and Correll, 1984; Jacobs and Gilliam, 1985).

The grazing of permanent grassland in areas adjacent to the stream network (which is common practice simply because this is where the permanent grassland is located) may be a major source of stream sediment and solute loads. It is unlikely that land some distance away from the stream will be an important primary contributor to stream water quality unless through partial area input. As a secondary source, it may be important, particularly in throughflow delivery which is not discussed in this chapter.

SUMMARY AND CONCLUSIONS

Infiltration capacity and bulk density were shown to vary with season (primarily rainfall input) and land use. The land-use distinction is greatest where grazing animals trample and compact the soil surface. This can result in up to an 80 per cent reduction in the infiltration capacity of the grazed area. This is in addition to the removal of the 'protective' crop cover which will increase the delivery in surface runoff of sediment and associated solutes. The results of the rainfall simulation experiments indicate the importance of an intact crop cover in reducing surface runoff and suspended sediment delivery from hillslope plots. The presence of grazing animals may also increase nitrogen and phosphorus delivery through increased organic inputs to the soil surface.

The runoff volume from heavily grazed permanent grassland is at least double that from lightly grazed areas, and nearly 12 times greater than that of ungrazed (temporary grassland) areas. This is comparable with the results of McColl (1979), who found that the runoff volume was seven times greater from grazed pasture when compared with ungrazed pasture. In terms of solute delivery, Ryden, Ball and Garwood (1984) report that nitrate leached from grazed grassland was 5.6 times greater than losses from a comparable ungrazed plot, and significantly greater than losses from arable land.

As a general indicator, heavily grazed land will result in high nitrogen, phosphorus, and suspended sediment delivery during storms. Although such areas are not extensive they may be critical in terms of stream water quality if the grazed land is located adjacent to the stream. Careful management of these land-use zones is required, in particular, limiting numbers of grazing animals and avoiding poaching around watering points.

REFERENCES

Bergstrom, L. (1987). Leaching of nitrate and drainage from annual and perennial crops in tile-drained plots and lysimeters. *J. Environmental Quality*, **16**, 11–18.

Betson, R. P. (1964). What is watershed runoff? *J. Geophysical Research*, **69**, 1541–52.

Bowyer-Bower, T. A. S. and T. P. Burt (1989). Rainfall simulators for investigating soil response. *Soil Technology*, **2**, 1–16.

Burt, T. P. (1978). Three simple and low cost instruments for the measurement of soil moisture properties. *Huddersfield Polytechnic, Department of Geography Occasional Paper No. 8.*

Burt, T. P. (1987). Slopes and slope processes. *Progress in Physical Geography*, **11**, 598–611.

Burt, T. P., D. P. Butcher, N. Coles and A. D. Thomas (1983). Hydrological processes in the Slapton Wood catchment. *Field Studies*, **5**, 731–52.

Burt, T. P., B. P. Arkell, S. T. Trudgill and D. E. Walling (1988). Stream nitrate levels in a small catchment in southwest England over a period of 15 years (1970–1985). *Hydrological Processes*, **2**, 267–84.

Crooke, W. M. and W. E. Simpson (1971). Determination of ammonium in kjeldahl digests of crops by an automated procedure. *J. Sci. Fd. Agric.*, **22**, 9–10.

Dunne, T. (1983). The relation of field studies and modelling in the prediction of storm runoff. *J. Hydrology*, **65**, 25–48.

Dunne, T. and R. D. Black (1970). Partial area contributions to storm runoff in a small New England watershed. *Water Resources Research*, **6**, 1296–1311.

Heathwaite, A. L., P. J. Johnes and T. P. Burt (forthcoming). The simultaneous determination of total nitrogen and phosphorus in freshwaters using a microwave digestion procedure. *Water Research*.

Henriksen, A. and A. R. Selmer-Olsen (1970). Automated methods for determining nitrate and nitrite in water and soil extracts. *Analyst*, **95**, 514–18.

Jacobs, T. C. and J. W. Gilliam (1985). Riparian losses of nitrate from agricultural drainage waters. *J. Environmental Quality*, **14**, 472–8.

Johnes, P. J. and O'Sullivan (in press). Nitrogen and phosphorus losses from the catchment of Slapton Ley—An export coefficient approach. *Field Studies*.

Kirkby, M. J. (1978). Implications for sediment transport. In M. J. Kirkby (ed.), *Hillslope Hydrology*, John Wiley, Chichester, pp. 325–63.

Lowrance, R. R., R. L. Todd and L. E. Asmussen (1984). Nutrient cycling in an agricultural watershed: I. Phreatic movement. *J. Environmental Quality*, **13**, 22–7.

Lowrance, R. R., R. Leonard and J. Sheridan (1985). Managing riparian ecosystems to control non-point pollution. *J. Soil and Water Conservation*, **40**, 87–91.

McColl, R. H. S. (1979). Factors affecting downslope movement of nutrients in hill pastures. *Prog. in Water Technology*, **11**(6), 271–85.

Mercer, I. D. (1966). The natural history of Slapton Ley nature reserve: Introduction and morphological description. *Field Studies*, **2**, 385–485.

Murphy, J. and J. P. Riley (1962). A modified single solution method for the determination of phosphate in natural waters. *Anal. Chim. Acta*, **27**, 31–6.

Oakes, D. B., C. P. Young and S. S. D. Foster (1981). The effects of farming practices on groundwater quality in the United Kingdom. *The Science of the Total Environment*, **21**, 17–30.

Peterjohn, W. T. and D. L. Correll (1984). Nutrient dynamics in an agricultural watershed: Observations on the role of a riparian forest. *Ecology*, **65**, 1466–75.

Ratsey, S. (1975). The climate of Slapton Ley. *Field Studies*, **4**, 191–206.

Ryden, J. C., P. R. Ball and E. A. Garwood (1984). Nitrate leaching from grassland. *Nature*, **311**, 51–3.

Sklash, M. G. and R. N. Farvolden (1979). The role of groundwater in storm runoff. *J. Hydrology*, **43**, 45–65.

Troake, R. P., L. E. Troake and D. E. Walling (1976). Nitrate loads of South Devon streams. In *Agriculture and Water Quality*. *MAFF Technical Bulletin*, **32**, 340–51.

Trudgill, S. T. (1983). The soils of Slapton Wood. *Field Studies*, **5**, 835–40.

Walling, D. E. and M. R. Peart (1980). Some quality considerations in the study of human influence on sediment yields. *Proceedings of the Helsinki Symposium*, June 1980, IAHS-AISH Publ. No. 130, pp. 293–302.

Warwick, J. and A. R. Hill (1988). Nitrate depletion in the riparian zone of a small woodland stream. *Hydrobiologia*, **157**, 231–40.

13 The Hydrological Implications of Heath Vegetation Composition and Management in the New Forest, Hampshire, England

A. M. GURNELL, P. A. HUGHES and P. J. EDWARDS
Departments of Geography and Biology,
University of Southampton

SUMMARY

This chapter describes field observations of hydrological processes derived from nested drainage basins and small plots of heathland of different age (since management by burning) in the New Forest, Hampshire, England. The field data are described and analysed in order to: (1) assess the influence of woodland and heathland cover on storm runoff from New Forest drainage basins ranging in area from 1.2 to 98.9 km^2; (2) evaluate the relationship between heath vegetation and soil water regime; and (3) consider the influence of management practices upon the hydrology of dry heathland. As a result of these analyses, it appears that heath areas generate more storm runoff than woodland areas, that the wet heath and mire areas, with their consistently near-surface water tables, are probably particularly significant in generating this enhanced runoff, and that management of dry heath has an effect on its hydrology which may impact upon the routing of storm water and thus the hydrology of downslope wet heath areas. As a result of the very close relationship between heath vegetation and the soil water regime, it is hypothesized that changes in the hydrology of heath hillslopes associated with repeated dry heath management may result in changes of vegetation composition, particularly within the wet heath zones.

Vegetation and Erosion
Edited by J. B. Thornes
©1990 John Wiley & Sons Ltd

INTRODUCTION

During the 25 years since the publication of Penman's (1963) book *Vegetation and Hydrology*, an increasing body of evidence has demonstrated the complexity of interactions between vegetation and hydrological processes. The enhanced water losses experienced in afforested catchments have been a major theme of this research and have been the subject of many catchment experiments (Courtney, 1981; Bosch and Hewlett, 1982; Trimble, Weirich and Hoag, 1987). In Britain, the best known catchment experiment concerned with the relationship between vegetation and hydrology has been undertaken by the Institute of Hydrology in central Wales (Newson, 1985). This experiment was used to compare hydrological processes in the afforested upper Severn catchment with those in moorland of the upper Wye; it is now being used to assess the impact on hydrology of deforestation and of the different methods of harvesting timber.

This chapter compares storm runoff from afforested and heath catchments at different spatial scales within the low-precipitation environment (approximately 800 mm per annum) of the New Forest, Hampshire, England. However, the main focus of the chapter is the hydrological processes which lead to enhanced runoff from heath areas in comparison with adjacent woodlands.

BACKGROUND TO VEGETATION AND HYDROLOGY STUDIES IN THE NEW FOREST

A major problem in investigating the hydrological characteristics of drainage basins in the New Forest is that river flows are extremely responsive to rainfall. It is very difficult to obtain accurate flood flow estimates even for quite high-frequency flood events because of overbank flows. The problem of monitoring flood flows is greatest in the larger catchments. As a result, it is necessary to consider a variety of types of hydrological information in addition to river flows, and gathered at different spatial and temporal scales, in order to synthesize the hydrological characteristics of areas under differing vegetation cover and to infer the likely outcome of vegetation management practices.

In order to attempt this synthesis, data are drawn from two geographical areas and from various experimental systems. Large-area data are gathered from the Lymington River catchment, which drains nearly 100 km² of the central area of the New Forest to a gauging station at Brockenhurst. Small-plot studies come from both within the Lymington River catchment and from a site near Beaulieu Road station in another part of the New Forest.

The information on storm runoff and baseflow comes from nested drainage basins of different size and vegetation cover. A review of observations of vegetation composition and near-surface water tables in a single subcatchment supplies data on the role of wet heathland and mire areas in runoff generation. Information on the interception, overland flow, infiltration and soil moisture content of a burnt and unburnt plot in the same subcatchment and of four plots

of different age near Beaulieu Road station, indicates the hydrological characteristics of dry heath areas and the impact of management on those characteristics.

VEGETATION AND HYDROLOGY—THE INFLUENCE OF WOODLAND COVER IN DRAINAGE BASINS OF DIFFERENT SIZE

The Lymington River and its tributaries above gauging station 6 (Figure 13.1)
drain 98.9 km² of the New Forest. All of the gauging stations have rated
sections apart from gauging station 2, which has a thin-plate V-notch weir, and
gauging station 6, which has a compound V-notch and rectangular thin-plate weir.
The network of six gauging stations provides an excellent basis for investigating
contrasts in runoff with differences in basin size and vegetation cover.

The catchment is underlain by Tertiary sands and clays. There is a contrast
between the northern part of the basin (upstream and to the west of gauging
station 5) where Barton Clay is exposed in the valley bottoms, and the southern
part (downstream and to the east of gauging station 5) where the more permeable
Barton Sand is exposed (sometimes with valley gravels) below the plateau gravels
on the interfluves. This contrast is reflected in the classification of winter rainfall
acceptance potential (WRAP), a catchment characteristic based on hydrological
properties of the soils, which is used in Britain to estimate flood characteristics

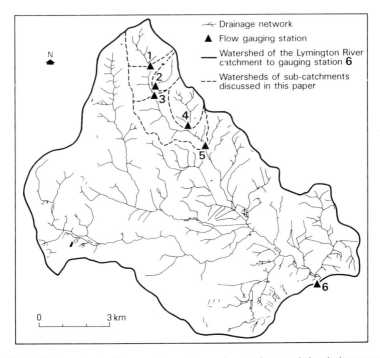

Figure 13.1 The Lymington River catchment: gauging stations and the drainage network

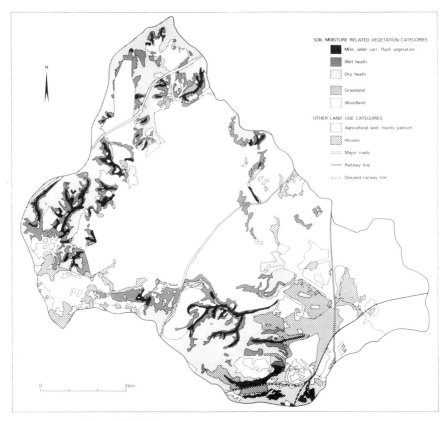

Figure 13.2   The Lymington River catchment: vegetation

for ungauged sites (NERC, 1975). The northern part of the Lymington River basin is mapped as WRAP class 4 (out of the five available classes), which implies that only a relatively small proportion of rainfall in winter storms will infiltrate the soil; in contrast the southern and eastern parts of the basin are mapped as class 1, which implies that a relatively high proportion of rainfall in winter storms will infiltrate the soil (Farquharson *et al.*, 1978). The junction between the two classes occurs about 0.5 km downstream of gauging station 5.

Figure 13.2 shows the distribution of vegetation categories within the Lymington River catchment based upon a visual interpretation of colour, tone and texture variations of a false colour composite of SPOT simulation data. On the basis of Gimingham's (1972) classification of heath vegetation, the nature of the vegetation categories mapped in Figure 13.2 are briefly as follows. Dry heath is dominated by the dwarf shrub *Calluna vulgaris* (which often exceeds 90 per cent cover), a species which can tolerate a wide range of soil moisture regimes but cannot survive prolonged waterlogging. Therefore, as the soil moisture regime becomes wetter in heath areas, the percentage cover of *Calluna vulgaris* tends to decrease, giving way to other species including *Erica tetralix, Myrica gale, Juncus* spp. and *Sphagnum* spp. This transition covers a range of recognizably distinct communities mapped as wet heath in Figure 13.2 but described in more detail by Gimingham as humid and wet heaths and terminates in mire communities, which are dominated by *Sphagnum* spp. on permanently water-logged sites. The acid grasslands of the New Forest are subjected to heavy grazing by ponies and cattle and so are locally referred to as 'lawns' (Tubbs, 1987).

Figure 13.2 provides little evidence of any north–south change in the vegetation to correspond with variations in rock and soil type or winter rainfall acceptance potential class. Areas of mire and wet heath typically occur close to the drainage network throughout the catchment where there is no woodland cover. Dry heath is most commonly located upslope of lawn, wet heath and mire areas. Figure 13.2 provides a base for subdividing the vegetation in each of the gauged catchments shown in Figure 13.1 into areas under woodland cover and areas under shorter vegetation including lawn, wet and dry heath. The percentage of woodland cover in each of the basins is shown in Table 13.1.

Table 13.1 provides percentage runoff statistics for three samples of storms from three scales of drainage basin. The statistics relate to simple, single-peaked, within-channel storm hydrographs. The first sample of only 16 storms, for catchments up to the scale of the whole basin, is a subset of the 52 storms in the second sample (for sub-basins within catchment 5). These samples of storms occurred at all seasons of the year over a three-year period of records, and include those storms which did not result in overbank flows or in a complex hydrograph at gauging station 5 (52-storm sample) or 6 (16-storm sample). The third sample of only 12 storms is for three of the headwater basins and is restricted in size because only one year of records was available for gauging station 4. The three sets of summary statistics describe the percentage runoff from different catchments (or the difference between catchments) in response to the same precipitation events. The 'wood' and 'heath' estimates of percentage runoff are derived by assuming simple mixing between runoff from areas of woodland and heathland in subcatchments 3-1 and 5-3 (see Gurnell and Gregory, 1987). For each storm the volume of storm runoff was estimated by employing a straight-line separation joining the base of the rising limb of the hydrograph to the point at which the falling limb of the hydrograph diverged from the station's baseflow recession curve. Depth of storm runoff was then calculated from each of the separated storm hydrographs.

The requirement for within-channel, single-peaked hydrographs at all of the gauging stations to ensure accurately gauged hydrographs yielded a sample of storms with low runoff amounts. Nevertheless, Table 13.1 clearly illustrates the strong association between vegetation and storm runoff from subareas of the catchment, with the woodland areas yielding considerably less runoff than the heath areas. It also indicates that this pattern extends over the whole catchment to gauging station 6 and appears to be independent of the change in winter

184

Table 13.1 Descriptive statistics for percentage storm runoff from subareas of the Lymington River catchment

Gauging station or area between two stations	Heath	2	3–1	3	1	7	5	4	5–3	Wood
Catchment area (km²)	—	1.3	2.7	5.4	2.8	98.9	11.6	2.1	6.1	—
% Woodland	0	4	13	21	28	54	58	91	92	100
WHOLE CATCHMENT (TO GAUGING STATION 7) ($n=16$, 4 years of records)										
Median	23	—	20	13	3	8	6	—	3	1
Maximum	48	—	45	33	26	28	28	—	24	21
Minimum	1	—	1	1	0	1	1	—	−9	−13
HIGHLAND WATER SUBCATCHMENT (TO GAUGING STATION 5) ($n=52$, 4 years of records)										
Median	27	—	25	16	6	—	10	—	4	2
Maximum	88	—	79	58	37	—	47	—	48	45
Minimum	1	—	1	1	0	—	0	—	−9	−13
HEADWATER CATCHMENTS (TO GAUGING STATIONS 1, 2, 4) ($n=12$, 1 year of records)										
Median	—	29	—	—	7	—	—	9	—	—
Maximum	—	88	—	—	42	—	—	33	—	—
Minimum	—	9	—	—	1	—	—	0	—	—

Precipitation / Percentage Runoff

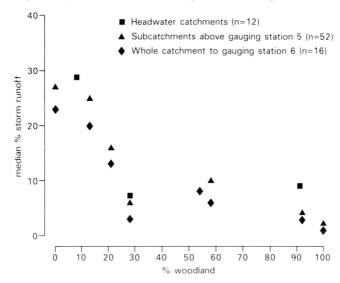

Figure 13.3 The relationship between median percentage storm runoff and percentage woodland for subareas of the Lymington River catchment

rainfall acceptance potential category from 4 in the northern part of the basin to 1 in the southern part. Figure 13.3 plots the median percentage storm runoff estimates from the three samples in Table 13.1 against the percentage woodland cover. There are insufficient data to estimate a reliable relationship between percentage runoff and percentage woodland but the data appear to describe a curve, with a rapid decline in the percentage runoff up to a woodland cover of about 30 per cent. Imprecision in the storm runoff–percentage woodland relationship probably results from interference from other factors. For example, the runoff from the basins represented in Figure 13.3 may be affected by: (1) scale factors (larger basins might be expected to have a larger proportion of flood plain to generate storm runoff); (2) the distribution of the woodland within the basin (woodland near the stream network might be expected to reduce storm runoff more than woodland on the interfluves); and (3) the character and water-relations of the areas of shorter vegetation (waterlogged areas might be expected to generate more storm runoff than areas with lower soil moisture content).

For relatively small storm events such as those presented in Table 13.1 and Figure 13.3, much of the difference between storm runoff from woodland and heathland areas is likely to be the result of the difference in the interception capacity of the two types of vegetation canopy. However, there appear to be other factors in addition to interception capacity influencing the difference in runoff from catchments under the two types of vegetation cover.

Within the area of the catchment in winter rainfall acceptance category 4, where the valley bottoms are underlain by Barton Clay, the area draining to the river channel between gauging stations 1 and 3 is predominantly under heath vegetation while that between stations 3 and 5 is under woodland cover (Table 13.1). Differences in baseflow runoff between these gauging stations indicate variation in soil moisture conditions and thus in the likely extent of areas that may respond rapidly to rainfall under the heath and woodland vegetation cover. Baseflow per unit area preceding the 52 storms analysed for the basin above gauging station 5 was higher from subareas of the basin with a greater percentage heath cover. The mean baseflow for the heath area (between gauging stations 1 and 3) was $17 \, \text{l} \, \text{s}^{-1} \, \text{km}^{-2}$, over 40 per cent higher than the $12 \, \text{l} \, \text{s}^{-1} \, \text{km}^{-2}$ observed from the woodland area (between gauging stations 3 and 5). The 52 storms were derived from four years of records and from all seasons of the year; the only storms with a higher baseflow from the woodland than the heath occurred in the periods April to June though this was not true of all storms at this time of year and the differences were not large. These results suggest that within the heath area there was generally higher soil moisture storage, especially in winter when the contrast in baseflow according to vegetation cover was found to be most marked. This is particularly relevant to the generation of storm runoff for large storm events, since differences in soil moisture storage will affect both the area of the catchment generating runoff (either as saturation overland flow or as rapid throughflow), and the amount and intensity of storm runoff production in those areas.

There are a number of possible explanations for the differences in soil moisture storage between the different vegetation types. The first of these is the differences in water losses between the vegetation types. Such differences in losses are largely a result of differences in the interception capacity of the vegetation; however, the presence of negative storm runoff estimates (i.e. storm hydrographs which decrease in size as they progress downstream through the woodland; Table 13.1) for some small summer storms suggests the possible influence of transpiration losses increasing soil moisture deficits and so inducing seepage into the river banks in the woodland at a greater rate than occurs within the heath. A second, and probably much more significant reason for contrasts in soil moisture storage between the heath and woodland areas is that within the heath there are mire and wet heath areas which have a water table at or near the ground surface for most of the year.

THE RELATIONSHIP BETWEEN HEATHLAND VEGETATION AND SOIL WATER REGIME

Rutter (1955) was the first to show that heath vegetation composition was closely related to the soil water regime. Such a relationship has been evaluated in more detail through the analysis of a year's records of water table levels from 168 shallow wells distributed across the heath areas of the catchment draining to gauging station 5 (Gurnell and Gregory 1986, 1987). This showed that the species composition of the heath areas can be quantitatively related to the level and variability of the near-surface water table. Indeed, it was possible, using a six-fold classification of vegetation composition to estimate water table level exceedance probabilities from the vegetation composition (see Gurnell *et al.*, 1985).

Analysis of water table levels recorded along downslope transects within the wet heath area of subcatchment 2 (Gurnell, 1981) also indicated the close relationship between vegetation composition and water table levels. Two factors were found to influence the level and variability of the water table. First, the permeability of superficial deposits caused variations in water table level down individual hillslopes. Second, beneath these variations, there were underlying seasonal trends, with water table levels rising nearer to the soil surface downslope in winter and upslope in summer. These trends result from the interaction of seepage of water into the superficial deposits at the junction between the Barton Clay and the overlying Barton Sand and gravel near the top of the slope, and the drainage from the superficial deposits into the stream at the base of the slope. Seepage probably occurs in most areas of the New Forest between the free-draining plateau gravels and the less permeable rock types beneath but seepage is greatest where heavy clays underly the gravels. Above the seepage zone, the gravels are covered by dry heath and have water table levels which remain below 60 cm depth from the soil surface (Hughes, 1984).

In summary, therefore, non-wooded hillslopes within the Lymington River catchment typically have dry heath at the top of the slope with wet heath, mire and lawn areas downslope (Figure 13.2). The dry heath area is subject to relatively deep water table levels throughout the year; water from this zone seeps downslope, often concentrating at a seepage zone or step where the lithology changes, to provide moisture for the vegetation communities downslope. The vegetation composition downslope reflects the near-surface water table regime which is itself a function of the water balance and the permeability of the soil and superficial deposits on the lower part of the slope. Waterlogged areas at the base of the slope or at seepage zones can result in the development of mires. This close link between vegetation and hydrology means that the composition of the vegetation can be used to map areas of high water table under varying hydrological conditions and thus areas likely to contribute to storm runoff through saturation overland flow, rapid throughflow and near-surface groundwater flow (e.g. Ward, 1984).

Since the supply of soil moisture to the wet heath and mire areas comes partly from seepage from the upslope dry heath areas, any change in the water relations of the dry heath may have an impact on the downslope soil water regime. This in turn may lead to changes in both the vegetation and the size and variability of the area contributing to storm runoff.

Heathland, particularly the dry heath, in the New Forest is managed by cutting or burning to improve the grazing (Tubbs, 1987; Webb, 1986). The analysis presented here suggests that such management practices, if used too frequently, could have long-term consequences for the hydrology of the entire hillslope.

THE HYDROLOGY OF DRY HEATH AREAS AND THE IMPACT OF MANAGEMENT

The hydrology of four areas of dry heath of varying age was investigated at a level site near Beaulieu Road railway station. The field data were collected in the summers of 1979 and 1980 on plots which had last been burnt in 1974 (plot 1), 1970 (plot 2), 1966 (plot 3) and 1962 (plot 4). In addition, an experimental burn in 1979 of an area of dry heath within the Withybed subcatchment draining to gauging station 2 permitted observations on a sloping site, where the adjacent unburnt plot was last burnt in 1965.

Table 13.2 shows the biomass of vegetation and litter on the study plots at the Beaulieu Road site. The results show that the dry weight of both litter and vegetation increase with age until at least 18 years after burning. It is likely that if cutting were applied to the dry heath vegetation, there would be a negligible effect on the litter and a more rapid recovery of the biomass of the vegetation with increasing age from cutting.

Table 13.3 summarizes the hydrological variables that were monitored at the two sites and provides details of the methods and measurement frequencies employed. Figures 13.4 and 13.5 show observations, from the four plots at Beaulieu Road station, of mean soil moisture content at three depths and soil moisture tension at one depth throughout the summers of 1979 and 1980; Figure 13.5 also shows variations in the mean infiltration rate monitored using single-ring, falling-head infiltrometers during the summer of 1980. All of these variables were monitored manually on the sampling dates marked on the horizontal axes of Figures 13.4 and 13.5. These figures illustrate the very high volumetric water contents that can be achieved in the surface peat horizon (0–5 cm) and the underlying sand horizon with high peat content (5–10 cm) during periods when these layers are at or near field capacity. These high volumetric soil moisture

Table 13.2 Summary of dry heath biomass observations, Beaulieu Road station

| Site | Date burnt | Biomass (g m^{-2} to nearest 10 g; June 1980) | | | |
| | | Standing crop ($n = 5$) | | Litter ($n = 10$) | |
		Mean	Standard error	Mean	Standard error
1	1974	510	60	180	50
2	1970	810	70	350	40
3	1966	1370	40	540	60
4	1962	1440	30	1020	110

Table 13.3 Summary of the dry heath, field data collection programme

Variable	Techniques and sample sizes
Soil moisture content	*Beaulieu Road* A mean of gravimetric determinations from six soil samples taken at random from each of three soil depths (0–5 cm, 5–10 cm, 10–15 cm) on each of 4 heather stands of different age. These were initially estimated as weight of water per unit dry soil weight and were then corrected to volumetric water content using bulk density estimates derived from twelve soil samples from each of the three depths (bulk density varied very considerably between soil layers but not between plots for the same soil depth). Sampling carried out on 18 occasions during summer 1979 and 15 occasions during summer 1980.
Soil moisture tension	*Beaulieu Road* Readings from one tensiometer on each of 4 heather stands on 18 occasions during summer 1979. Mean of readings from three tensiometers on each of 4 heather stands on 15 occasions during summer 1980.
Interception	*Beaulieu Road* (i) 4 (0.5×0.5 m, sand-graded, bitumen sealed) interception plots on each of 4 heather stands provided net precipitation estimates for 6 natural, simple, rainfall events in 1980. (ii) 10 high-intensity artificial storms were applied to 2 interception plots in each of 4 heather stands during 1980. A 0.75 m boom was attached to a watering can to distribute precipitation evenly over each plot and two adjacent raingauges.
Infiltration	*Beaulieu Road* Single ring, 15.4 cm diameter, falling-head infiltrometers were used to record the time taken to infiltrate 500 ml water at 8 sites, on each of 4 heather stands, on 12 occasions during summer 1980. Mean infiltration rates were calculated on each occasion for each heather stand. Double-ring, 10 and 15.4 cm diameter, constant-head infiltrometers were used to record a stable infiltration rate after prewetting on 2 sites on each of 4 heather stands on up to 9 occasions during summer 1980.
Overland flow	*Withybed* Two 7 m^2 areas on a 20° slope on each of the burnt and unburnt plots. Overland flow was limited upslope by a diverting trench and was trapped downslope in a Gerlach trough and was monitored using a tipping bucket mechanism. 15 natural rainfall events and 2 artificial watering events with different antecedent moisture conditions were recorded.

contents can be compared with the even higher 91–98 per cent moisture contents quoted by Ivanov (1981) for peat deposits located below the water table in mire systems.

Paired t-tests applied to the soil moisture data showed that in 1979 the peat of the top 5 cm of the soil profile (soil level 1) had a significantly higher soil

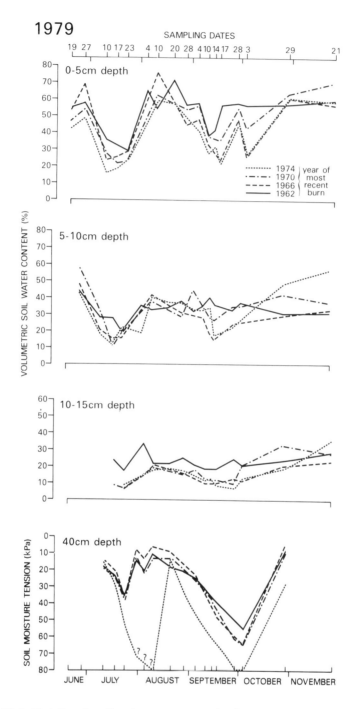

Figure 13.4 Variations in soil moisture content and soil moisture tension under dry heath canopies of different age; summer 1979

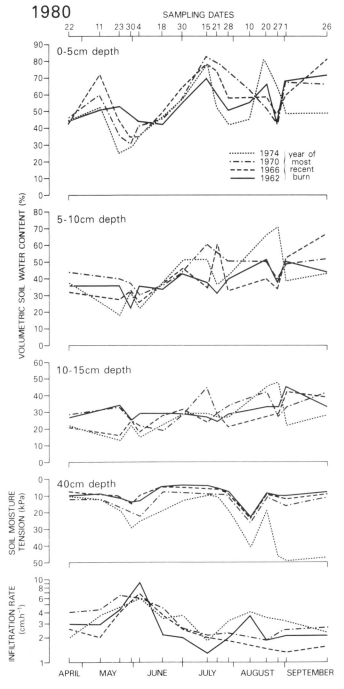

Figure 13.5 Variations in soil moisture content, soil moisture tension and infiltration rates under dry heath canopies of different age; summer 1980

moisture content on plot 4 (the oldest site) than on plots 1 and 2 ($P<0.05$) and plot 3 had a significantly higher moisture content than plot 1. At soil level 2 (5–10 cm, a sand layer with high peat content) there was no significant difference between sites but at soil level 3 (10–15 cm, a sand and gravel layer) plot 4 had a significantly higher soil moisture content than all other plots. In 1980, which was a wetter summer than 1979, there was no significant difference in the soil moisture content between plots. Thus, at times when a soil moisture deficit develops, the soil moisture content at the older sites remains higher than at the younger sites. In addition the soil moisture tension at 40 cm was higher on the younger than on the older sites, particularly during the drying cycles experienced in 1979. The estimates of initial infiltration derived using single ring infiltrometers to infiltrate 500 ml of water were found to be inversely correlated with soil moisture content, with the infiltration rates on the four plots in the order $1>2>3>4$ ($P<0.05$) and 1 and $2>3$ and 4 ($P<0.01$). In order to provide a more accurate assessment of infiltration rates, a smaller number of double-ring, constant-head infiltrometers were also used on the four sites in 1980 to monitor stable infiltration rates after prewetting of the soil. The results showed that infiltration rates after prewetting varied according to time since burning, with paired t-tests indicating $1>3$, 4 and $2>4$ ($P<0.05$). These infiltration observations are in accordance with the view that 'the most important controls are generally agreed to be: soil texture and structure, initial moisture content and some aspect of the vegetation density and organic content of the surface' (Dunne, 1983), since soil moisture content, vegetation and litter biomass are observed to be different on the four plots but the character of the underlying soils is similar.

The interception characteristics of the sites at Beaulieu Road station were investigated using observations of natural rain storms, artificial watering

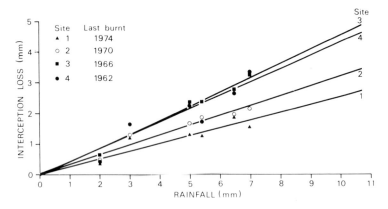

Figure 13.6 Interception loss during natural rainfall events under heather canopies of different age; summer 1980, Beaulieu Road station

experiments in the field, and laboratory experiments. Figure 13.6 illustrates that interception loss during natural rainfall events increased with dry heath age. Simulated rainfall events were applied at such high 'rainfall' intensities that interception loss resulting from evaporation during the storms was negligible. As a result, plots of net precipitation against gross precipitation had steep slopes (approaching unity) and the intercepts on the gross precipitation axis were assumed to be indicative of the interception capacity of the vegetation canopies (Table 13.4). In addition, laboratory experiments were carried out using 210 shoots of heather cut from the Withybed Bottom site; these were weighed, dipped in water, allowed to drip for 30 seconds and reweighed to assess their maximum water-holding capacity. A regression relationship was established between the weight gain of the shoot from retained water (y) and the shoot dry weight (x) $(\log y = 0.165 + 0.754 \log x; n = 210; R^2 = 0.951)$. In a related field experiment, 20 shoots from each of the four plots at Beaulieu Road station were weighed and placed within the vegetation canopy. The canopy was then thoroughly watered and the wet shoots were reweighed to assess the increase in weight resulting from intercepted water. The shoots were subsequently dried at 80° C to a constant weight so that the weight gain resulting from intercepted water could be related to shoot dry weight.

Estimates of interception capacity of the dry heath canopies of different age at Beaulieu Road station were derived from these two sets of observations (Table 13.5). The mean of the dry weight of the 20 cut shoots at each site was calculated and divided into the canopy biomass (Table 13.2) to derive an

Table 13.4 The relationship between net precipitation (y) and gross precipitation (x) (expressed in mm) during high-intensity artificial rainstorms applied to selected interception plots in four dry heath sites of different age

Site	Regression relationship		Estimated canopy interception capacity (x at $y=0$)	n	R^2
1 (1974 burn)	$y = 0.140 + 0.768x$	$x = -0.182$		10	0.986
	$y = -0.258 + 0.901x$	$x = 0.286$		8	0.992
		average =	0.052 mm		
2 (1970 burn)	$y = -0.937 + 0.875x$	$x = 1.071$		11	0.996
	$y = -0.780 + 0.884x$	$x = 0.882$		9	0.978
		average =	0.977 mm		
3 (1966 burn)	$y = -1.113 + 0.857x$	$x = 1.299$		11	0.978
	$y = -1.002 + 0.984x$	$x = 1.018$		12	0.994
		average =	1.159 mm		
4 (1962 burn)	$y = -1.768 + 0.856x$	$x = 2.065$		6	0.943
	$y = -1.563 + 1.038x$	$x = 1.506$		7	0.970
		average =	1.786 mm		

Table 13.5　Estimates of canopy interception capacity at Beaulieu Road station based on shoot watering experiments

Site	1 Average shoot weight (g)	2 Estimated number of shoots per m^2	3 Predicted shoot interception capacity (g)	4 Observed mean shoot interception capacity (g)	5 6 Canopy interception capacity (mm) estimated estimated from 3 from 4	
1	1.39	365	1.87	2.00	0.68	0.73
2	2.05	394	2.51	2.30	0.99	0.91
3	4.23	323	4.34	4.10	1.40	1.32
4	6.50	222	6.00	6.29	1.33	1.40

Column 1 represents the mean weight of 20 cut shoots.
Column 2 is the canopy biomass for the site (Table 13.2) divided by the mean shoot weight.
Column 3 predictions are derived from $\log y = 0.165 + 0.754 \log x$ ($n = 210$, $R^2 = 0.951$) where x is the dry shoot wt (g) and y is the increase in shoot weight from wetting in the laboratory (g).
Column 4 is the mean increase in weight of 20 shoots/site inserted within the watered canopy in the field.

approximate number of shoots per square metre on each site. An estimate of interception capacity per shoot was derived firstly from the observed interception when the shoots were placed in the sprayed canopy (Table 13.5, column 4) and secondly from the regression relationship for weight gain per shoot derived from the laboratory experiments using 210 shoots from the Withybed site (Table 13.5, column 3). Both these techniques produced similar estimates of canopy interception capacity (columns 5 and 6).

Additionally, litter interception capacity was estimated by drying 20 samples of litter (three from each of the plots at Beaulieu Road station and four from each of the burnt and unburnt plots in the Withybed catchment). The litter was placed on a 210 μm sieve, thoroughly sprayed with water, allowed to drain for five minutes and the increase in weight noted. A regression relationship was established between the increased weight resulting from the addition of water and the dry weight of litter which were combined with the dry weight of litter on each plot (Table 13.2) to estimate the interception capacity of the litter on the Beaulieu Road plots. Table 13.6 lists the estimates of the interception capacity of the dry heath vegetation and litter on the Beaulieu Road plots using the different field and laboratory techniques. The sharp increase in interception capacity with increasing dry heath age is evident regardless of the method of estimation.

Differences in overland flow were monitored on two 7 m^2 plots on each of the burnt and unburnt plots in the Withybed catchment. In response to 15 natural rainfall events from 1 to 39.5 mm precipitation, the burnt plots yielded an average of 2.3 times as much overland flow as the unburnt plots. Two high-intensity, artificial precipitation events were also applied to the plots. In both cases the soil moisture content on the unburnt plots was markedly higher than

Table 13.6 Summary table of interception estimates (in mm) for dry heath sites of different age at Beaulieu Road station

Source of estimate	Site 1	Site 2	Site 3	Site 4
Canopy interception capacity				
Field artificial watering experiments	0.05	0.98	1.16	1.79
Immersion of individual shoots (Table 13.5)	0.68	0.99	1.40	1.33
Spraying of shoots placed in the canopy (Table 13.5)	0.73	0.91	1.32	1.40
Mean litter interception capacity	0.42	1.01	1.65	3.29

on the burnt plots and yet the latter yielded substantially more overland flow. These results indicate that the influence of the high interception capacity of the unburnt site more than compensates for the lower soil moisture content (and probably higher infiltration capacity) of the burnt site, resulting in infiltration excess overland flow. Indeed, during very dry periods, the exposed litter and peat layer of the younger sites may become sufficiently dry that it actually repels water at the beginning of rainfall events, so ensuring a high level of overland flow.

Table 13.7 summarizes the results of all of the observations of hydrological variables on dry heath sites of different age. The time since burning of the sites results in strong contrasts in their hydrological characteristics. Recently burnt sites have higher infiltration rates than unburnt sites but their decreased

Table 13.7 Summary of the hydrological characteristics of dry heath stands of different age

Variable	Withybed	Beaulieu Road
Soil moisture content		Wet antecedent conditions—no significant difference dry antecedent conditions— 4 > 3 > 2 > 1
Soil moisture tension		1 > 2 > 3 > 4
Infiltration rate for 500 ml water (falling-head infiltrometer)		1 > 2 > 3 > 4
Interception capacity (canopy)		4 > 3 > 2 > 1
Interception capacity (litter)		4 > 3 > 2 > 1
Overland flow	A > B	

Dates of most recent burn: site A–1979; B–1965; 1–1974; 2–1970; 3–1966, 4–1962. Thus in 1980 the age of the heath on each site was: A, < 1 year; B, 15 years; 1, 6 years; 2, 10 years; 3, 14 years; 4, 18 years.

vegetation and litter interception capacity and decreased surface roughness override these soil moisture contrasts and result in higher overland flow.

DISCUSSION

The research results presented in this chapter can be used to give an indication of the complexity of hydrological processes within drainage basins in the New Forest. Whilst the chapter reports the result of simple analyses of field observations, it underlines the pivotal role of vegetation in influencing hillslope hydrology and thus the importance of developing plot (e.g. Hanks, 1984), hillslope (e.g. Kirkby and Neale, 1987) and basin (e.g. Schulze and George, 1987) scale hydrological models which emphasize the role of vegetation. 'It is claimed that vegetation is the most important and multifarious influence upon infiltration . . . yet there is not a rigorous theory describing the influence of vegetation and its interaction with other variables, and there are no detailed measurements of the physical parameters controlling flow into a vegetated surface' (Dunne, 1983). The field data presented in this chapter make a contribution towards relating properties of the vegetation canopy to variability in the underlying hydrology.

It is apparent from the field data presented here that during small storm events, heath areas generate considerably more runoff than woodland areas. The contrast in runoff decreases with increasing precipitation amount for these storms. This suggests that a significant proportion of the difference in runoff is due to differences in the interception capacity of the vegetation canopies. These contrasts in storm runoff are superimposed upon higher baseflows from the heath than from the woodland areas, which suggest higher soil moisture storage in the heathland.

The heath areas can be subdivided into dry heath (which usually occurs near the top of hillslopes), wet heath, mire and lawn areas. The wet heath and mire areas are subject to near-surface water tables which favour the generation of saturation overland flow and rapid throughflow. The link between the heath and mire vegetation composition and the level and variability of the water table is so close that it is possible to estimate water table exceedance probabilities from vegetation composition.

It appears that the runoff from dry heath sites may take different routes according to the degree to which the vegetation has been managed and the time since the site was last managed. Mature dry heathland maintains a higher and more stable soil moisture content, has higher interception losses and generates less overland flow than recently burnt dry heathland. This implies that recently managed sites will deliver moisture downslope in more exaggerated pulses rather than supplying attenuated seepage of soil moisture and groundwater. Such differences in the routing of the water may have an impact on the hydrology of the downslope area. For example, if the downslope area develops a high-density ephemeral drainage network during storm events (such as in the humid heath and lawn areas observed by Gurnell, 1978), then overland flow from recently managed dry heath could connect into this ephemeral drainage network and be efficiently and rapidly drained from the slope, presumably resulting in a complementary decrease in the soil moisture content and near-surface groundwater levels on the slope; in contrast, if the water were delivered by attenuated seepage it would be more likely to remain in soil moisture storage or percolate to groundwater downslope. Thus changes in the proportion of overland flow from managed areas could lead to differences in water routing downslope and, because of the sensitivity of the heath vegetation communities to the soil water regime, this could result in vegetation composition change downslope. These suggestions need to be tested by field observation but they illustrate a potentially complex hillslope response to management of the dry heath according to the type and recency of the management and the nature of the hydrology of the wet heath downslope.

The data presented here are from sites which have been burnt; similar but more subdued contrasts might be predicted on sites managed by cutting, since the destruction of the vegetation (and especially the litter) is less complete and regeneration is likely to be more vigorous (Gimingham, 1972). Nevertheless, if the dry heath is repeatedly managed by burning or cutting, it is conceivable that the soil water regime will be changed downslope and that the vegetation will respond by a change in its composition.

ACKNOWLEDGEMENTS

NERC are very gratefully acknowledged both for the provision of a research grant for a part of the Lymington River study and also for the award of a research studentship to P. A. Hughes.

REFERENCES

Bosch, J. M. and J. D. Hewlett (1982). A review of catchment experiments to determine the effect of vegetation changes on water yield and evapotranspiration. *Journal of Hydrology*, **55**, 3–23.

Courtney, F. M. (1981). Developments in forest hydrology. *Progress in Physical Geography*, **5**, 217–41.

Dunne, T. (1983). Relation of field studies and modelling in the prediction of storm runoff. *Journal of Hydrology*, **65**, 25–48.

Farquharson, F. A. K., D. Mackney, M. D. Newson and A. J. Thomasson (1978). *Estimation of run-off potential of river catchments from soil surveys*. Soil Survey Special Survey No. 11, Rothamsted Experimental Station, Harpenden, Herts.

Gimingham, C. H. (1972). *Ecology of Heathlands*, Chapman & Hall, London.

Gurnell, A. M. (1978). The dynamics of a drainage network. *Nordic Hydrology*, **9**, 293–306.

Gurnell, A. M. (1981). Heathland vegetation, soil moisture and dynamic contributing area. *Earth Surface Processes and Landforms*, **6**, 553–70.

Gurnell, A. M. and K. J. Gregory (1986). Water table level and contributing area: The generation of runoff in a heathland catchment. In S. M. Gorelick (ed.), *Conjunctive Water Use*, International Association of Hydrological Sciences Publication 156, pp. 87–95.

Gurnell, A. M. and K. J. Gregory (1987). Vegetation characteristics and the prediction of runoff: Analysis of an experiment in the New Forest, Hampshire. *Hydrological Processes*, **1**, 125–42.

Gurnell, A. M., K. J. Gregory, S. Hollis and C. T. Hill (1985). Detrended correspondence analysis of heathland vegetation: The identification of runoff contributing areas. *Earth Surface Processes and Landforms*, **10**, 343–51.

Hanks, R. J. (1984). Soil water modelling. In M. G. Anderson and T. P. Burt (eds), *Hydrological Forecasting*, Wiley, Chichester, pp. 15–36.

Hughes, P. A. (1984). *Effects of management by burning on heathland hydrology*. Unpublished PhD thesis, University of Southampton.

Ivanov, K. E. (1981). *Water Movement in Mirelands*. Translated from the Russian by A. Thomson and H. A. P. Ingram, Academic Press, London.

Kirkby, M. J. and R. H. Neale (1987). A soil erosion model incorporating seasonal factors. In V. Gardiner (ed.), *International Geomorphology 1986, II*, Wiley, Chichester, pp. 189–210.

NERC (1975). *Flood Studies Report* (5 vols), Natural Environment Research Council, London.

Newson, M. D. (1985). Forestry and water in the uplands of Britain—the background of hydrological research and options for harmonious land use. *Quarterly Journal of Forestry*, **79**, 113–20.

Penman, H. L. (1963). *Vegetation and Hydrology*, Technical Communication 53. Commonwealth Bureau of Soils Harpenden, Commonwealth Agricultural Bureau.

Rutter, A. J. (1955). The composition of wet-heath vegetation in relation to the water table. *Journal of Ecology*, **43**, 507–43.

Schulze, R. E. and W. J. George (1987). A dynamic, process-based, user-oriented model of forest effects on water yield. *Hydrological Processes*, **1**, 293–307.

Trimble, S. W., F. H. Weirich and B. L. Hoag (1987). Reforestation and the reduction of water yield on the Southern Piedmont since circa 1940. *Water Resources Research*, **23**, 425–37.

Tubbs, C. (1987). *The New Forest: A Natural History*, Collins, New Naturalist Series, London.

Ward, R. C. (1984). On the response to precipitation of headwater streams in humid areas. *Journal of Hydrology*, **74**, 171–89.

Webb, N. (1986). *Heathlands*, Collins, New Naturalist Series, London.

14 The Impact of Agricultural Landuse Changes on Soil Conditions and Drainage

IAN REID
Birkbeck College, University of London

ROB PARKINSON, STEPHEN TWOMLOW
Seale-Hayne Faculty of Agriculture, Food and Landuse, Polytechnic South West, Newton Abbot

and **ANGELA CLARK**
Birkbeck College, University of London

SUMMARY

Agriculture is undoubtedly one of the largest single determinants of current geomorphic process rates. Especially important in advanced agricultural economies is the way that changes in landcover and the seedbed preparation that these entail affect the disposal of water, particularly where underdrainage techniques are employed. Field experiments on a number of different soil types show that seedbed preparation and deep soil loosening decrease storm drainage by as much as 40 per cent and 63 per cent, respectively, at least in the short term. This is contrary to expectation, and leads to soil water contents that are increased by 12 per cent and soil shear strengths that are reduced by 23 per cent, so that the soil remains plastically deformable for longer. The upshot of untimely land management is deterioration of soil structure and an increase in the likelihood of permanent soil damage. An unexpected benefit that arises from the changes in ploughland hydrology is a reduction in the discharge of underdrainage. This signals reduced flood hazard in the arterial waterways of drainage basins where soil loosening is extensively practised.

INTRODUCTION

Agriculture is arguably the most significant biogeomorphological process. Much more energy is expended through the use of a plough in a single preparation

Vegetation and Erosion
Edited by J. B. Thornes
©1990 John Wiley & Sons Ltd

of the seedbed than can be provided by the annual fall of rain, and the conscious rearrangement of earth surface materials has often gone far beyond turning over the topsoil (Lutwick and Hobbs, 1964). Man is not the first substantial earth-mover, of course, and far humbler organisms have been shown to shift considerable quantities of material (Darwin, 1881; Nye, 1955; Barley, 1959). However, the inversion of a 20 cm plough layer with tractor and plough is accomplished 10^4 times quicker than can be achieved by a healthy earthworm population and 10^5 times faster than can be managed by termites. Such an assault on the soil is bound to have adverse side-effects. In fact, it has been argued (Hallsworth, 1987) that the problems of accelerated soil erosion almost entirely post-date the rise of agriculture, and it is now realized that important organic bonding agents such as the polysaccharides are rapidly decomposed once the soil has been disturbed by ploughing, so that soil structure is at risk (Greenland, Lindstrom and Quirk, 1962; Tisdall and Oades, 1982).

However, modification of the soil profile extends beyond inversion of the topsoil. It has long been recognized that crop productivity can be increased by controlling the soil water regime (Trafford and Oliphant, 1977). In advanced agriculture this can be achieved most conveniently by installing piped underdrainage. Green (1979) and Robinson and Armstrong (1988) have shown just how extensively the method has been used where agriculture is heavily mechanized, such as in the United Kingdom. But the method achieves only limited success in poorly draining soils when used by itself, and extensive work has demonstrated that a secondary treatment is necessary to provide the flowpaths that conduct water to the permanent pipe network (Trafford and Rycroft, 1973; Leeds-Harrison, Spoor and Godwin, 1982). These consist of either producing a number of closely spaced soil pipes at depth with a mole-plough, or the heaving and shattering of the soil with a subsoiler. The application of one or other of these secondary treatments depends on soil properties such as clayeyness, but both are temporary measures and need to be repeated at between a few and ten years (Nicholson, 1934).

While it is commonly expected that the installation of an underdrainage system and the modification of the soil profile by moling and subsoiling will ensure evacuation of soil water, it is becoming clear that the efficiency of these artificial treatments is less than has been anticipated. It is also becoming clear that a change in cultivation following the installation of a drainage scheme has a marked impact on its performance. In both cases, topsoils are left in a state that makes them more prone to structural damage than has been anticipated hitherto. As a consequence, there is a greater risk of encouraging the problems that eventually attend structural degradation such as those associated with higher runoff ratios. Besides this, there are the inevitable economic disbenefits of reduced crop yields.

SOIL LOOSENING AND THE WATER REGIME

The use of soil loosening is recommended where low clay content precludes successful deployment of a mole plough. Soils are usually structure-poor, if not actually structureless. As a result, hydraulic conductivity is almost inevitably low prior to treatment.

The results of a field experiment that was established at Seale-Hayne College farm in Devon, England (Figure 14.1), allow us to assess the hydrological consequences of soil loosening. The soil is a member of the Sportsman Series (Findlay *et al.*, 1984) and is developed on Devonian slate. Texturally, it is a clay loam (Table 14.1), and may be taken to represent soils on similar parent materials throughout the United Kingdom.

Figure 14.1 Location of the drainage experiments in the British Isles. 1—Seale-Hayne College Farm, Sportsman Series (Dystric Cambisol); 2—Rectory Farm, Windsor Series (Eutric Cambisol); 3—Brooksby Hall, Ragdale Series (Gleyic Luvisol); 4—Langabeare Farm, Hallsworth Series (Dystric Gleysol)

Table 14.1 Climate and soil properties of the experimental sites

Soil	Soil	Climate		Soil texture*		
		Rain (mm a^{-1})	Evaporation (mm a^{-1})	% sand 63 μm–2 mm	% silt 2–63 μm	% clay <2 μm
Seale-Hayne College	Sportsman Series	974	527	25	53	22
Rectory Farm	Windsor Series	646	521	5	55	40
Brooksby Hall	Ragdale Series	622	510	30	36	34
Langabeare Farm	Hallsworth Series	1051	468	10	41	49

* Subsoil, fine earth basis (<2 mm)

Six straight plastic laterals were installed at 0.75 m depth and 20 m spacing in 1985 and their outfalls were monitored using weir boxes and water-stage recorders. After a period of one year in which the entire site was checked for uniformity of rainfall-drainflow response, half of it was loosened to a depth of 0.4 m using a five-leg McConnel 'Shakaerator'; the other half was left as an 'undisturbed' control. The whole field was then ploughed and put down to winter wheat. Various soil physical parameters were monitored continually.

Figure 14.2 shows the impact of loosening to 0.4 m depth. Subsoil bulk density is reduced by approximately 15 per cent. Superimposed on this is the counter-effect of topsoil seedbed preparation which reduces the difference between treatment and control down to a depth of 0.2 m. However, even this subsequent treatment fails to remove the effect of the loosening where this effect is most

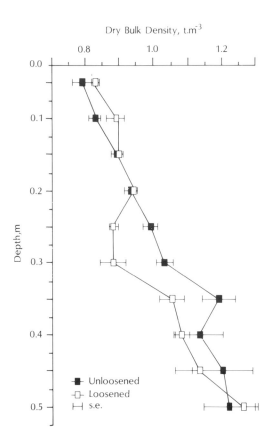

Figure 14.2   Impact of soil loosening to 0.4 m and subsequent seedbed preparation on the bulk density of the Sportsman Series at Seale-Hayne College Farm, Devon

important—the macroporosity. Figure 14.3 shows not only that the fraction of the pore space drained at low tensions (soil water potentials above − 1 kPa) has been substantially increased—especially around 0.2 m where a plough pan had previously developed—but also that the seedbed preparation has left the macropores that were created by loosening largely intact.

Curiously, the effect of loosening on the water regime is not that which would be expected by the farmer. The loosened soil is *wetter* in winter than its unloosened counterpart. The winter mean water content has been shown to be a good diagnostic parameter of the soil water regime in dynamic equilibrium (Reid and Parkinson, 1987). It is a parameter that can be approximated with 'field capacity'. In the case of the Sportsman Series, the winter mean of the soil profile to 0.8 m depth is 6.3 per cent higher where it has been loosened (Figure 14.4).

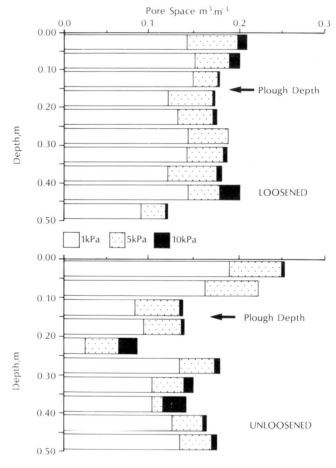

Figure 14.3 Pore space in unloosened and deep loosened Sportsman Series soil drained at three matric potentials

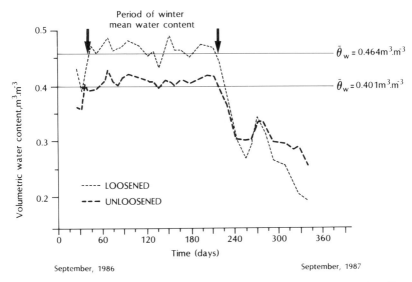

Figure 14.4 Water regimes of unloosened and loosened Sportsman Series Soil. $\bar{\Theta}_w$ is the value of the winter mean water content in the treated and control plots

One of the reasons for the higher water content is undoubtedly the shift in the pore size distribution (Figure 14.3), and the inevitable consequence that this has for soil water retention. In a winter situation where rainfall inputs are continual, the winter mean water content of this soil corresponds to a matric potential of about -1 to -3 kPa. In these circumstances, the loosened soil has a higher fraction of its volume occupied by pores of appropriate size and so remains wetter. As soon as the water balance favours evaporation and these comparatively large pores have been evacuated, the water contents of the loosened and unloosened soils approach each other (Figure 14.4). In the mean time, the fact that the winter water content of the treated soil is higher than its unloosened equivalent is contrary to any farmer's expectations and indicates that the widespread use of subsoiling to alleviate adverse soil water conditions may be inappropriate. Leeds-Harrison *et al.* (1986) had previously predicted this increase in wetness for cracking clay soils using a simulation model of water redistribution processes. The empirical results from Seale-Hayne not only confirm the effect, but broaden the geographical extent of the farm management problems that it brings by including structure-poor silty soils.

SOIL LOOSENING AND DRAINAGE

As might be anticipated, the shift in the pore size distribution that is brought about by deep loosening alters the hydraulic conductivity (K) of the soil.

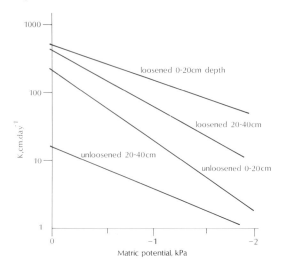

Figure 14.5 Hydraulic conductivity (*K*) of the topsoil and subsoil of an unloosened and deep loosened Sportsman Series soil over a limited range of matric potential

Figure 14.5 shows the results of determinations of vertical conductivity on large 0.2 m diameter cores from both the unloosened and loosened plots using the gypsum crust method of Bouma, Dekker and Wosten (1978). Although there is an inevitably large standard error (Twomlow, 1989), the least-squares trend lines show that loosening has brought about a ten-fold increase in *K* in the subsoil. The impact on the topsoil is less in relative terms and varies throughout the range of water potential over which the test was carried out. Nevertheless, the effect is similarly dramatic and in the direction that has led to positive government recommendations concerning subsoiling in poorly draining soils (MAFF, 1981).

However, as with the winter soil water contents, the results of monitoring the drainage of this Sportsman Series soil are at odds with conventional 'wisdom'. In reality, loosening the soil does *not* hasten the evacuation of newly added rainfall (Parkinson, Twomlow and Reid, 1988). This is illustrated by two storms (Figure 14.6). In both cases, the time between the rainfall centroid and peak drainflow (time of concentration, t_c) is *greater* in the case of the loosened soil. Furthermore, the peak discharge is *much less*. In fact, when we take into account all of the simple winter storms that were monitored, drainflow from the loosened plot peaks, on average, just under half-an-hour later than the unloosened soil. In addition, drainage from the loosened plot reaches a peak discharge that is, on average, only 60 per cent that of its unloosened counterpart. In the short term—say within 24 hours of rainfall, which is the period conventionally taken as a standard for assessing drain performance, and beyond which serious crop damage may occur due to lack of root aeration—the *loosened*

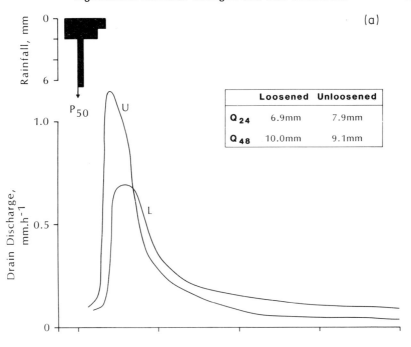

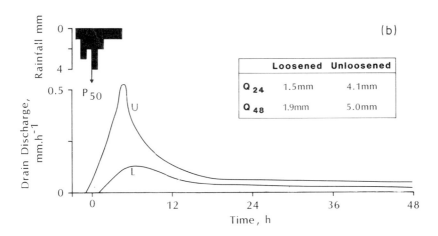

Figure 14.6 Drain outfall hydrographs from unloosened (U) and loosened (L) Sportsman Series soils in response to simple winter storms. P_{50} is the rainfall centroid; Q_{24} and Q_{48} are the total drainage at 24 and 48 hours, respectively, after the initial rise of the hydrograph

soil not only starts wetter (Figure 14.4), it also retains more of the storm water. The 24-hour efficiency of underdrainage (expressed as drainage as a percentage of rainfall without a secondary treatment such as soil loosening averages 38 per cent. With a deep loosening treatment designed to improve drainage, efficiency does not improve. Indeed, the average value actually drops, albeit marginally, to 37 per cent.

The slower and lesser drainage response of the loosened soil is due mainly to a change in flowpaths after application of the secondary treatment. The saturated hydraulic conductivity of the subsoil before loosening is as low as 2×10^{-6} m s^{-1}, or 7 mm h^{-1}. At a matric potential of -1 kPa, this falls to 6×10^{-7} m s^{-1}, or 2 mm h^{-1}. These values do not apply to the plough pan itself, which can be expected to have much lower conductivities. Even so, in order to appreciate why these soils suffer from winter water problems, the subsoil conductivities should be set against the average maximum half-hour rainfall intensity for winter storms. This was 3.5 mm h^{-1} during the period of field observations (Twomlow, 1989), or 1.75 times higher than K at a level of matric potential that is generous for winter conditions. Further reductions in matric potential antecedent to a storm, however small, produce even greater disparity between rainfall intensity and the subsoil's hydraulic conductivity.

Parkinson et al. (1988) have argued that comparatively low hydraulic conductivity below the plough layer (Figure 14.5), together with the rapid increase of water potential that occurs as newly arrived rain is 'perched' in the plough layer, encourages lateral flow in the unloosened soil. In fact, plough layer interflow is now well documented for a number of agricultural soils (Harris et al., 1984; Parkinson and Reid, 1986), and it is responsible for the first rise in the drain hydrograph. On average, this rise occurs half-an-hour after the centroid of simple winter storms in the case of the unloosened silty soil at Seale-Hayne. In fact, plough layer interflow is the main evacuation pathway where the soil has not been loosened. In contrast, the first flush of drainage from the loosened plot takes another half-an-hour to appear. Parkinson et al. (1988) have argued that this is because water is no longer perched in the plough layer to the same extent as in the unloosened soil. Water is now dissipated through a soil column of 0.4 m thickness, i.e. down to the depth of subsoil loosening. Matric potentials do not rise as quickly because of this dissipation, and so the hydraulic gradient causing flow towards the drains is smaller. Drainage is therefore less 'flashy' (lower peak discharge; Figure 14.6) and the system is less efficient (diminutive outfall over 24 hours).

CULTIVATION AND DRAINAGE

In some respects, cultivation appears to cause a similar hydrological response as deep loosening, if only because it brings about a similar, if shallow, shattering of the soil. However, the vast permutation of climatic and land management factors has meant that considerable uncertainty has surrounded the isolation of its effects from those of others. Reid and Parkinson (1984) reported a drop in the 40-hour efficiency of a tile-cum-mole drainage system from 50 to 30 per cent after a grass ley had been ploughed out and the field put down to winter wheat. This pattern had been mooted by Nicholson (1934) but on the basis of a much scantier database. On the other hand, Harris (1977) had reported no measurable change in drainage efficiency with a change from grass to cereals on a soil of the Hallsworth Series.

The storm drainage of the Windsor Series sites investigated by Reid and Parkinson (Figure 14.1) is given in Figure 14.7. The soil is a heavy silty clay (Table 14.1). The outfall of tile Lateral 1 is used as a control because the land it serves remained under grass ley during the period of observations. On the contrary, tile Network 2 was turned from ley to cereals in the last year of record. The large permutation of factors governing soil hydrology is reflected in the wide scatter of the ratio of the storm drainage outfalls of the two systems. However, quite clear is the dramatic shift in the mean level of drainage issuing from Network 2 after ploughing, from approximately equal to about 60 per cent of that of the individual lateral.

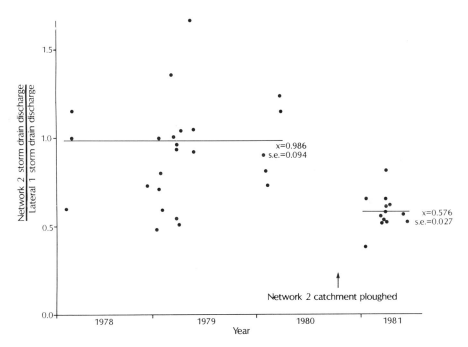

Figure 14.7   Storm drainage of a Windsor Series soil at Rectory Farm on Enfield Chase for simple winter storms in four successive drainage seasons. Each 40-hour storm drainage of tile Network 2 is ratioed with that of tile Lateral 1 to indicate the dramatic fall in efficiency that occurs when the catchment of Network 2 is ploughed out before the winter of 1980–81

Clearly, the ploughed soil retains more water—at least in the short term. The reasons are complex. Undoubtedly, the loosening of the plough layer increases the opportunity for absorption of water by the soil peds, and this has been used as an argument by Leeds-Harrison and Jarvis (1986) in a theoretical modelling of drainage. There is also the fact that ploughing shifts the relationship between matric potential and water content by changing the pore size distribution so that more water can be held in the soil, at least temporarily. Besides these factors, Reid (1979) has shown the significant part played by the microtopography of a ploughed soil surface in promoting depression storage. This is particularly encouraged in clay soils where the development of a surface seal causes a dramatic reduction in infiltration capacity. Inspection of any arable field after winter rainfall will inevitably reveal surface stored water and this storage will help to reduce drainage efficiency.

In probing the effect that cultivation has on soil hydrology still further, a search has been made of the Ministry of Agriculture Fisheries and Food, Field Drainage Experimental Unit's data archive. It was important to choose successive years in which storm sizes were reasonably similar. Besides this, the deterioration of drainage systems dictates that comparison should be made only of data derived for periods falling roughly within the same lapse of time since mole drains were last drawn at each site. After satisfying these conditions (the most important among a host of others), there had also to be a change in cropping from grass to cereals.

Out of this selection procedure, two experimental sites emerged with an adequate amount of experimental data. These are at Brooksby Hall in the English Midlands where the soil is a clay loam of the Ragdale Series, and Langabeare Farm in Devon (Figure 14.1, Table 14.1). The soil at Langabeare Farm is a clay of the Hallsworth Series and the site is the same as that reported on by Harris (1977). Each site had a tile-cum-mole system installed, with permanent pipes at 0.9 m depth and 40 m spacing and mole channels drawn roughly orthogonal at 0.6 m depth and at 2 m spacing to connect with the permeable fill covering the tiles. This is identical to the drainage treatment at the Windsor Series site of Reid and Parkinson (1984) already discussed.

Ideally, in deriving levels of drainage efficiency, only simple winter storms that are isolated in time from previous events would be selected from the database of each site in order to remove yet other unwanted sources of variability. For a number of reasons, including instrument failure and the nature of the rainfalls recorded as producing a drainflow, there are few such storms in the records. However, the four simple storm drainflows recorded in each of the successive years at Brooksby Hall (Ragdale Series) indicate a reduction in 24-hour drainage efficiency of 34 per cent with a change in landuse from grass ley to cereals. This is the same order of magnitude as the 40 per cent reduction in 40-hour efficiency reported for the Windsor Series at Rectory Farm

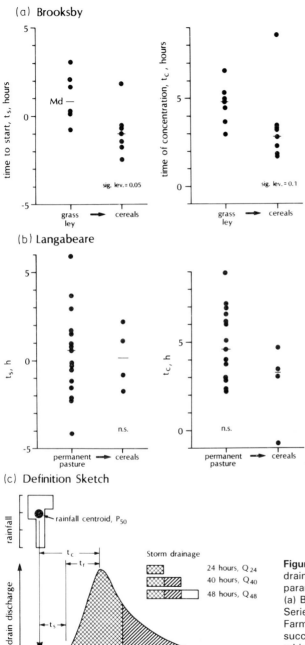

Figure 14.8 Temporal drainflow hydrograph parameters for winter storms at (a) Brooksby Hall (Ragdale Series) and (b) Langabeare Farm (Hallsworth Series) in two successive years between which landuse changed from either grass ley or permanent pasture to winter cereals. The parameters are defined in (c)

and seems to confirm the fact that seedbed preparation reduces the speed at which water leaves the soil.

The small number of storms in the records of Brooksby Hall and Langabeare Farm that fit the strict criteria for selection and comparison of drainage efficiencies meant that other measures of rainfall-drainflow response had to be used in detecting the impact of landuse change on soil hydrology. The temporal parameters of the drainflow hydrograph (as defined in Figure 14.8c) provide a set of such measures and have been shown to be sufficiently sensitive to detect changes in the flowpaths of water in the soil (Reid and Parkinson, 1984). Important for present purposes is that, by taking the effects of the first burst of rain only in complex, multiple-burst storms, the data-set can be enlarged relative to that used in calculating drainage efficiency.

The Ragdale Series (Brooksby Hall; Figure 14.8a) shows a well-defined shift in the first flush of drainage water, with the median value of the time of start parameter, t_s, being brought forward by 1.8 h. The time of concentration, t_c, also shifts forward, this time by 2 h, so that peak drainflow occurs earlier in this soil with a change in landuse from grass ley to winter cereal. The pattern for the steeper and wetter (Table 14.1) Hallsworth Series soil at Langabeare Farm (Figure 14.8b) is less clear cut and none of the annual patterns are significantly different from each other, statistically, confirming Harris's (1977) cautious assessment, albeit from a smaller and less selective sample of storms at the same site, that there was no change in drainage. Nevertheless, the median levels of the temporal hydrograph parameters do mimic those of the Ragdale Series at Brooksby Hall, with a suggestion of faster response once the land had been ploughed.

The temporal hydrograph parameters offer no instruction about changes in the *efficiencies* of the drainage systems at these two sites after a change in crop, but they do indicate a change in the *nature* of the flowpath taken by rain on its passage through the soil. The faster response after seedbed preparation suggests that the hydraulic conductivity of the plough layer has been increased. This has facilitated a more rapid initial interflow response. However, it might be speculated that ploughing has also produced a smear at the base of the plough layer as is the case in most clay soils. In this circumstance, newly added water might be expected to perch above the smear, so increasing matric potential as well as providing the hydraulic head necessary to actually promote interflow. There is no suggestion of a change in the flowpath itself.

In this respect, the ploughed soils appear to be behaving similarly to the unloosened Sportsman Series soil at Seale-Hayne. On the other hand, the shift in the efficiency index derived for the Windsor Series soil and mooted in the data for the Ragdale Series site suggests that landuse changes involving ploughing cause greater short-term retention of storm water because of the alteration in pore size distribution and the increase in depression storage that they entail.

CONCLUSIONS

Agriculture involves frequent and often cyclical changes in landcover. In preparing a seedbed, the deployment of minimal cultivation techniques offers a way of reducing soil disturbance (Goss, Howse and Harris, 1978). However, common practice still involves ploughing and other secondary treatments, and in certain soils, artificial loosening is extended deeper in order to facilitate root penetration and control the water regime.

With respect to soil drainage, the effects have been shown to counter expectation. The evacuation of water from the soil profile after seedbed preparation and deep loosening is less efficient, at least in the short term. Besides this, a shift in the pore size distribution of the loosened soil leads to an increase in the winter water content. This in turn contributes to a reduction in the shear strength of the soil below the plough layer from approximately 90 to 80 kPa (Twomlow, Parkinson and Reid, 1988), so that the soil is more susceptible to the structural damage of animal poaching or farm machine traffic.

Ironically, the agricultural practice that increases the risk of soil degradation appears to decrease the risk of another hydrogeomorphic hazard. Flooding in arterial waterways has been variously attributed to 'improvements' in field drainage within a river catchment (Newson and Robinson, 1983; Robinson, Ryder and Ward, 1985; Parkinson and Reid 1987). However, it is clear that the loosening of the soil whether as a function of seedbed preparation or as a deep modification of the soil profile actually reduces both the peak and short-term total fluxes of water to the arterial system. This is not the result of soil loosening that is expected by the farmer who has an interest in maintaining as equable a root environment as possible by expediting soil drainage. Nevertheless, it carries the unexpected benefit that this particular farm operation cannot be blamed for any increase in flooding.

ACKNOWLEDGEMENTS

We are indebted to Doug Castle, Director of the Field Drainage Experimental Unit (Ministry of Agriculture, Fisheries and Food) for access to the records of the drainage experiments at Brooksby Hall and Langabeare Farm, and to Adrian Armstrong for his advice and collaboration. Stephen Twomlow was in receipt of a MAFF Research Training Award and Angela Clark was in receipt of a NERC Research Training Award.

REFERENCES

Barley, K. P. (1959). Earthworms and soil fertility. *Australian Journal Agricultural Research*, **10**, 171–85.

Bouma, J., L. W. Dekker and J. H. M. Wosten (1978). A case study on infiltration into dry clay soil. II Physical measurements. *Geoderma*, **20**, 41–51.

Darwin, C. (1881). *The Formation of Vegetable Mould Through the Action of Worms*, John Murray, London.

Findlay, D. C., G. J. N. Colbourn, D. W. Cope, T. R. Harrod, D. V. Hogan and S. J. Staines (1984). *Soils and Their Use in South West England*. Soil Survey of England and Wales, *Bulletin* 14.

Goss, M. J., K. R. Howse and W. Harris (1978). Effects of cultivation on soil water retention and water use by cereals in clay soils. *Journal Soil Science*, **29**, 475–88.

Green, F. H. W. (1979). Field underdrainage and the hydrological cycle. In G. E. Hollis (ed.), *Man's Impact on the Hydrological Cycle in the United Kingdom*, GeoAbstracts, Norwich, pp. 9–17.

Greenland, D. J., G. R. Lindstrom and J. P. Quirk (1962). Organic materials which stabilize natural soil aggregates. *Soil Science Society America Proceedings*, **26**, 366–71.

Hallsworth, E. G. (1987). *The Anatomy, Physiology and Psychology of Erosion*, John Wiley & Sons, Chichester.

Harris, G. L. (1977). An analysis of the hydrological data from the Langabeare experiment. *Field Drainage Experimental Unit Technical Bulletin* 77/4.

Harris, G. L., M. J. Goss, R. J. Dowdell, K. R. Howse and P. Morgan (1984). A study of mole drainage with simplified cultivation for autumn-sown crops on a clay soil. Soil water regimes, water balances and nutrient loss in drain water, 1978–80. *Journal Agricultural Science, Cambridge*, **102**, 561–81.

Leeds-Harrison, P. B. and N. J. Jarvis (1986). Drainage modelling in heavy clay soils. In J. Saavalainen and P. Vakkilainen (eds.), *Proceedings of International Seminar on Land Drainage, Helsinki*, pp. 198–220.

Leeds-Harrison, P. B., C. J. P. Shipway, N. J. Jarvis and E. G. Youngs (1986). The influence of soil macroporosity on water retention, transmission and drainage in a clay soil. *Soil Use and Management*, **2**, 47–50.

Leeds-Harrison, P. B., G. Spoor and R. J. Godwin (1982). Water flow to mole drains. *Journal Agricultural Engineering Research*, **27**, 81–91.

Lutwick, L. E. and E. H. Hobbs (1964). Relative productivity of soil horizons, singly and in mixture. *Canadian Journal Soil Science*, **44**, 145–50.

MAFF (1981). Subsoiling as an aid to drainage. *Drainage Leaflet* 730, HMSO.

Newson, M. D. and M. Robinson (1983). Effects of agricultural drainage on upland streamflow: Case studies in mid-Wales. *Journal of Environmental Management*, **17**, 333–48.

Nicholson, H. H. (1934). The role of field drains in removing excess water from the soil. 1. Some observations on rates of flow from outfalls. *Journal Agricultural Science, Cambridge*, **24**, 349–67.

Nye, P. H. (1955). Some soil forming processes in the humid tropics. 4. The action of the soil fauna. *Journal Soil Science*, **6**, 73–83.

Parkinson, R. J. and I. Reid (1986). Effect of local ground slope on the performance of tile drains in a clay soil. *Journal Agricultural Engineering Research*, **34**, 123–32.

Parkinson, R. J. and I. Reid (1987). Field drainage, soil water management and flood hazard. *Soil Use and Management*, **3**, 133–8.

Parkinson, R. J., S. Twomlow and I. Reid (1988). The hydrological response of a silty clay loam following drainage treatment. *Agricultural Water Management*, **14**, 125–36.

Reid, I. (1979). Seasonal changes in microtopography and surface depression storage of arable soils. In G. E. Hollis (ed.), *Man's Impact on the Hydrological cycle in the United Kingdom*, GeoAbstracts, Norwich, pp. 19–30.

Reid, I. and R. J. Parkinson (1984). The nature of the tile-drain outfall hydrograph in heavy clay soils. *Journal Hydrology*, **72**, 289–305.

Reid, I. and R. J. Parkinson (1987). Winter water regimes of clay soils. *Journal Soil Science*, **38**, 473–81.

Robinson, M. and A. C. Armstrong (1988). The extent of agricultural field drainage in England and Wales, 1971–80. *Transactions Institute British Geographers*, New Series, **13**, 19–28.

Robinson, M., E. L. Ryder and R. C. Ward (1985). Influence on streamflow of field drainage in a small agricultural catchment. *Agricultural Water Management*, **10**, 145–58.

Tisdall, J. M. and J. M. Oades (1982). Organic matter and water-stable aggregates in soils. *Journal Soil Science*, **33**, 141–63.

Trafford, B. D. and J. M. Oliphant (1977). The effect of different drainage systems on soil conditions and crop yield of a heavy clay soil. *Experimental Husbandry*, **32**, 75–85.

Trafford, B. D. and D. W. Rycroft (1973). Observations on the soil-water regimes in a drained clay soil. *Journal Soil Science*, **24**, 380–91.

Twomlow, S. (1989). *Soil loosening and drainage efficiency of silt soils*. Unpublished PhD Thesis, Birkbeck College, University of London.

Twomlow, S., R. J. Parkinson and I. Reid (1988). Water retention and soil physical properties following the deep loosening of a silty clay loam. *Proceedings 11th Conference International Soil Tillage Research Organization, Edinburgh*, 165–70.

15 Vegetation Patterns in Relation to Basin Hydrogeomorphology

CLIFF R. HUPP
*US Geological Survey,
Reston, Virginia*

SUMMARY

Persistent distribution patterns of woody vegetation within the bottomland forests of the Passage Creek basin, in northern Virginia, were related to hydrogeomorphic characteristics, including fluvial landforms, channel geometry, streamflow characteristics, stream gradient, and stream order. Vegetation patterns were determined from species presence as observed in transects and traverses on landforms developed along the stream. The species-hydrogeomorphic relations were analyzed both across-valley and up-valley into basin heads. Distinct species distributional patterns occur on four common fluvial-geomorphic landforms: depositional bar, channel shelf, floodplain, and terrace. Independent hydrologic characteristics (flow duration and flood frequency) were determined for each landform. Vegetation data were analyzed through multivariate statistical analyses. Results suggest that certain species are significantly associated with each landform. Flood disturbance may be an important factor in maintaining vegetation patterns. Floodplains and their characteristic species are restricted upstream above gradients exceeding 0.150. Channel shelves and their characteristic species are particularly well developed along high-gradient reaches and may persist to basin heads. Stream order and landform are significantly related to species distribution up-valley. Channel gradient strongly affects channel morphology and, thus, species distribution. Streamflow characteristics are probably the most important factors maintaining the vegetation patterns characteristic of bottomlands. Furthermore, water, in general, may be the most influential agent affecting both

Vegetation and Erosion
Edited by J. B. Thornes
©1990 John Wiley & Sons Ltd

fluvial and interfluvial vegetation patterns at the scale of landforms. The interdisciplinary study of vegetation patterns may provide both utilitarian and theoretical approaches to problems associated with plant-ecological and geomorphic endeavor.

INTRODUCTION

Studies emphasizing environmental explanations of plant distributional patterns have developed under the aegis of plant ecology. Many of the earliest theoretical statements in ecology came from plant ecology, particularly in the relations between species distributions and geomorphic features (Cowles, 1901). Bottomland-vegetation patterns have been studied by many plant ecologists and a strong relation among bottomland vegetation, fluvial landforms, and hydrogeomorphic processes has been documented (Wistendahl, 1958; Sigafoos, 1961; Ellenberg, 1963; Hosner and Minckler, 1963; Robertson, Weaver and Cavanaugh, 1978; Hupp and Osterkamp, 1985; Girel and Pautou, 1985; Pautou and Girel, 1985; Simon and Hupp, 1987; Hupp, 1988). Bottomlands develop characteristic fluvial landforms, each with associated distinct hydrogeomorphic processes (Wolman and Leopold, 1957; Richards, 1982). Flow duration, flood frequency, flood intensity, depositional environment, channel gradient, and variation within each parameter on different geomorphic levels within a bottomland are hydrogeomorphic processes that help shape fluvial landforms and the vegetation they support (Hupp, 1988). The purpose of the present chapter is to synthesize several works by the author and close colleagues, pertaining to vegetation-distributional patterns as they relate fluvial-hydrogeomorphic process and form, into a systematic basin-wide perspective. This presentation will begin with an analysis of the vegetation-geomorphic relations developed across the valley section, perpendicular to the trunk-stream axis. This will be followed by an analysis of the relations as they change, up-valley, along the trunk-stream axis, including an analysis of the relations developed in the vicinity of the low-order basin heads.

This research pertains to the vegetation and fluvial features that develop along relatively high-energy streams common in most humid climates; it does not include coastal-plain streams under tidal influences. The basin selected for intensive study is drained by Passage Creek, located in northwestern Virginia (Figure 15.1); part of the Valley and Ridge physiographic province. Passage Creek is representative of most streams in this region, exhibiting a variety of hydrogeomorphic conditions including alluvial meandering reaches, pool-riffle reaches, and bedrock-controlled reaches.

Passage Creek has been gaged for a relatively long period and displays a variety of flow regimes along its different reaches (there are dramatic shifts from gentle to steep gradients owing to variety in bedrock; Hupp, 1982). The stream

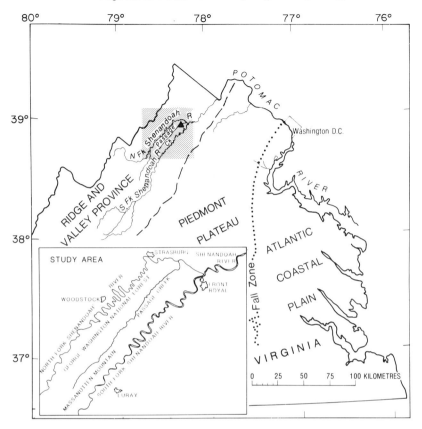

Figure 15.1 Location of Passage Creek in northern Virginia; detail of Passage Creek is inset. (From Hupp and Osterkamp, 1985.)

has a basin area of 277 km² above the gage. Mean discharge through the period of record (1932 to present) is 1.93 m³/s with a peak discharge of 595 m³/s on 15 October 1942. Much of Passage Creek and most of its tributaries drain and flow through George Washington National Forest where the bottomland forests have been undisturbed for at least 50 years; in some places, over 100 years. Nearly all perennial reaches of Passage Creek were sampled including many tributaries through to their basin head.

Fluvial Landforms

The term 'bottomland', in the present chapter, refers to all fluvially generated landforms and their vegetation. These landforms occur as terraces high in the valley section and, in order of descending elevation, proceed through floodplains,

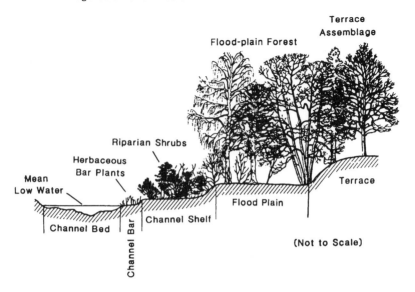

Figure 15.2 Diagram of generalized cross-section, along Passage Creek, showing the relative position of bottomland geomorphic features and associated vegetation types. (From Hupp, 1986.)

various riparian features, channel bars down to the channel bed (Figure 15.2). The term 'riparian' refers only to bank features and vegetation and not the floodplain as is commonly done in many ecological papers. The landforms described are defined by independent-hydrologic parameters including flow duration (percent of time, annually, that a level is reached or exceeded by streamflow) for surfaces below the floodplain, and flood frequency (recurrence interval of flooding) for the floodplain and terrace levels (Figure 15.2).

The channel bed of a perennial stream (Figure 15.2) is the surface that is wholly or partly covered by flows below mean discharge. Thus, at least part of the channel bed of a perennial stream is inundated at all times. Channel or depositional bars (Figure 15.2) occur in the active-channel part of the valley section and are the lowest prominent geomorphic feature higher than, but within, the active-channel bed. The level of the channel bar generally corresponds to a stage slightly higher than the low-flow water stage, about the 40 per cent flow duration (Hedman, More and Livingston, 1972).

The channel shelf, a riparian feature, is a horizontal to gently sloping surface (Figure 15.2) that normally extends the short distance between the break in the relatively steep-bank slope and the lower limit of persistent woody vegetation that marks the channel-bed edge (Osterkamp and Hupp, 1984). The channel shelf is best developed along relatively steep-gradient reaches (Hupp, 1986). The level of the channel shelf of many perennial streams corresponds to the stage of the average discharge; this level is typically reached by streamflow between the 5 and 25 per cent flow duration.

Floodplains are generally flat surfaces flooded, on the average, once every 1 to 3 years. In many cases, areas that fit this definition represent a minor part of the valley section (Hack, 1957). Because flood events typically last only a few days or less, floodplains are inundated a very small percentage of time each year, much less than channel shelves. Floodplains are commonly delimited channelward by a bank.

Terraces generally represent former floodplains and occur at various levels above modern floodplains (Figure 15.2). The likelihood of inundation for the terrace is always lower than that for the floodplain, greater than 3 years. Low terraces may be flooded slightly less frequently than floodplains, whereas flooding may be extremely improbable for topographically high terraces.

Channel bars, channel shelves, floodplains, and terraces are common features along Passage Creek and some of its larger tributaries. However, any one reach may have one or more of these missing; apparently the processes that promote extensive floodplain development are not conducive to extensive channel-shelf development and vice versa. Furthermore, the relative proportion and presence of these fluvial landforms changes, up-valley, from the main channel to the basin heads.

Vegetation Patterns

It is beyond the scope of the present chapter to provide an exhaustive literature review. However, extensive reviews are provided in the papers upon which this work is based (Hack and Goodlett, 1960; Hupp, 1986; Hupp and Osterkamp, 1985; Olson and Hupp, 1986). Other papers that have related vegetation distribution to fluvial landforms and hydrogeomorphic processes include Wistendahl (1958), Sigafoos (1961), Teversham and Slaymaker (1976), and Osterkamp and Hupp (1984). Several authors (Hefley, 1937; Hickin and Nanson, 1975; Nanson and Beach, 1977) have acknowledged a relation with elevation (thus to some extent landform), but stress that age of surface may be the most important factor (thus implying a strong plant-successional component). Depositional environment and sediment size, which does not exclude hydrogeomorphic importance, has been suggested to be of primary importance in vegetation patterns (Everitt, 1968, 1980; Wolfe and Pittillo, 1977; Hupp and Simon, 1986). A review of the commonly suggested agents for bottomland vegetation patterns is provided in Hupp (1988).

Although many factors may affect species distribution patterns across a bottomland, the particular landform with its attendant hydrogeomorphic characteristics appears to be the most influential (Hupp and Osterkamp, 1985). The species included in any patterns of bottomland vegetation distribution are, of course, determined in part by regional floristics and climatic variation.

It is stressed that clear process-oriented definitions of landforms and use of the proper spatial scale are necessary for the interpretation of vegetation patterns. The vegetation studies, detailed in the present chapter, adopted the physiographic approach to plant geography, which presupposes collaboration with geology, geomorphology, soil science, and hydrology, and which seeks to place vegetation in the context of a dynamic physical landscape (Zimmermann and Thom, 1982). This approach uses the scale of landforms which relies heavily on topographic mapping and generally avoids species-plot data in deference to simple presence–absence data. Presence–absence data are better suited for work at the scale of landforms (Hack and Goodlett, 1960).

METHODS

Physical Environment

Along the main stem of Passage Creek, 17 valley cross-sections were established to determine fluvial landform–vegetation variation across-valley. At each section measurement included surveying for cross-sectional geometry, collection of sediment samples, and identification of the geomorphic surfaces. Detailed sediment analyses, beyond the scope of this chapter, were conducted; these techniques, results, and interpretations are reported in Osterkamp and Hupp (1984).

At all sections, identification of geomorphic surface was consistent with the published definitions or descriptions of these features provided earlier. In no case was vegetation used to help identify a geomorphic surface. Flow duration and flood frequency were estimated for each geomorphic surface; these estimates are for the lower elevational limit of each surface. Hydrologic estimations were determined by relating the stage–discharge curves for the streamflow gaging station to the fluvial landform near the gaging site. Crest-stage gages were established at several of the valley cross-sections. Crest-stage information and numerous visits in the basin during low and flood stages allowed for confirmation throughout the basin of the discharge–landform relations, determined at the gage site.

The analysis of fluvial landform–vegetation variation in the up-valley direction also included the determination of stream order, lithology, subbasin area, streambed characteristics, and stream gradient. These data were obtained using topographic and geologic maps and ground traverses. Results from the across-valley study were used to determine fluvial landforms for the up-valley study. This parallel-to-stream study was done in 18 first- and second-order intermittent montane basins in or adjacent to the Passage Creek basin; all intervening stream orders below fifth order were similarly analyzed.

Vegetation Analysis

Presence–absence data were collected for all woody species identified on 20 to 40 m transects positioned, along the 17 cross-sections, parallel to the stream and extending the width of the fluvial landform. Generally, a species was of reproductive size and age before listing as present. Botanical nomenclature followed Radford, Ahles and Bell (1968). This same procedure was used in the up-valley study except that entire reaches were traversed, stopping at 20- to 50-m intervals for notation of species present per fluvial landform. This sampling interval corresponds to the minimum resolution mappable on standard US Geological Survey 7½ minute topographic quadrangles. Basin heads were similarly mapped for vegetation corresponding to the interfluvial landforms, including nose (divergent, convex-upward) slopes, side (linear) slopes, and cove (convergent, concave-upward) slopes (Hack and Goodlett, 1960).

Several types of statistical analyses were applied to the vegetation–landform data sets. Binary discriminant analysis (BDA) is immensely suited for vegetation presence–absence data (Strahler, 1978). One of the outputs of BDA are standardized residuals (*D*-values) from $2 \times k$ contingency tables. *D*-values, ranging from about 7 to -7, may be interpreted as standard deviations away from 0, where a species would show no 'preference' (positive *D*-value) or 'avoidance' (negative *D*-value) for a particular environmental condition, such as a floodplain or sediment-size class. These *D*-values may then be used as inputs to one of numerous types of species-ordination procedures; notable among these is a multivariate-reciprocal averaging analysis developed at Cornell University termed detrended correspondence analysis (DCA; Gauch, Whittaker and Wentworth, 1977; Hill and Gauch, 1980). Details of these statistical procedures as applied to the present studies may be found in the citations given above and in the papers listed at the beginning of the preceding 'Vegetation patterns' section of this chapter. In short, the vegetation data, presented here, were assembled by landform and systematically analyzed through contingency analysis, BDA, and finally through ordination, typically DCA. These procedures ultimately provide quantitative values of vegetation–geomorphic relations and allow for their concise graphical depiction.

ACROSS-VALLEY VEGETATION–HYDROGEOMORPHIC RELATIONS

The fluvial landforms (channel bar, channel shelf, floodplain, and terrace) are found along most perennial reaches of Passage Creek; their hydrologic characteristics and relative sizes are illustrated in Figure 15.3. Of 43 species of woody plants identified on cross-section transects, 22 had significant ($P < 0.05$) distributions relative to fluvial landforms, based on chi-square evaluations from the contingency analysis (Table 15.1). Sediment-size class and depth to water

Table 15.1 Standardized residuals[†] and chi-square values from two binary discriminant analyses of plant species distribution near Passage Creek

Species	Species vs. fluvial landform					Species vs. sediment size class					
	DB	CS	FP	T	χ^2	Gravel	Sand	Low	Medium	High	χ^2
1. Acer negundo	-1.34	1.83	0.34	-1.59	5.91	-2.18	-1.51	4.50	0.28	-0.87	21.29
2. Acer rubrum	-1.77	1.10	1.10	1.25	5.60	-3.49	1.99	0.31	0.05	0.46	12.83
3. Alnus serrulata	0.24	4.68	-3.29	-1.99	24.46*	-1.42	3.04	-1.69	-1.35	-1.20	11.60*
4. Amelanchier arborea	-0.68	-1.32	-1.32	4.00	16.00*	-1.77	1.48	0.20	0.51	-0.78	1.41
5. Asimina triloba	-0.68	-1.32	2.42	-0.80	5.84	-1.21	-0.37	2.56	-0.60	-0.53	4.13
6. Betula lenta	-0.99	-1.03	2.62	-1.17	7.13	-1.15	1.22	-1.03	-0.60	-0.53	1.28
7. Carpinus caroliniana	-1.60	1.63	0.95	-1.89	7.49	-2.68	3.62	-1.14	-1.39	-1.24	18.00*
8. Carya cordiformis	-1.34	-2.63	4.80	-1.59	23.03*	-1.33	-0.95	0.71	2.44	1.12	6.95
9. Carya glabra	-1.08	-2.12	-2.12	6.40	40.98*	-2.50	-0.60	1.00	1.86	2.37	8.67
10. Carya tomentosa	-0.89	-1.75	-1.75	5.28	27.91*	-1.88	-0.48	1.41	0.28	1.92	5.87
11. Celtis occidentalis	-0.68	-0.08	1.17	-0.80	1.82	-1.15	1.92	0.31	-0.60	-0.53	3.62
12. Cercis canadensis	-0.79	-0.45	1.73	-0.93	3.34	-2.15	-1.13	2.32	2.33	0.28	12.42*
13. Chionanthus virginicus	-0.55	-1.07	1.95	-0.65	3.81	-1.15	1.92	-0.88	-0.60	-0.22	1.44
14. Cornus amomum	-0.14	3.72	-2.46	-1.49	14.73*	-1.71	2.53	-0.74	-1.21	-1.08	8.44*
15. Cornus florida	-1.60	0.27	0.27	0.27	2.76	-2.85	0.27	1.73	1.50	0.22	9.30
16. Diospyros virginiana	-0.79	-1.55	1.73	0.47	4.20	-1.15	-0.35	0.10	3.09	-0.53	5.94
17. Fraxinus americana	-0.68	-1.32	-1.32	4.00	16.00*	-2.50	0.27	1.65	0.81	0.02	5.82
18. Fraxinus pennsylvanica	-2.12	-2.02	1.40	-2.51	13.07*	-3.49	2.37	0.89	-0.88	-0.58	9.97*
19. Hamamelis virginiana	-1.51	1.22	-0.17	0.00	2.98	-2.59	3.04	-0.41	-1.35	1.20	13.36*
20. Ilex verticillata	-0.79	2.82	-1.55	-0.93	7.96*	-1.33	1.80	-0.07	-0.66	-0.59	0.77

Species											
21. Juglans nigra	−1.49	−2.80	5.10	−1.69	26.02*	−1.97	−1.21	1.22	2.62	0.44	10.58*
22. Lindera benzoin	−1.60	0.95	1.62	−1.89	7.49	−2.15	1.08	3.13	−0.18	−1.12	16.55*
23. Liriodendron tulipifera	−1.60	−1.77	2.99	−0.15	12.14*	−2.59	1.76	0.23	−0.31	−0.05	5.51*
24. Nyssa sylvatica	−1.34	−1.15	1.83	0.32	4.67*	−2.50	0.27	1.65	0.81	0.02	5.82
25. Physocarpus opulifolius	−0.79	2.82	−1.55	−0.93	7.96*	−1.27	2.12	−1.14	0.66	−0.59	3.84
26. Pinus strobus	−1.08	−2.12	2.16	0.91	7.62	−1.97	2.79	−1.02	−1.02	−0.91	9.20
27. Pinus virginiana	−0.38	−0.75	−0.75	2.26	5.11	−1.88	−1.51	1.41	1.53	0.53	7.27
28. Platanus occidentalis	−0.33	2.64	−0.45	−2.51	9.96*	−2.24	3.96	−2.38	−0.77	−1.53	21.39*
29. Populus deltoides	−0.68	1.17	−0.08	−0.80	1.82	−1.38	2.31	−1.24	−0.72	−0.64	5.09
30. Prunus serotina	−0.79	−1.55	1.73	0.47	4.20	−1.21	1.48	0.14	1.30	−0.53	0.25
31. Quercus alba	−1.34	−2.63	0.34	4.13	20.36*	−2.68	0.27	1.37	0.65	1.03	6.67
32. Quercus prinus	−0.79	−1.55	−1.55	4.67	21.82*	−1.33	2.31	−1.24	−0.72	−0.64	5.09
33. Quercus rubra	−1.51	−0.87	−0.17	2.68	8.51*	−2.68	0.69	1.37	0.65	−0.10	5.93
34. Quercus velutina	−0.89	−1.75	−1.75	5.28	27.91*	−1.97	−0.71	1.22	1.40	1.79	7.33
35. Robinia pseudoacacia	−0.55	0.44	0.44	−0.65	0.86	−1.77	−0.55	1.97	0.51	0.76	3.55
36. Salix nigra	3.39	0.23	−1.75	−1.06	13.01*	0.25	1.59	−1.69	−0.98	−0.37	3.26
37. Staphylea trifolia	−0.38	1.36	−0.75	−0.45	1.86*	−1.33	1.80	−0.07	−0.66	−0.59	0.77
38. Tilia heterophylla	−0.99	−0.11	0.80	0.00	0.55	−1.88	2.10	−0.14	−0.98	−0.87	5.23
39. Tsuga canadensis	−0.55	−1.07	0.40	1.29	2.85	−1.45	1.29	0.79	−0.72	−0.64	0.86
40. Ulmus americana	−1.26	−2.46	4.49	−1.49	20.20*	−2.15	1.23	0.19	−1.12	−1.00	2.41
41. Ulmus rubra	−0.89	3.19	−1.75	−1.06	10.18*	−1.78	1.91	0.00	−0.93	−0.83	4.15
42. Viburnum acerifolium	−0.79	−1.55	2.82	−0.93	7.96*	−2.06	3.44	−1.85	−1.07	−0.96	6.96
43. Viburnum dentatum	−0.89	3.19	−1.75	−1.06	10.18*	−2.85	3.12	−0.72	−0.49	−1.32	14.96*

†A positive residual indicates frequent occurrence on the particular landform or sediment size; negative residuals indicate that the species is rarely found there.

*$P<0.05$.

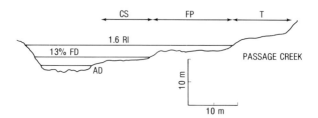

Figure 15.3 Cross-section of Passage Creek near Shenandoah River. CS, FP, and T represent channel shelf, floodplain, and terrace, respectively. Note that the recurrent interval (RI) for the flood plain level and flow duration (FD) for the channel shelf is shown. AD is the level of the average discharge. (From Hupp and Osterkamp, 1985.)

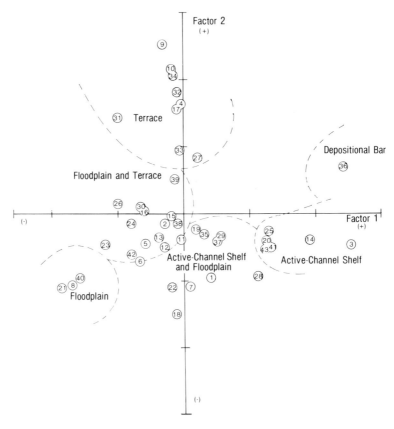

Figure 15.4 Woody species plotted by factor scores, from BDA, on Factors 1 and 2 from the species versus fluvial landform ordination. Species names and numbers are given in Table 15.1. Association zones are indicated and delineated by dashed lines. (From Hupp and Osterkamp, 1985.)

table, when tested similarly, did not show nearly as strong a relation with vegetation distribution (only 12 species had significant distributions for these parameters). Results of the species ordination (Figure 15.4) show the associations of the 43 species with the various fluvial landforms. Species associated with channel shelves are shown on the right side of the plot; those on the left side are associated with floodplains and terraces. Species in the upper part of Figure 15.4 are associated with only terraces; species centrally located do not show statistically significant associations.

Results suggest that some species may be site specific. *Alnus serrulata* (Figure 15.4; see Table 15.1), *Cornus amomum* (14), *Physocarpus opulifolius* (25), *Ilex verticillata* (20), *Ulmus rubra* (41), and *Viburnum dentatum* (43) are characteristic of Passage Creek channel shelves and the first four are virtually restricted to the channel shelf. Similarly, *Juglans nigra* (21), *Carya cordiformis* (8), and *Ulmus americana* (40) characterize and are restricted to floodplains (Figure 15.4; Table 15.1). Qualitative and statistical analyses indicate that many species may not discriminate between two adjacent geomorphic features, but may be largely excluded from a third. Species common on Passage Creek bottomland that can be routinely found on either the channel shelves or floodplains, or both (and thus usually had low chi-square values; Figure 15.4; Table 14.1), include: *Acer negundo* (1), *Carpinus caroliniana* (7), *Fraxinus pennsylvanica* (18), *Populus deltoides* (29), and *Staphylea trifolia* (37). Among terrace species, only *Quercus prinus* (32) was restricted to the terrace, most others being occasionally found on other surfaces. In contrast, both channel-shelf and floodplain surfaces supported many species that were not found on other landforms. Only *Salix nigra* (36), typically a channel-shelf species, occasionally grows on channel bars that are normally dominated by herbaceous aquatic plants (Figure 15.4; Table 15.1).

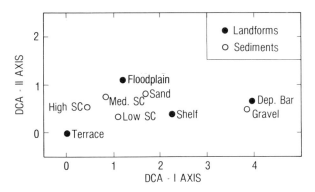

Figure 15.5 DCA ordination of species versus fluvial landforms and sediment types. Note that site parameters are plotted rather than species. SC refers to silt and clay fraction. Dep. Bar is depositional bar. (From Hupp and Osterkamp, 1985.)

DCA may ordinate environmental conditions as well as species. A combined DCA of species against both landforms and sediment-size class (Figure 15.5) illustrates the more important role of landforms rather than sediment size in bottomland-vegetation patterns; note the horizontal- and vertical-polar positions of landforms versus sediment sizes. Other statistical analyses support the primary role of landforms (Hupp and Osterkamp, 1985).

These species patterns are nearly identical to those identified along the larger Potomac River near Washington, DC (Sigafoos, 1961; Yanosky, 1982; Osterkamp and Hupp, 1984). These results also support the work of Everitt (1968), which describes a low and a high floodplain with distinct suites of vegetation. Analogous results were obtained by Wistendahl (1958) for the Raritan River in New Jersey, where bank, inner and outer floodplain, and terrace vegetation types were separated. All of these authors suggest that distinct variation in flow characteristics, across the various fluvial landforms, strongly affects the distribution of. most bottomland species.

The long and detailed records for flood magnitude and frequency available on many streams offer an opportunity to assess the effect of periodic-flood disturbance on vegetation. Disturbance theory has drawn increasingly more interest in modern plant ecology (White, 1979; Pickett, 1980); bottomland forests may be well adjusted to, if not maintained by, periodic-destructive flooding (Hupp, 1983). Apparently the observed plant distributions in this study are at least in part controlled by inundation frequency and the susceptibility of plants to damage by destructive floods. Thus, it is the hydrogeomorphic processes operating differently on different landforms that actually affect the plant patterns, not the landforms *per se*. As examples, two shrubs common on the channel shelf, *Alnus serrulata* and *Cornus amomum*, are relatively resistant to destruction by flooding due to small, highly resilient stems and the ability to sprout rapidly from flood-damaged stumps (Hupp, 1983). Conversely, *Cornus florida* and some species of *Quercus* and *Carya*, which commonly grow on terraces but rarely on lower surfaces, may be intolerant of repeated flood damage or inundation. Floodplain species, such as *Carya cordiformis* and *Juglans nigra*, are probably less tolerant of destructive flooding than channel-shelf species, but more tolerant of periodic inundation than are terrace species.

UP-VALLEY VEGETATION—HYDROGEOMORPHIC RELATIONS

Distinct trends in species distributions were noted both along decreasing stream order and between channel shelves and floodplains. Previous study has shown that the channel shelf develops significantly farther upstream than the floodplain (Hupp, 1986). Because many species are associated with one or the other of these geomorphic features, species distributional differences can be expected in the upstream direction. Floodplains do not develop along many of the

Figure 15.2 Distribution of bottomland woody species according to stream order and landform in the northern Massanutten Mountain area of Virginia. (CS in the channel shelf, FP is the flood plain.)

Species	Basin head	Stream order									
		1st		2nd		3rd		4th		5th	
		CS	FP	CS	FP	CS	FP	CS	FP	CS	FP
Betula Lenta	X										
Quercus rubra	X										
Tilia heterophylla	X		X		X		X		X		X
Liriodendron tulipifera	X				X		X		X		X
Hamamelis virginiana	X	X		X	X	X	X	X	X	X	X
Lindera benzoin	X	X	X	X	X		X		X	X	X
Asimina triloba	X		X		X		X		X		X
Platanus occidentalis		X		X		X		X	X	X	
Ilex verticilata		X		X		X					
Viburnum alnifolium		X		X		X					
Fraxinus pennsylvanica				X	X	X	X	X	X	X	X
Ulmus rubra				X		X	X	X	X	X	X
Viburnum dentatum				X		X		X			
Juglans nigra			X		X		X		X		X
Carya cordiformis			X		X		X		X		X
Salix nigra						X		X		X	
Cornus amomum						X		X		X	
Vitus riparia						X		X		X	
Acer negundo						X	X	X	X	X	X
Populus deltoides							X		X		X
Ulmus americana							X		X		X
Viburnum acerifolium							X		X		X
Staphylea trifolia							X	X	X		X
Carpinus caroliniana							X	X	X		X
Cephalanthus occidentalis								X			
Physocarpus opulifolius								X			
Celtis occidentalis											X
Acer saccarhinum											X
Total number of species	7	5	5	8	9	11	15	13	16	9	17

montane first-order basins. Species distribution by stream order and landform is shown in Table 15.2. Note that some species like *Hamamelis virginiana* are present from the basin heads through large-order streams and usually on both the floodplain and banks, whereas other species, such as *Acer negundo*, are limited upstream regardless of landform. Still other species, such as *Alnus serrulata* or *Juglans nigra*, are not limited upstream but generally are restricted to one or the other geomorphic surfaces (Table 15.2).

Species typically present on riparian surfaces (below the floodplain) had variable distributions along decreasing stream orders (Table 15.2). *Hamamelis virginiana*, *Lindera benzoin*, *Alnus serrulata*, and *Ilex verticilata* are the most ubiquitous bank species; note that *Alnus* and *Ilex* are restricted to channel shelves, except that *Alnus* may become part of the subcanopy in wet, rocky, basin heads. *Salix nigra*, *Cornus amomum*, *Cephalanthus occidentalis* and *Physocarpus opulifolius* along with the above species characterize the riparian-shrub forest of channel shelves (Hupp, 1982; Hupp and Osterkamp, 1985) on third- through fifth-order streams, especially along high-channel gradients. *Salix*, *Cornus*, *Cephalanthus*, and *Physocarpus* become limited toward the stream head where the forest canopy closes over the increasingly smaller stream channel, thus suggesting that light also limits at least some channel-shelf species. Additionally, *Cephalanthus* and *Physocarpus* appear to be limited downstream where gentler gradients preclude the relatively coarse channel and bank deposits commonly associated with channel shelves.

Species typically present on floodplains also showed analogous variation in distributions along decreasing stream orders (Table 15.2). Of species restricted to floodplains, only *Juglans nigra* and *Carya cordiformis* were found upstream of third-order reaches. Most first-order and many second-order streams have no floodplain development due to their high gradients (Hupp, 1986). The only first-order streams with appreciable floodplains and associated floodplain species are those that flow on erodible shales, which create sufficiently gentle channel gradients. *Acer saccarhinum* and *Celtis occidentalis* are the most restricted floodplain species; they are present only along fifth-order and larger streams (Table 15.2).

The basin head is generally not considered a fluvial feature, but is included in this research, because it represents the terminal landform above the point where fluvial processes begin. Because of the inherently moist nature of the basin head, some bottomland species, chiefly subcanopy (i.e., *Lindera benzoin*), are present there along with species typical of cove forests (Hack and Goodlett, 1960; Hupp, 1984). The ubiquitous species *Acer rubrum* and *Nyssa sylvatica* grow along most reaches and on most fluvial landforms, as well as on many upland areas; they were not considered in the analyses.

The polar ordination of sites categorized by stream order and landform (Figure 15.6) reveals strong relations among bottomland-species distribution, fluvial landform, and stream order. The ordination, when tested for correlation

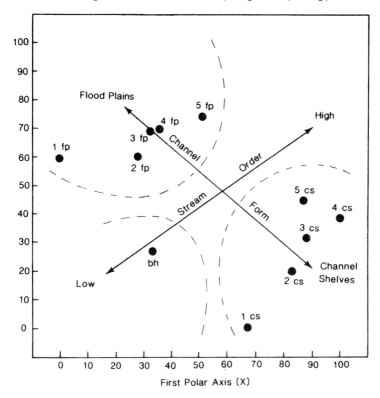

Figure 15.6 Plot of polar ordination of site types. Numeral before landform abbreviation is stream order, fp = flood plain, cs = channel shelf, bh = basin head. Arrows indicate landform and stream-order gradients. Note that stream order generally increases from lower left to upper right. (From Hupp, 1986)

and significance (Mueller-Dombois and Ellenberg, 1974), had a correlation coefficient (r) of 0.96. A t-test indicated a highly significant correlation between dissimilarity indicates and ordination intervals ($t = 10.29$, $P < 0.01$). This ordination, based entirely on species presence, accurately reflects the fluvial–geomorphic characteristics of the area. Note that floodplains and channel shelves are clearly separated in the ordination, and the distinct nature of the basin head is reflected in its isolated position (Figure 15.6). Furthermore, within each landform cluster a relatively distinct stream-order gradient is borne out by the ordination. These results support the contention that bottomland species distributions are strongly affected by fluvial geomorphic form and process.

Trends in species richness are illustrated in Table 15.2. Floodplain forests reach maximum species diversity along the highest-order stream (fifth order and greater). This condition probably results from the gentler gradient that promotes floodplain development; they would be less affected by the 'island effect' that may be associated with small localized floodplains along narrow upstream reaches. It is not felt that the range of elevations (180 to 390 m) represent a significant environmental gradient. Riparian-shrub communities reach maximum diversity on third- and fourth-order reaches (Table 15.2). This vegetation would be affected less from the 'island effect' upstream than floodplain forests, because channel shelves are typically present farther upstream. However, the channel gradients that promote channel-shelf development usually occur locally on large streams where they flow over particularly resistant bedrock. Bottomland-species richness usually decreases upstream where upland vegetation may grow to the channel edge along the low-order streams.

Channel gradient seems to be the most important factor controlling the distribution of floodplains and channel shelves. A channel-gradient comparison of eight first-order streams, four on sandstone and four on shale, demonstrates the relative importance of gradient compared to basin area (Hupp, 1986). All of the eight streams had basin areas less than 100 ha. The basins on shale slopes had floodplains and associated vegetation along parts of their lower reaches and had average gradients less than 0.150. The streams on sandstone had no floodplains or associated vegetation above their confluence with another stream and had average gradients above 0.150. Landform, bottomland vegetation, and

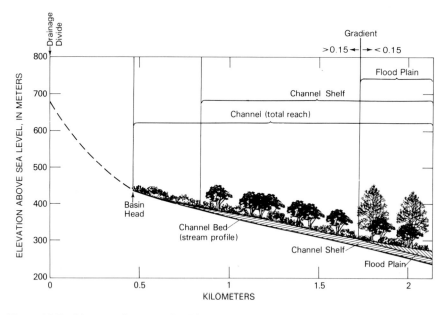

Figure 15.7   Diagram of a generalized low-order channel profile. Tall trees indicate flood-plain vegetation, shrubby trees indicate channel-shelf vegetation. Note the greater up-valley extent of channel shelves and associated vegetation, and gradient limits. (From Hupp, 1986.)

channel-gradient relations are illustrated in Figure 15.7 along a generalized low-order stream from the floodplain upstream to basin head.

A minimum threshold drainage area, dependent on gradient, is necessary for channel development and fluvial deposition. Channel gradient appears to be the 'fine tuning' that determines the shift between channel dynamics dominated by floodplain processes and those dominated by channel-shelf processes (Figure 15.4). These two suites of processes are not mutually exclusive, but one or the other appears to be more dominant along any given reach. The shift to predominantly channel-shelf processes occurs headward above average gradients of about 0.150 and along reaches of large streams at about the same channel gradient.

The importance of channel gradient in most aspects of fluvial hydrogeomorphic processes has been indicated by Hack (1957, 1973), Lane (1957), and Osterkamp (1978). Kilpatrick and Barnes (1964) suggested that channel gradient had a great influence on the character, elevation, and extent of 'benches' along southern Piedmont streams. They found that deposition, as an agent in the formation of floodplains, was strongly related to channel gradient. Furthermore, they found that steep-gradient streams contained the mean-annual flood within the floodplain banks, which would promote channel-shelf development. However, the floodplains of gentle-gradient streams commonly were inundated by the mean annual flood.

The basin heads of several first-order streams have been mapped in the vicinity of the Passage Creek basin (Hupp, 1984; Olson and Hupp, 1986); three of these are provided here (Figure 15.8) to illustrate vegetation–geomorphic relations in the vicinity of basin heads. Five general forest types are illustrated that have distinct hydrogeomorphic affinities, largely related to water availability. From dry (xerophytic) to wet (hydrophytic), the forest types are ordered pitch pine–bear oak, chestnut oak, mixed oak–hickory, northern hardwood, and floodplain (Figure 15.8).

Interfluvial-convergent slopes tend to support relatively moist northern-hardwood forests where surface- and groundwater flow converges. The northern-hardwood forest routinely forms the dominant vegetation type in basin heads, from which the rudiments of a stream channel emanate downslope. Ultimately these first-order streams, either along their course or after confluence with others, develop floodplains with their attendant distinctive vegetation type. Divergent slopes where surface- and groundwater flow diverges are dry interfluvial slopes, particularly when south facing, and typically support pitch pine–bear oak forests. Areas with intermediate moisture conditions, linear or side slopes, typically support the common mesophytic, mixed oak–hickory forest (Figure 15.8). Any slope with a coarse-sandstone colluvium commonly supports the singularly present chestnut oak.

Mapping provides large amounts of data, which can be interpreted without the pitfalls involved in the exclusive use of plot data. However, unlike plot data,

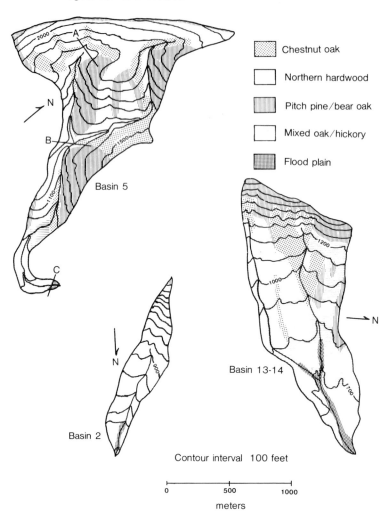

Chestnut oak

Northern hardwood

Pitch pine/bear oak

Mixed oak/hickory

Flood plain

Basin 5

Basin 13-14

Basin 2

Contour interval 100 feet

0   500   1000

meters

Figure 15.8 Vegetation maps of three low-order basins in the vicinity of the Passage Creek basin. Note the greater extent of flood-plain forest on the more gentle gradient basin 13–14 compared to the steep gradient basin and north-facing basin 2. The xerophytic pitch pine-bear oak forest is largely limited to south-facing slopes, nose slopes, and ridge crests. (From Hupp, 1984.)

map data, until recently, have not allowed for extensive quantification or for substantial data reduction. If for no other reason, vegetation mapping at the scale of landforms (about 1:20,000) has not found a place in current plant-ecological methodology. Mapping at this scale is a very promising way to relate plants to their physical environment and may offer many new insights in the future, including the mapping of vegetation type to identify variable source areas of runoff production and sites of groundwater recharge.

Results of the present study suggest that hydrogeomorphic form and process have a marked influence on the distributional characteristics of bottomland-forest species. The importance of hydrogeomorphic forms is not a new concept; Cowles (1901) described vegetation patterns on floodplains and tied these patterns to geomorphic forms and processes. He noted a trend in bottomland vegetation composition from xerophytic to mesophytic, which coincided with a down-valley physiographic trend from initial ravines (steep gradient) to mature streams (gentle gradient).

Clear definitions of fluvial landforms and stream-ordering systems are necessary for interpreting most patterns of bottomland vegetation. The hydrologic definitions should be couched in such terms as flow duration and flood frequency, providing independent parameters consistent with all streams that facilitate accurate comparisons among different streams. A process-oriented interdisciplinary explanation of bottomland-species distribution improves our ability to understand bottomland ecology and fosters practical application of vegetation studies in other physical sciences.

REFERENCES

Cowles, H. C. (1901). The physiographic ecology of Chicago and vicinity. A study of the origin, development and classification of plant societies. *Bot. Gazette*, **31**(2), 73–182.

Ellenberg, H. (1963). *Vegetation Mitteleuropas mit den Alpen*, Eugen Ulmer, Stuttgart.

Everitt, B. L. (1968). Use of cottonwood in an investigation of the recent history of a floodplain. *Am. J. of Science*, **266**, 417–39.

Everitt, B. L. (1980). Ecology of saltcedar—a plea for research. *Environ. Geol.*, **3**, 77–84.

Gauch, H. G., R. H. Whittaker and T. R. Wentworth (1977). A comparative study of reciprocal averaging and other ordination techniques. *J. of Ecol.*, **65**, 157–74.

Girel, J. and G. Pautou (1985). Associations vegetales et types de cultures dans les sections a forte pente du Haut-Rhone Francais. Colloques phytosociologiques, *Vegetation et Geomorphologie*, **13**, 669–89.

Hack, J. T. (1957). Studies of longitudinal stream profiles in Virginia and Maryland. *US Geol. Surv. Prof. Pap.*, **294-B**, 1–97.

Hack, J. T. (1973). Stream profile analysis and stream-gradient index, *US Geol. Surv. Jour. of Research*, **1**(4), 421–9.

Hack, J. T. and J. C. Goodlett (1960). Geomorphology and forest ecology of a mountain region in the Central Appalachians. *US Geol. Surv. Prof. Pap.*, **347**, 1–66.

Hedman, E. R., D. O. More and R. K. Livingston (1972). Selected streamflow characteristics as related to channel geometry of perennial streams in Colorado. *US Geol. Surv. open-file report*.

Hefley, H. M. (1937). Ecological studies on the Canadian River flood plain in Cleveland County, Oklahoma. *Ecol. Mono.*, **7**, 346–402.

Hickin, E. J. and G. C. Nanson (1975). The character of channel migration on the Beatton River, Northeast British Columbia, Canada. *Geol. Soc. Am. Bull*, **86**, 487–94.

236   Vegetation and Erosion

Hill, M. O. and H. G. Gauch (1980). Detrended correspondence analysis: An improved ordination technique. *Vegetatio*, **42**, 47–58.

Hosner, J. G. and L. S. Minckler (1963). Bottomland hardwood forests of southern Illinois. *Ecology*, **44**, 29–41.

Hupp, C. R. (1982). Stream-grade variation and riparian-forest ecology along Passage Creek, Virginia. *Bull. Torrey Bot. Club*, **109**, 488–99.

Hupp, C. R. (1983). Vegetation patterns on channel features in the Passage Creek gorge, Virginia. *Castanea*, **48**, 62–72.

Hupp, C. R. (1984). *Forest ecology and fluvial geomorphic relations in the vicinity of the Strasburg Quadrangle, Virginia*. Unpublished PhD thesis, The George Washington Univ.

Hupp, C. R. (1986). The headward extent of fluvial landforms and associated vegetation on Massanutten Mountain, Virginia. *Earth Surf. Proc., and Landforms*, **11**, 113–26.

Hupp, C. R. (1988). Plant ecological aspects of flood geomorphology and paleoflood history. In V. R. Baker, R. C. Kochel and P. C. Patton (eds.), *Flood Geomorphology*, Wiley, pp. 335–56.

Hupp, C. R. and W. R. Osterkamp (1985). Bottomland vegetation distribution along Passage Creek, Virginia, in relation to fluvial landforms. *Ecology*, **66**(3), 670–81.

Hupp, C. R. and A. Simon (1986). Vegetation and bank-slope development. *Proc. 4th Fed. Interagency Sed. Conf.*, **2**(5), 83–91.

Kilpatrick, F. A. and H. H. Barnes, Jr. (1964). Channel geometry Piedmont streams are related to frequency of floods. *US Geol. Surv. Prof. Pap.*, **422-E**, 1–10.

Lane, E. W. (1957). A study of the shape of channels formed by natural streams flowing in erodible materials. US Army Engineer Div., Missouri River, *M.R.D. Sediment Ser. No. 9*.

Mueller-Dombois, D. and H. Ellenberg (1974). *Aims and Methods in Vegetation Ecology*, Wiley.

Nanson, G. C. and H. F. Beach (1977). Forest succession and sedimentation on a meandering river floodplain, northeast British Columbia, Canada. *J. Biogeogr.*, **4**, 229–51.

Olson, C. G. and C. R. Hupp (1986). Coincidence and spatial variability of geology, soils, and vegetation, Mill Run Watershed, Virginia. *Earth Surf. Proc. and Landforms*, **11**, 619–29.

Osterkamp, W. R. (1978). Gradient, discharge, and particle size relations of alluvial channels in Kansas, with observations on braiding. *Am. J. of Sci.* **278**, 1253–68.

Osterkamp, W. R. and C. R. Hupp (1984). Geomorphic and vegetative characteristics along three northern Virginia streams. *Bull. of the Geol. Soc. of Am.*, **95**(9), 501–13.

Pautou, G. and J. Girel (1985). Le role des processus allogeniques dans le deroulement des successions vegetales: L'exemple de la plaine alluviale du Rhone entre Geneve et Lyon. Colloques phytosociologiques, *Vegetation et Geomorphologie*, **13**, 655–68.

Pickett, S. T. A. (1980). Non-equilibrium coexistence of plants. *Bull. Torrey Bot. Club*, **107**, 238–48.

Radford, A. E., H. W. Ahles and C. R. Bell (1968). *Manual of the Vascular Flora of the Carolinas*, Univ. of N.C. Press.

Richards, K. (1982). *Rivers, Form and Process in Alluvial Channels*, Methuen.

Robertson, P. A., G. T. Weaver and J. A. Cavanaugh (1978). Vegetation and tree species patterns near the north terminus of the southern flood-plain forest. *Ecol. Mono.*, **48**, 249–67.

Sigafoos, R. S. (1961). Vegetation in relation to flood frequency near Washington, DC. *US Geol. Surv. Prof. Pap.*, **424-C**, 248–9.

Simon, A. and C. R. Hupp (1987). Geomorphic and vegetative recovery processes along modified Tennessee streams: An interdisciplinary approach to disturbed fluvial systems. *Proc. Int'l. Assoc. of Hydrol. Sci.*, Vancouver, British Columbia, August 1987, IAHA-AISH, **167**, 251–62.

Strahler, A. H. (1978). Binary discriminant analysis: A new method for investigation species-environment relationships. *Ecology*, **59**(1), 108–16.

Teversham, J. M. and O. Slaymaker (1976). Vegetation composition in relation to flood frequency in Lillooet River Valley, British Columbia. *Catena*, **3**, 191–201.

White, P. S. (1979). Pattern, process and natural disturbance in vegetation. *Botanical Rev.*, **45**, 229–99.

Wistendahl, W. A. (1958). The flood plain of the Raritan River, New Jersey. *Ecol. Mono.*, **28**, 129–52.

Wolfe, C. B. and J. D. Pittillo (1977). Some ecological factors influencing the distribution of *Betula nigra* L. in western North Carolina. *Castanea*, **42**, 18–30.

Wolman, M. G. and L. B. Leopold (1957). River flood plains; some observations on their formation. *US Geol. Surv. Prof. Paper 282-C.*

Yanosky, T. M. (1982). Effects of flooding upon woody vegetation along parts of the Potomac River flood plain. *US Geol. Surv. Prof. Paper*, **1206**.

Zimmermann, R. C. and B. G. Thom (1982). Physiographic plant geography. *Prog. in Phys. Geog.*, **6**, 45–59.

16 Historic Riparian Vegetation Development and Alluvial Metallophyte Plant Communities in the Tyne Basin, North-east England

MARK G. MACKLIN
Department of Geography, University of Newcastle upon Tyne.
and
ROGER S. SMITH
Department of Agricultural and Environmental Science, University of Newcastle upon Tyne

SUMMARY

One of the most important ecological legacies of historic metal mining in many of Britain's base-metal orefields is a characteristic group of uncommon plants, typically including *Thlaspi alpestre* and *Minuartia verna*, that grow mainly or entirely on recent alluvial terraces severely contaminated by heavy metals. Such metallophyte plant communities are exceptionally well developed in the South Tyne and Tyne valleys. north-east England, where they have been investigated in detail at six sites that experienced mining and flood-related channel transformations during the late nineteenth century. Floodplain sedimentation has been documented over the last 130 years using serial topographic maps, aerial photographs and lichenometry. These have provided age estimates for vegetation development by indicating the time elapsed since sites were last reworked by the river. The present floristic composition of metallophyte plant communities in the South Tyne and Tyne valleys appears to be related to river terrace age, the degree to which alluvial soils are contaminated by heavy metals and their current management as farmland.

INTRODUCTION

Since 1950 extensive river bed incision has occurred in the Tyne basin, north-east England, in response to climatically related changes in flood frequency and

Vegetation and Erosion
Edited by J. B. Thornes
© 1990 John Wiley & Sons Ltd

cessation of metal mining in the catchment (Macklin and Lewin, 1989; Macklin and Newson, 1988). During this period parts of the valley floor that were sites of active alluvial sedimentation during the heyday of metal mining (c. 1850–1930) have been transformed into a series of low terraces. Subsequent vegetation development on these has produced a number of plant communities that seem to be related in some way to river terrace age, the degree to which alluvial soils are contaminated by heavy metals and their current management as farmland. Mining-age river terraces provide a habitat for a number of rare metallophyte plant species specifically adapted to contaminated soils; these are *Minuartia verna* (Spring Sandwort), *Thlaspi alpestre* (Alpine Penny-cress) and *Epipactis leptochila* (Narrow-lipped Helleborine). *Armeria maritima* (Thrift) is also found as an ecotype distinctly different from those on saltmarshes (Richards *et al.*, 1989). Their distribution on recent alluvial terraces is patchy with alluvium deposited since the mid-nineteenth century presently being covered with infertile grassland, scrub or woodland, or by metallophyte communities. It was initially unclear whether these represented a polyclimax dependent upon edaphic and management factors or plant communities at different stages along the same sere. Therefore, a key question for effective conservation management of current metallophyte vegetation types is whether or not they represent climax communities that have survived through adaptation to high levels of heavy metals in alluvial soils.

THE STUDY AREA

The River Tyne (Figure 16.1) has a drainage area of 2927 km^2 and is fed by two major tributaries, the Rivers North and South Tyne. The River South Tyne drains the northern part of the North Pennine Orefield, formerly the most productive lead and zinc mining area in Britain (Schnellmann and Scott, 1970). The principal metal ores exploited were galena (PbS) and sphalerite (ZnS), both of which contain significant concentrations of cadmium, copper and silver (Young, Bridges and Ineson, 1987). The peak of lead mining in the Northern Pennines occurred between the end of the Napoleonic wars and 1880 (Figure 16.2). The heyday of zinc extraction (1880–1920) was somewhat later and more short-lived. The last major metal mine in the Tyne basin closed in the early 1950s.

Metallophyte plant communities developed on metal-contaminated alluvium are found at more than 30 sites along the South Tyne and Tyne valleys (Richards *et al.* 1989). They range from above Alston downstream to Newburn (8 km west of Newcastle upon Tyne) on the tidal reach of the River Tyne; a total river distance of more than 80 km. Six of these sites, five on the River South Tyne and one in the Tyne valley, were chosen for detailed study (Figure 16.1). Together they cover the full geographical and geomorphological range of alluvial

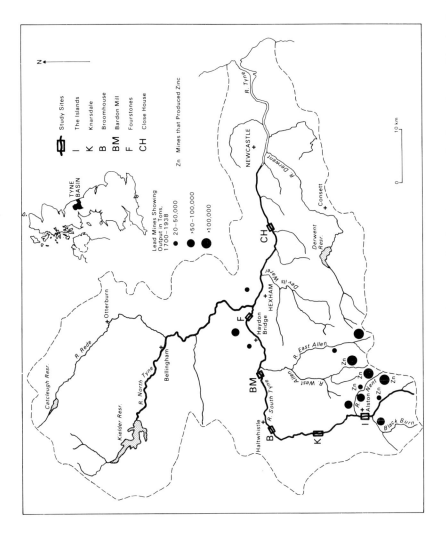

Figure 16.1 The Tyne basin, north-east England, showing locations on study sites, mines and lead output

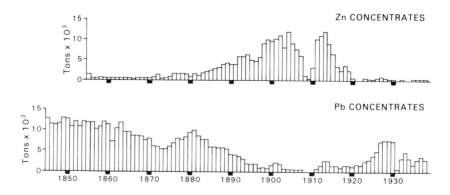

Figure 16.2   Zinc and lead production in the Tyne basin, 1845–1938 (after Dunham, 1944)

environments that provide habitats for metallophyte and associated riverside plants in the Tyne basin.

The two most upstream sites at the Islands (NY 716445), near Alston, and Knarsdale (NY 680535), are situated within extensive alluvial basins (1.7–1.3 km long and 0.25–0.35 km wide) in the upper reaches of the River South Tyne. Both sites have experienced accelerated transformation of channels from a single thread to a multithread braided pattern together with channel bed aggradation and subsequent incision over the last hundred years or so (Macklin and Lewin, 1989).

River development since 1860 at the Broomhouse (NY 686623) and Beltingham/Bardon Mill (NY 785642) sites, in the middle part of the South Tyne valley, has occurred by progressive lateral migration and sedimentation of high-sinuosity channels punctuated by channel cutoffs. The height of late nineteenth-century mining-related coarse alluvial sediment at Broomhouse are some 1.0 m lower than at the Islands or at Knarsdale. At Beltingham/Bardon Mill, however, alluvial gravel extraction in the 1960s caused the channel to incise 2.5 m below the level of late nineteenth- and early twentieth-century bed material sedimentation.

The fifth site at Fourstones (NY 888676) is located 3 km upstream of the confluence of the North and South Tyne rivers, with the most downstream site at Close House (NZ 131563) situated in the tidal reach of the Tyne. Historic maps and more recent aerial photographs show very little lateral movement of the channel bordering both sites; however, considerable river bed incision (accentuated by gravel extraction and dredging) has occurred since the cessation of mining. Fine-grained metal-contaminated alluvium currently lies between 7 and 9 m above the contemporary river bed and more than 2 m above the overbank sedimentation levels associated with the 30-year flood recorded on 26 August 1986 (Macklin and Dowsett, 1989).

METHODS

Six-inch Ordnance Survey maps (1860–1952) and aerial photographs (1947–85), in conjunction with lichenometric growth patterns from the species *Huilia tuberculosa* (Macklin, 1986) and trace metal analyses (Knox, 1987; Lewin and Macklin, 1987), were used to date alluvial sediments deposited in the South Tyne and Tyne valleys over the last 130 years. These techniques provided maximum age estimates for the development of plant communities by indicating the time elapsed since an alluvial valley floor site was last reworked by the channel. Alluvial sediments deposited before 1860 generally have higher concentrations of lead than zinc; while zinc levels are higher than those of lead in alluvium laid down between 1880 and 1920, during the period of large-scale zinc mining in the Allendale and Alston Moor mining fields (Macklin, 1986; Macklin and Lewin, 1989). Each dated alluvial unit, and its associated plant community, was assigned to one of six age classes: pre-1860, 1860–80, 1881–1900, 1901–20, 1921–52 and 1953–87.

The vegetation of the dated alluvial units was sampled with randomly positioned 1 m² quadrats. The percentage cover of each vascular plant species was visually estimated in these quadrats. The fine sediment fraction (less than 2 mm), collected from the top 10 cm of the soil profile, was also sampled in each quadrat. These samples were air dried, sieved through a 2 mm mesh and analysed to determine their pH, organic matter content, phosphorus and heavy metal concentrations. Levels of organic matter were taken as the loss of weight on ignition after 24 hours at 430 °C; pH was measured in a 2.5:1 water to soil/sediment mixture. Phosphorus content was measured by extraction with sodium bicarbonate solution and spectrophotometric measurement at 880 nm of the blue complex formed on reaction with ammonium molybdate. 'Total' heavy metal concentrations were determined as the amount of these elements brought into solution by digesting sediment samples in 25 per cent nitric acid. 'Plant available' metals were estimated in extractant solutions using 0.5 M acetic acid. Concentrations of cadmium, copper, iron, lead, manganese, silver and zinc were determined using atomic absorption spectrophotometry (air/acetylene flame).

Vegetation data were classified by two-way indicator species analysis (TWINSPAN) using Hill's default pseudospecies cut levels (Hill, 1979a), after removing two atypical woodland quadrats from Close House. Vegetation types were identified using the scaling system advocated by Gauch and Whittaker (1981). Relationships between plant species and quadrats were assessed with indirect ordinations by detrended correspondence analysis (DECORANA) (Hill, 1979b), using Hill's 'typical' species data transformation, downweighting the rare species and detrending by second-order polynomials as recommended by Ter Braak (1988). Such indirect ordinations arrange species and quadrats along 'environmental' gradients derived indirectly from the species composition of

Table 16.1 Frequency of the main species in each vegetation type

| | Habitat and vegetation types | | | | | | | |
| | Festuca–Agrostis grassland | | | | | Tall grassland | | |
Vegetation types[1]	4	2	3	1	5	6	7	8
Number of quadrats	30	8	7	9	28	6	5	6
Betula spp.	7				21	33**		
Briza media					21*	17		
Lotus corniculatus			29		43*	33		
Leontodon hispidus					46*		20	
Thymus praecox	7		57*		75*			
Ameria maritima	53	25			57			
Euphrasia officinalis	13		14	11	46*			
Minuartia verna	57				57			
Polygala vulgaris		13	43*		18			
Nardus stricta			29**					
Juncus effusus			29**					
Festuca ovina	100*	88	86	33	53			
Thlaspi alpestre	73*	63	29		43			
Agrostis capillaris	63	100*	100*	100*	57			
Galium saxatile			29**	11				
Deschampsia flexuosa		13	57**					
Campanula rotundifolia	40	63*	100*		32	33		17
Deschampsia cespitosa		25	71*		25	17	40*	
Hieracium pilosella	3		71**		7		20	
Anthoxanthum odoratum	3	13	71**	22	14	33		
Cerastium fontanum	37	38		89*	43	17	40	17
Luzula campestris	13		71*	33	25	17		
Trifolium repens	10	38	14	100**	14	17	20	
Cynosurus cristatus				22**				
Lolium perenne				33**				
Poa trivialis				44**			20	
Ranunculus acris	7		29*	44*	4		20	
Ranunculus repens				44*				17
Plantago lanceolata	7	13		44	39	50*		33
Avenula pratensis	7				43*	67*	40	17
Leontodon autumnalis	13			33*	11	50	20	
Festuca rubra				22	43*	83*	60*	
Holcus lanatus		50*		89*	14	17	80*	33
Arrhenatherum elatius		13			25	83*	20	100**
Centaurea nigra			14		11	33*	20	83**
Crepis paludosa								33**
Festuca arundinacea								50**
Filipendula ulmaria								33**
Knautia arvensis								50**
Vicia cracca			14		7	17	20	83**
Galium cruciata						17	40**	33*
Galium mollugo	3					33*	60**	50*

(continued)

Table 16.1 *(continued)*

	Habitat and vegetation types							
	Festuca–Agrostis grassland					Tall grassland		
Vegetation types[1]	4	2	3	1	5	6	7	8
Number of quadrats	30	8	7	9	28	6	5	6
Holcus mollis						33*		
Rubus spp.							40**	
Silene dioica							80**	17
Veronica chamaedrys							40**	30**
Hieracium spp.	3		29**		11	17	20	
Achillea millefolium		13	43*	11	21	17		83**
Species richness (species/m^2)	5.8	6.3	11.4	8.6	11.5	8.8	8.6	8.3

[1]Vegetation type names are given as binomials with the first species being the most frequent and the second being the most characteristic. The latter is defined by the highest value of $(O-E)^2/E$, where O is the observed frequency in the vegetation type and is greater than E the expected value, taken as the mean value over all quadrats. This index is based on chi-square and is used to identify characteristic species, i.e. those with values greater than or equal to 20 (*) or 200 (**).
1. *Agrostis capillaris/Poa trivialis*
2. *Agrostis capillaris/Holcus lanatus*
3. *Agrostis capillaris/Deschampsia flexuosa*
4. *Festuca ovina/Thlaspi alpestre*
5. *Thymus praecox/Leontodon hispidus*
6. *Festuca rubra/Arrhenatherum elatius*
7. *Holcus lanatus/Silene dioica*
8. *Arrhenatherum elatius/Festuca arundinacea*

the community. The relationships that such gradients have with true environmental factors can be interpreted through the use of stepwise multiple regression, the environmental factors being used as independent variables.

Constrained ordinations position species and sample quadrats along gradients derived directly from the environmental data. Canonical correspondence analysis (CANOCO) (Ter Braak, 1986, 1987, 1988; Ter Braak and Prentice, 1988) develops ordination axes by optimizing the level of correlation between vegetation axes and supplied environmental variables. The vegetation axes are linear combinations of these environmental variables. The technique leads to an ordination diagram (species–environment biplot) in which the points represent species and sites, and the vectors represent environmental variables. The vectors (arrows) show the direction in which each environmental factor appears to be operating. The longest arrows indicate the most important environmental variables. The position of nominal variables (age classes, presence of grazing livestock) are denoted by points. These empirical approaches to vegetation–environment relationships enable the data to be summarized and hypotheses to be generated.

RESULTS

Vegetation Characteristics

Of the 104 vascular plant species found in 101 quadrats, the most frequent were
Festuca ovina (61 per cent), *Agrostis capillaris* (59 per cent), *Thlaspi alpestre*
(41 per cent), *Cerastium fontanum* (38 per cent), *Campanula rotundifolia* (36
per cent), *Armeria maritima* (34 per cent) and *Minuartia verna* (33 per cent).
Eight vegetation types were recognized from the TWINSPAN output and were
given binomial names from their most frequent and most characteristic species
(Table 16.1). These fall into two main categories, on the basis of the presence
or absence of *Festuca ovina* and *Agrostis capillaris*. Both species are absent from
tall grassland vegetation types 6, 7 and 8, but are generally present in vegetation
types 1 to 5, which make up the bulk of the data set. These *Festuca–Agrostis*
grasslands can be subdivided into two groups. The first group is characterized
by a high cover of *Trifolium repens*, *Agrostis capillaris*, *Holcus lanatus* and
Cerastium fontanum (types 1 and 2). The second group consists of grassland
which has relatively small amounts of these species (types 3, 4 and 5). Type
3 is a dry, acid grassland with few heavy metal species, while types 4 and 5 are
the main heavy metal grasslands in which more than 50 per cent of the quadrats
were located. The vegetation of individual alluvial deposits is often very uniform
with respect to these types.

   The main trend in the vegetation is the separation, along the first ordination
axis, of the tall grassland vegetation (types 6–8) from the *Festuca–Agrostis*
grassland (vegetation types 1–5) (Figures 16.3 and 16.4). The other trend along
the second axis moves from pasture (type 1) to dry, acid grassland. The main
metallophyte communities (types 4 and 5) are situated at the ends of both axes.

Chemical and Physical Characteristics of Alluvial Soils

Historic alluvial soils in the South Tyne and Tyne valleys are generally slightly
acid (Table 16.2), although pH ranges from acid to neutral. Phosphorus and
organic matter levels are generally low and most soils are grossly contaminated by
cadmium, lead and zinc. Mean levels of these metals exceed those of uncontaminated soils in the region (Aspinall, Macklin and Openshaw, 1988), and elsewhere
in the UK (Archer and Hodgson, 1987), by more than an order of magnitude.

   Age classes of alluvial sediments deposited since the mid-nineteenth century
are not uniformly distributed between sites (Figure 16.5). River deposits that
date from before 1880 are found only at The Islands and Knarsdale in the upper
reaches of the South Tyne. At the four sites investigated downstream of
Knarsdale, mining-related alluvium deposited before 1880 appears either to have
been eroded as the result of lateral channel migration (e.g. Broomhouse and
Beltingham/Bardon Mill), or buried by younger fine-grained sediments (e.g.
Fourstones and Close House).

Table 16.2 Chemical and physical data for mining contaminated alluvial soils (93 samples) in the South Tyne and Tyne valleys

	Mean	Standard deviation	Median	Range		Derived mean*
				Min.	Max.	
pH	6.09	0.78	6.30	4.14	7.22	—
Phosphorus (mg kg^{-1})	1.48	1.34	0.88	0.08	7.80	1.02
Organic matter (%L.O.I.)	7.8	5.5	6.0	1.3	35.8	6.5
Total lead (mg kg^{-1})	2834	2060	2270	410	9798	2193
Available lead (mg kg^{-1})	685	500	512	4	2158	597
Total zinc (mg kg^{-1})	5504	3394	4640	590	16520	4560
Available zinc (mg kg^{-1})	888	466	976	98	2023	—
Total copper (mg kg^{-1})	57	48	43	8	384	47
Available copper (mg kg^{-1})	7.2	6.4	5.7	<0.1	22.8	—
Total silver (mg kg^{-1})	3.1	2.3	2.6	<0.1	15.0	2.8
Total cadmium (mg kg^{-1})	14.0	14.1	10.3	2.3	116.9	11.1
Available cadmium (mg kg^{-1})	5.0	3.8	4.3	<0.1	15.0	4.2
Total iron (mg kg^{-1})	38716	14120	37790	3670	85790	35975
Available iron (mg kg^{-1})	362	401	184	16	1850	192
Total manganese (mg kg^{-1})	2430	930	2370	690	5518	2249
Available manganese (mg kg^{-1})	487	335	476	23	1222	427

*These are based upon transformations used to bring their distributions to approximate normality. Log$_{10}$ transformations were used for phosphorus, organic matter, total lead, zinc, copper, iron and manganese and available iron. Square root transformations were used for available lead, cadmium and manganese and total silver.

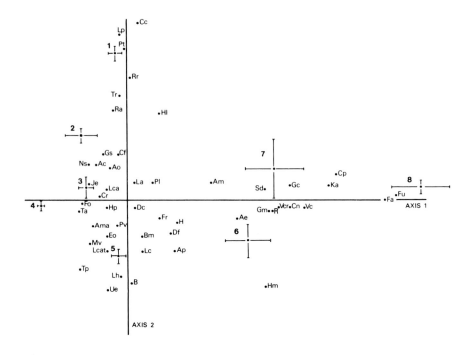

Figure 16.3 Indirect ordination of the vegetation types and species. Vegetation types are numbered in Table 16.1

Ac	*Agrostis capillaris*	Ae	*Arrhenatherum elatius*
Am	*Achillea millefolium*	Ama	*Armeria maritima*
Ap	*Avenula pratensis*	Ao	*Anthoxanthum odoratum*
B	*Betula* spp.	Bm	*Briza media*
Cc	*Cynosurus cristatus*	Cf	*Cerastium fontanum*
Cn	*Centaurea nigra*	Cp	*Crepis paludosa*
Cr	*Campanula rotundifolia*	Dc	*Deschampsia cespitosa*
Df	*Deschampsia flexuosa*	Eo	*Euphrasia officinalis*
Fa	*Festuca arundinacea*	Fo	*Festuca ovina*
Fr	*Festuca rubra*	Fu	*Filipendula ulmaria*
Gm	*Galium mollugo*	Gc	*Galium cruciata*
Gs	*Galium saxatile*	H	*Hieracium* spp.
Hl	*Holcus lanatus*	Hm	*Holcus mollis*
Hp	*Hieracium pilosella*	Je	*Juncus effusus*
Ka	*Knautia arvensis*	La	*Leontodon autumnalis*
Lc	*Lotus corniculatus*	Lca	*Luzula campestris*
Lcat	*Linum catharticum*	Lh	*Leontodon hispidus*
Lp	*Lolium perenne*	Mv	*Minuartia verna*
Ns	*Nardus stricta*	Pl	*Plantago lanceolata*
Pv	*Polygala vulgaris*	Pt	*Poa trivialis*
R	*Rubus* spp.	Ra	*Ranunculus acris*
Rr	*Ranunculus repens*	Sd	*Silene dioica*
Ta	*Thlaspi alpestre*	Tp	*Thymus praecox*
Tr	*Trifolium repens*	Ue	*Ulex europaeus*
Vc	*Veronica chamaedrys*	Vcr	*Vicia cracca*

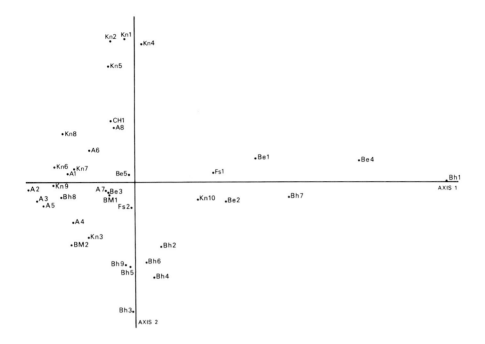

Figure 16.4 Indirect ordination of the alluvial deposits. Alluvial deposits are numbered sequentially within each of the following sites:
A — Alston/the Islands
Bh — Broomhouse
BM — Bardon Mill
Be — Beltingham
CH — Close House
Fs — Fourstones
Kn — Knarsdale

At upstream sites the pH of mining-contaminated alluvial deposits is lower and available phosphorus concentrations are generally higher, although the Close House site is anomalous in both respects. Nevertheless, it is clear that soils become increasingly acidic with age and that concentrations of phosphorus tend to be higher in the older alluvial deposits. Sediment metal concentrations generally decrease downstream, although the rates and patterns of decline vary between metals and also between their total and available forms. Thus sediment-associated cadmium, silver and zinc appear to have been transported further away from the Alston Moor and Allendale mining fields than either copper or lead, and available forms of all metals attenuate less rapidly than metals associated with primary and secondary minerals ('total' metal concentrations).

250

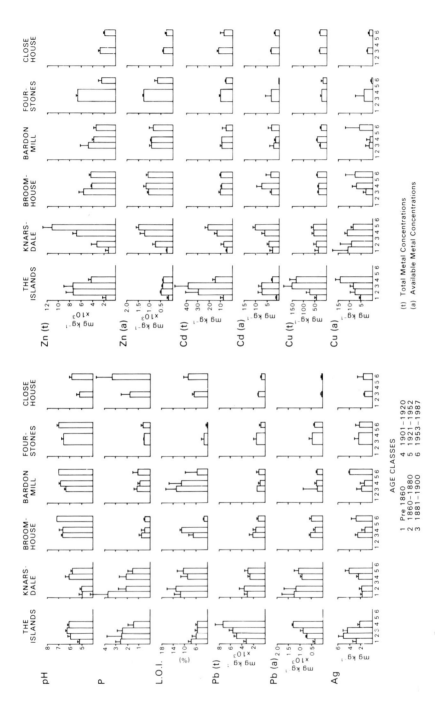

Figure 16.5 pH, extractable phosphorus, organic matter (L.O.I.), total and available metal concentrations plotted against age classes of alluvial deposits at the study sites

Vegetation and Soil/Habitat Relationships

The vegetation trends recognized in the first two indirect ordination axes have been interpreted by using the sample coordinates on these axes as dependent variables in a stepwise multiple regression against the independent environmental variables. The equations developed from these analysed were as follows:

$$\text{AXIS } 1 = 645 - 135\log_{10}\text{Tot.Pb} + 66\text{pH} - 126\log_{10}\text{Tot.Zn}$$
$$(r^2 = 0.398, \ P < 0.001)$$

$$\text{AXIS } 2 = 769 - 51\text{pH} - 120\log_{10}\text{Tot.Mn} + 71\log_{10}\text{P} + 45\log_{10}\text{Av.Fe}$$
$$(r^2 = 0.610, \ P < 0.001)$$

where AXIS 1 is the trend towards tall grassland with species indicative of damp conditions and AXIS 2 is the trend from dry, acid grassland with the beginnings of scrub invasion to permanent pasture. Tot.Mn, Tot.Pb and Tot.Zn are, respectively, total manganese, lead and zinc. Av.Fe is available iron and P is extractable phosphorus.

The trend towards tall grassland is associated with decreasing levels of both total lead and zinc but increasing pH. Total lead appears to be the most important environmental variable, accounting for 27.2 per cent of the variation along the first axis. The trend from acid grassland to permanent pasture is associated with decreasing pH and total manganese, and increasing levels of phosphorus and available iron. pH appears to be the most important environmental variable, accounting for 44.5 per cent of the variation along the second axis, with soils being generally more acidic under the pasture.

The species–environment biplot (Figure 16.6) includes only the most frequent species and those continuous variables to which the species react significantly when they alone are used to constrain the axes. Total and available copper, total silver and available iron did not meet this criterion and were not included in the diagram. Nominal variables (age classes and the presence of grazing livestock) are included, as are the mean positions of each of the vegetation types. The resultant biplot suggests that, in terms of the first two constrained axes, the number of vegetation types can be reduced to five by amalgamating types 2 and 3, types 5 and 6 and types 7 and 8. The species composition of these vegetation types is reflected in the positions of the individual species. The close association of vegetation type 4 with the three metallophyte species is of particular note. Total lead, cadmium, zinc and manganese are important in separating this from other vegetation types (Table 16.3). Metallophyte species appear to be most closely associated with alluvial deposits laid down between 1920 and 1952.

Vegetation types 1, 2 and 3 are found on the oldest alluvial deposits (pre-1860 and 1860–80) which have the highest levels of phosphorus and organic matter and are presently grazed. Types 2 and 3 contain some metallophyte species are found

Table 16.3 Alluvial soil physical and chemical characteristics for each vegetation type (**means and standard errors**)

Vetetation types	1	2	3	4	5	6	7	8
No of samples	9	7	7	28	24	4	5	5
pH	**4.88**	**5.09**	**5.47**	**6.10**	**6.60**	**6.56**	**6.83**	**6.88**
	0.13	0.17	0.20	0.11	0.07	0.28	0.06	0.18
Phosphorus	**2.61**	**2.68**	**2.16**	**1.74**	**0.59**	**0.87**	**0.89**	**0.72**
(mg kg^{-1})	0.58	0.71	0.17	0.30	0.09	0.28	0.14	0.17
Organic matter	**15.2**	**9.7**	**7.0**	**7.3**	**6.8**	**6.7**	**5.4**	**5.6**
(%L.O.I.)	2.9	1.7	0.7	0.8	1.1	1.2	1.5	1.6
Total lead	**3260**	**2052**	**3472**	**4039**	**1959**	**1601**	**882**	**1262**
(mg kg^{-1})	513	438	304	424	270	352	195	169
Available lead	**1241**	**327**	**431**	**912**	**554**	**385**	**277**	**464**
(mg kg^{-1})	242	164	91	84	58	66	14	70
Total zinc	**2746**	**3514**	**2471**	**7853**	**6260**	**4918**	**3700**	**5024**
(mg kg^{-1})	602	627	662	687	678	438	210	636
Available zinc	**560**	**537**	**266**	**939**	**1193**	**1194**	**1030**	**1099**
(mg kg^{-1})	109	143	49	96	65	119	91	53
Total copper	**46.1**	**44.0**	**44.9**	**86.2**	**47.1**	**41.0**	**30.6**	**46.2**
(mg kg^{-1})	7.7	5.2	4.5	13.8	7.2	5.1	3.0	8.7
Available	**7.8**	**5.9**	**5.0**	**10.1**	**4.6**	**6.7**	**9.2**	**6.4**
copper (mg kg^{-1})	2.7	2.0	0.4	1.1	1.3	2.9	4.0	2.8
Total silver	**2.1**	**1.9**	**3.1**	**3.5**	**3.2**	**2.1**	**3.6**	**4.8**
(mg kg^{-1})	0.3	0.5	0.5	0.3	0.6	1.2	0.9	2.1
Total cadmium	**7.2**	**10.6**	**9.4**	**23.3**	**12.3**	**10.7**	**7.7**	**11.1**
(mg kg^{-1})	1.3	1.2	2.6	4.2	1.2	1.7	1.4	0.4
Available cadmium	**3.1**	**3.6**	**1.9**	**7.3**	**4.8**	**5.9**	**4.9**	**3.4**
(mg kg^{-1})	0.8	0.6	0.7	0.7	0.9	1.6	1.5	1.4
Total iron	**33428**	**27571**	**24426**	**41258**	**44822**	**42508**	**37542**	**40820**
(mg kg^{-1})	3013	4243	693	2705	3382	4258	1313	3070
Available iron	**373**	**68**	**35**	**278**	**576**	**247**	**680**	**567**
(mg kg^{-1})	100	22	6	67	97	28	235	179
Total manganese	**1581**	**1510**	**1927**	**3050**	**2639**	**2370**	**2016**	**2402**
(mg kg^{-1})	233	257	219	146	196	305	220	186
Available manganese	**292**	**242**	**84**	**430**	**722**	**671**	**767**	**704**
(mg kg^{-1})	50	128	10	62	57	31	123	54

on alluvium where levels of total cadmium, zinc and manganese are relatively high. Vegetation type 1 is primarily found at Knarsdale and is located at the high ends of the organic matter, phosphorus and lead gradients, and at the low ends of the pH, cadmium, zinc and manganese gradients. Vegetation types 5 and 6 occur in the centre of the ordination and are generally developed on alluvium deposited between 1880 and 1920 at Beltingham/Bardon Mill and Broomhouse. Vegetation types 7 and 8 are most closely associated with the youngest (post-1952) alluvial deposits at Broomhouse, Beltingham and Fourstones (Figure 16.7) which have a high pH and relatively low levels of heavy metals.

DISCUSSION AND CONCLUSIONS

The dating techniques used to age historic alluvial deposits in the South Tyne and Tyne valleys were only sufficiently accurate to place sites within broadly defined age classes. As a consequence physical and chemical soil data predominate, with the emphasis on the heavy metal content of the sites. This may be one reason why their present floristic composition cannot be entirely explained by the age of the substrate on which the communities have developed. However, seral change is not necessarily unidirectional. Furthermore, the initial soil environments in each of the age classes evidently varied considerably, particularly with respect to their heavy metal content. This was determined primarily by the magnitude and type of mining activity in the catchment. Variation in this initial environment would influence the rate and pattern of vegetation development, creating a range of vegetation types. Some of these may represent true climaxes, others plagioclimaxes consequent upon grazing by domestic livestock. Much would also depend upon the location of the sites in the catchment, particularly the potentially greater number of plant species available for colonization in the lower parts of the Tyne valley. The influence of grazing livestock is greatest at sites upstream of Haltwhistle, whereas downstream, alluvial deposits dating from the mid-nineteenth century are generally fenced against farm livestock to prevent them drowning in the river. This has enabled tall herb and woodland vegetation to develop on sites less contaminated and further downstream from the mines.

The association between age class and vegetation type is, therefore, likely to be a consequence of the different initial soil environments and their subsequent grazing history. The tall grassland occurs on ungrazed recent deposits (post-1953) relatively uncontaminated by heavy metals. Heavy metal contamination (except for lead) is similarly low on the oldest deposits (pre-1860) and these show intact grazed pastures with no metallophyte species. If grazing ceased it is likely that these pastures would rapidly scrub over. Deposits laid down between 1880 and 1952, particularly those that date from 1920–52, are characterized by metallophyte plant communities that have developed under grazed conditions.

It has been suggested that the original source of the metallophyte colonists of river alluvium were the montane populations (Hajar, 1987). These evolved metal tolerance to inhabit sites that were otherwise very similar to their montane homes. The shoots of *Thlaspi alpestre* have been found to accumulate five times as much zinc and twice as much cadmium as those of *Minuartia verna* (Hajar, 1987). *T. alpestre* is, therefore, a much more extreme metallophyte than *M. verna* and both are restricted by their inability to successfully compete with more aggressive species on less contaminated sites. Their position at the extremity of the cadmium and zinc vectors (Figure 16.6) reinforces this.

The wildlife conservation interest in the vegetation of these sites resides in their populations of *Thlaspi alpestre*, *Minuartia verna* and *Armeria maritima*. Maintenance of metallophyte species populations, therefore, depends on the continued existence of metal-contaminated alluvial deposits which today persist as river terraces well above the influence of the river. Alluvial gravel extraction and poorly planned channel regulation schemes together pose the greatest threat

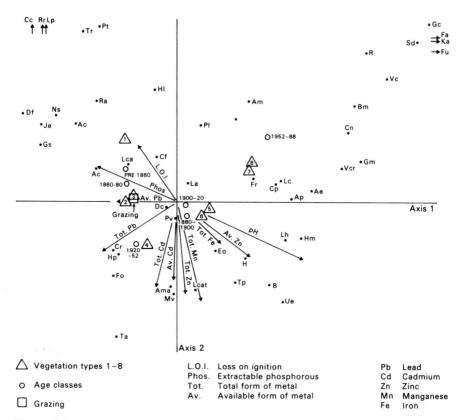

Figure 16.6 Species–environment biplot from CANOCO. Species codes are given in Figure 16.3; codes for vegetation types 1–8 are given in Table 16.1

to these rare plant communities, and it is very unlikely that this habitat will be destroyed by future river erosion. Natural succession to scrub and woodland is, however, a possibility but this is likely to be prevented by the combination of grazing and especially adverse soil conditions.

Metallophyte species are distributed throughout the North Pennine orefield on both old spoil heaps and river alluvium in the Tees and Wear valleys, and a number of riverside inundation communities have been described by Graham (1988). The six sites surveyed here are but a small sample of a more extensive resource. Surveys of dated alluvial deposits from a wider range of contaminated and uncontaminated sites are needed to refine our proposed scheme of seral vegetation development on metal-contaminated alluvium. This should enable key conservation sites to be identified and management prescriptions developed for the conservation of the metallophyte communities.

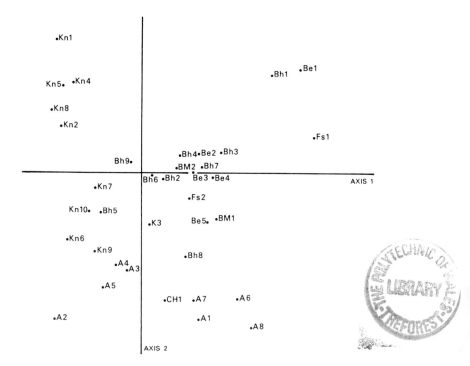

Figure 16.7   Constrained ordination of alluvial deposits. Site codes are given in Figure 16.4

ACKNOWLEDGEMENTS

The authors wish to especially thank Dr John Richards and Angela Nicholson for assistance in the collection of the vegetation data, and also Watts Stelling and Carol Campsell for undertaking metal and nutrient analyses. We also thank the two anonymous referees for their most helpful comments.

REFERENCES

Archer, F. C. and I. H. Hodgson (1987). Total and extractable trace element contents of soils in England and Wales. *Journal of Soil Science*, **38**, 421–31

Aspinall, R. J., M. G. Macklin and S. Openshaw (1988). Heavy metal contamination in soils of Tyneside: A geographically-based assessment of environmental quality in an urban area. In J. M. Hooke (ed.), *Geomorphology in Environmental Planning*, Wiley, Chichester, pp. 87–102.

Dunham, K. C. (1944). The production of galena and associated minerals in the northern Pennines; with comparative statistics for Great Britain. *Trans. Instn. Min. Metall.*, **53**, 181–252.

Gauch, H. G. and R. G. Whittaker (1981). Hierarchical classification of community data. *Journal of Ecology*, **69**, 537–57.

Graham, G. G. (1988). *The Flora and Vegetation of County Durham*, The Durham Flora Committee and the Durham County Conservation Trust, Durham.

Hajar, A. S. M. (1987). The comparative ecology of *Minuartia verna* and *Thlaspi alpestre* in the Southern Pennines with particular reference to heavy metal tolerance. PhD Thesis, University of Sheffield.

Hill, M. O. (1979a). TWINSPAN: A FORTRAN program for arranging multivariate data in an ordered two-way table by classification of the individuals and attributes. Ecology and Systematics Department, Cornell University, New York.

Hill, M. O. (1979b). DECORANA: A FORTRAN program for Detrended Correspondence Analysis and Reciprocal Averaging. Ecology and Systematics Department, Cornell University, New York.

Knox, J. C. (1987). Historical valley floor sedimentation in the Upper Mississippi valley. *Annals of the Association of American Geographers*, **77**, 224–44.

Lewin, J. and M. G. Macklin (1987). Metal mining and floodplain sedimentation in Britain. In V. Gardiner (ed.), *International Geomorphology 1986*, Wiley, Chichester, pp. 1009–27.

Macklin, M. G. (1986). Channel and floodplain metamorphosis in the River Nent, Cumberland. In M. G. Macklin and J. Rose (eds), *Quaternary River Landforms and Sediments in the Northern Pennines, England Field Guide*. British Geomorphological Research Group/Quarternary Research Association, pp. 19–33.

Macklin, M. G. and R. B. Dowsett (1989). The chemical and physical speciation of trace metals in fine grained overbank flood sediments in the Tyne basin, North East England. *Catena*, **16**(2), 135–51.

Macklin, M. G. and J. Lewin (1989). Sediment transfer and transformation of an alluvial valley floor, the River South Tyne, Northumbria, U.K. *Earth Surface Processes and Landforms*, **14**, 233–46.

Macklin, M. G. and M. D. Newson (1988). Man-related channel changes in the Tyne: Lessons for river basin management. *Northern Economic Review*, **16**, 34–41.

Richards, A. J., C.. Lefebvre, M. G. Macklin, A. Nicholson and X. Vekemans (1989). The population genetics of *Armeria maritima* (Mill.) Willd. on the River South Tyne, UK. *New Phytologist*, **112**, 281–293.

Schnellmann, G. A. and B. Scott (1970). Lead-zinc mining areas of Great Britain. In Jones (ed.), *Proceedings of the Ninth Commonwealth Mining and Metallurgical Congress 1969*, Vol. 2. Institute of Mining and Metallurgy, London, pp. 325–56.

Ter Braak, C. J. F. (1986). Canonical correspondence analysis: A new eigenvector technique for multivariate direct gradient analysis. *Ecology*, **67**(5), 1167–79.

Ter Braak, C. J. F. (1987). The analysis of vegetation–environment relationships by canonical correspondence analysis. *Vegetatio*, **69**, 69–77.

Ter Braak, C. J. F. (1988). CANOCO—A FORTRAN program for canonical community ordination by [partial] [detrended] [canonical] correspondence analysis, principal components analysis and redundancy analysis (version 2.1). Technical report: LWA-88-02. GLW, Postbus 100, 6700 AC Wageningen.

Ter Braak, C. J. F. and I. C. Prentice (1988). A theory of gradient analysis. *Advances in Ecological Research*, **18**, 271–317.

Young, B., T. F. Bridges and P. R. Ineson (1987). Supergene cadmium mineralisation in the Northern Pennine Orefield. *Proceedings of the Yorkshire Geological Society*, **46**(3), 275–8.

17 Seasonal Change in Aquatic Vegetation and its Effect on River Channel Flow

J. F. WATTS*
Cricklade College, Andover

and

G. D. WATTS
School of Biological Sciences, University of East Anglia

SUMMARY

This study investigates the effect of seasonal growth of aquatic vegetation on stream flow at a single cross-section. Measurements of the size and shape of the channel, velocity of the flow, and distribution, type and effective height of the bed vegetation were taken at regular intervals from March to September. Vegetation growth was clearly seen to disrupt the flow pattern, reducing velocity and leading to a doubling of water depth despite decreasing discharge. Standard resistance coefficients were calculated and Manning's n showed a change from 0.02 to 0.15. However, many roughness estimates may be inadequate to describe this situation since the hydraulic radius becomes almost meaningless.

INTRODUCTION

Aquatic vegetation is present in many rivers, streams and man-made channels throughout the world. In humid temperate climates, and particularly in lowland areas, the presence of vegetation on the bed of a watercourse is the rule rather

*now J. F. Birnie

Vegetation and Erosion
Edited by J. B. Thornes

than the exception. Yet in the work of fluvial geomorphologists, vegetation is rarely mentioned. Studies in the field have often been located where vegetation is absent, in semi-arid (e.g. Schumm, 1960) or mountain areas. Alternatively, any vegetation present is removed before the channel is studied. Aquatic vegetation contributes to channel roughness, reduces its capacity, and retards flow (Chow, 1981). For the fluvial geomorphologist some consideration of the influence of vegetation is desirable if the dynamics of flow in a natural channel are to be understood, and successful links made between process and form. This chapter demonstrates the potential significance of aquatic vegetation in natural channels by recording its seasonally changing influence at one point.

Channel roughness is not generally measured directly but by means of its effect. For example, Manning's n is a roughness estimate based on observed relationships between velocity, hydraulic radius and channel slope. There are published figures for a wide variety of channels, some of which are vegetated. For one drainage ditch the value of n increased from 0.033 to 0.099 as a consequence of vegetation growth (Chow, 1981). Such measurements are used as guidance when engineers plan or modify open channels, with the roughness of a reach often guessed by eye, on the basis of experience (Anglian Water Authority engineers, personal communication, 1976).

Direct measurement of roughness might be undertaken in laboratory conditions, but there have been few attempts to model aquatic vegetation in a flume. Kouwen and Unny (1973) simulated vegetation with flexible plastic strips and derived an equation for estimating vegetative roughness height. This recognized the common characteristic of many river plants to bend with the increased flow, so that the effective height (i.e. the height of the obstruction) is partially dependent on flow conditions. This makes vegetation a particularly difficult roughness factor to model. In addition, plastic strips only simulate one form of vegetation, omitting those types with, for example, branching stems, broad leaves or finely divided leaves. The equation has not found wide application.

Vegetation is the only channel roughness factor which may change seasonally (many aquatic plants die back in winter). It is this characteristic that has enabled its influence to be demonstrated in field studies. For example, Powell (1972, 1978) found summer peaks of Manning's n in a Lincolnshire river, with a maximum of 0.25, contrasting with a winter range of 0.04 to 0.021. Dawson (1978) recorded seasonal changes in biomass of *Ranunculus calcareus* in a small chalk stream. Dry weight increases from 1 to 350 g/m were accompanied by a change in n from 0.05 to 0.4. The use of biomass is a welcome attempt to quantify the roughness factor directly and is further developed in Dawson and Robinson (1985) and Westlake and Dawson (1988). Dawson and Charlton (1988) have recently published an extensive bibliography on the subject of the roughness of vegetated watercourses.

The object of the field study reported here was to examine the role of aquatic vegetation in detail at one cross-section, to calculate standard roughness estimates, and to consider whether these were the most appropriate way of accounting for the effect of vegetation on channel form and process. The cross-section examined was a reach of the River Yare, about three miles south-west of Norwich, Norfolk (OS TG212052). The site selected lay between a riffle and a pool in a compromise between the need for symmetry (since most channel form and flow theory makes the assumption of a trapezoidal channel), and the need for representative vegetation cover, which did not occur on the coarse bed material of the riffles. The chosen cross-section was about 9 m wide, 0.35 m deep, and with an initial mean velocity of $0.3\,\mathrm{m\,s^{-1}}$.

METHODS

At fortnightly intervals from 31 March to 2 September observations were made to record the type, distribution and height of aquatic vegetation; to construct velocity profiles; and to calculate commonly used coefficients of channel roughness. In addition, suspended sediment load was recorded in order that it could be certain that it was not affecting the outcome.

Depth was measured at half-metre intervals across the section, and flow velocity was recorded at the bed, and at 10 cm intervals from the bed to the surface, at each depth measurement. Velocity was measured with a simple flowmeter consisting of a hinged flap, 1.5×1.0 cm, on the end of a rod which could be inserted vertically into the water. The current lifted the flap, which could then be locked in position by a spring-loaded bar, operated by a lever at the top end of the rod. The flowmeter was then lifted out of the water and the angle of the flap read off an attached scale. At each point several velocity measurements were taken in this way, and the mean was recorded. The advantages of the flowmeter were that it was smaller than standard current meters, with no rotating blades, so it caused the minimum of disturbance to flow and was appropriate for use *within* clumps of vegetation. Prior to use it was calibrated by towing through still water at a known rate using a dynamics trolley. The level of precision was $\pm 0.05\,\mathrm{m\,s^{-1}}$.

The slope of the water surface (hydraulic gradient) was measured with a quick-set level over a 50 m reach in which the cross-section was the mid-point. Suspended sediment was sampled systematically at a number of points across the section. Discharge figures were obtained from the nearest upstream gauging station, operated by Anglian Water Authority. In the four-mile reach between the gauging station and the cross-section there were no major inputs to, or outflows from, the river. Obviously discharge could not be calculated from the measurements at the section because of the influence of the vegetation.

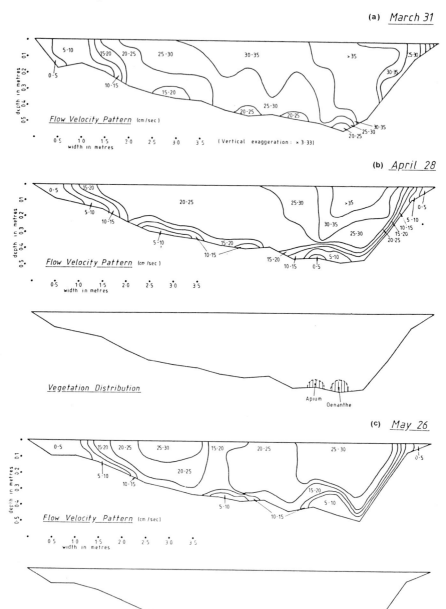

Figure 17.1 (a) to (c). Flow velocity pattern and vegetation distribution at the cross-section from March to September.

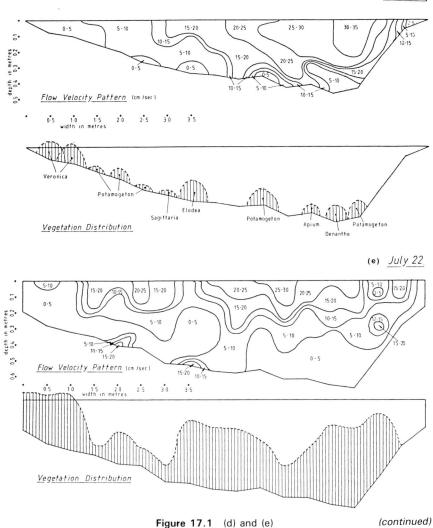

Figure 17.1 (d) and (e)   *(continued)*

Once vegetation began to appear, from 12 May onwards, its effective height at each half-metre point was recorded. Distribution of species, both vertically, in the section, and horizontally, upstream and downstream, was mapped.

RESULTS

Figure 17.1 (a) to (g) show the changing velocity profile and vegetation distribution at the cross-section. They show a clear disruption of the winter

(continued)

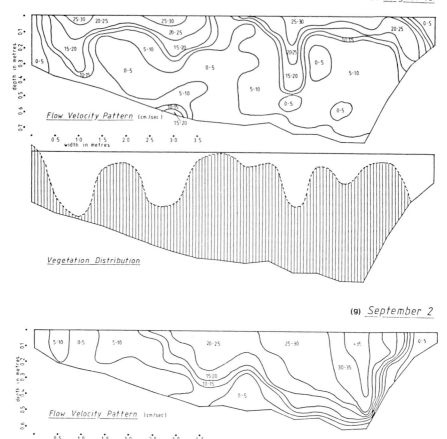

Figure 17.1   (f) and (g).

pattern as the region of maximum velocity was moved away from the bed near
the west bank of the channel during April and May. The pattern on 7 June
was particularly patchy, but a more even flow was restored on 20 June. During
July the region of indeterminate flow extended again, affecting over half the

channel depth and the entire width. In August the majority of water flow was confined to the upper 20 per cent of the cross-section, and in the centre the low flow area extended to the surface. On 2 September there was a dramatic change as the velocity distribution reverted to the winter pattern.

In addition to these general changes it was noted that the development of low velocities at the bed during May and June was not homogeneous but consisted of local areas of reduced flow separated by areas of higher velocity. Even during July and August, when reduced velocities were characteristic of the lower cross-section, there were still localized patches of higher-velocity flow at or near the bed.

On 12 May the low velocities associated with the clump of *Apium* and *Oenanthe* (Figure 17.1b) had led to the formation of a silt mound downstream of the cross-section which was 0.15 m high and 1.5 m in length. This was being colonized by the plants.

The appearance of aquatic vegetation began with this single clump of *Apium nodiflorum* and *Oenanthe fluviatilis* on the west wide of the bed. This had grown to over 20 cm by 7 June, and *Veronica anagallis-aquatica* was encroaching from the east bank. Between 7 and 20 June the main clump of *Apium* and *Oenanthe* was cut by Water Authority employees, but small patches of *Potamogeton perfoliatus* and other species were beginning to appear at intervals across the bed. By early August vegetation reached the surface in several places, with *P. perfoliatus* the dominant plant, covering clumps of less rapidly growing species. Between 18 August and 2 September weed cutting occurred, leaving plant stems and trailing debris less than 10 cm above the bed. Figure 17.2 gives a plan of vegetation cover just prior to cutting.

Figure 17.3 (a)–(d) shows discharge, mean depth, mean velocity and mean vegetation height for the study period. Width changed little as flow was below bankfull, and suspended sediment remained below 0.03 g/litre on all occasions except 18 August when it reached 0.1 g/litre. Roughness estimates are shown

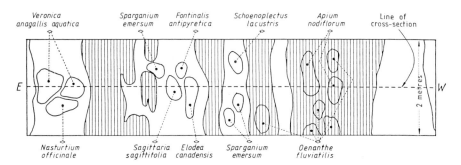

Figure 17.2 Belt transect of vegetation at the cross section, 18 August. Shaded areas represent *Potamogeton perfoliatus*, either rooted in the transect or trailing across it from anchorage upstream

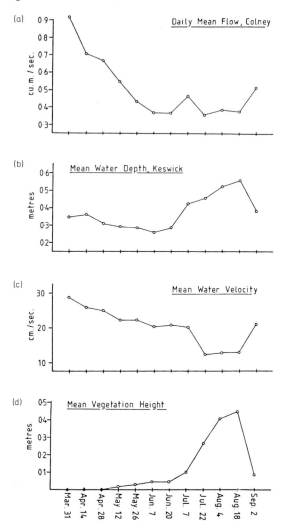

Figure 17.3 (a) Daily mean flow, Colney gauging station. (b) Mean water depth. (c) Mean water velocity. (d) Mean vegetation height

in Figure 17.4. A simple regression of mean velocity against vegetation height gave $r = 0.94$, which suggests a close association, although the small number of observations means the significance levels are low. The visual evidence from the graphs themselves is more convincing than the statistics in this case. Regressions of vegetation height against the roughness estimates did not show a close association, with the exception of the Darcy Weisbach friction factor in which velocity is more significant.

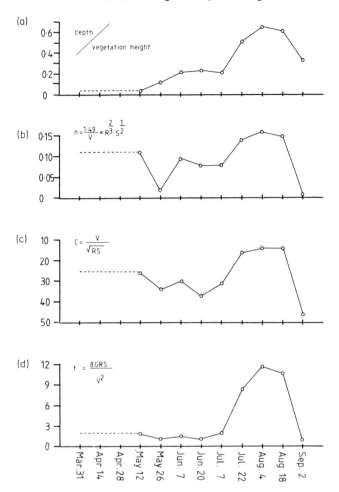

Figure 17.4 Roughness estimates. (a) Relative roughness. (b) Manning's n. (c) Chezy's C. (d) Darcy Weisbach f. Key: G, acceleration due to gravity; R, hydraulic radius; S, tangent of angle of water surface slope; V, mean velocity

DISCUSSION

There can be little doubt that in some instances aquatic vegetation has a greater effect on the flow of water in natural channels than any other single source of flow resistance, as demonstrated in this study by the scale of the seasonal change in the standard roughness estimate, Manning's n. The direct measurements of flow in the field study show that changes in the pattern of velocity distribution within the channel were quite clearly related to the distribution and type of aquatic vegetation. The distribution of sediment erosion and deposition within the reach was apparently modified, although this was not specifically studied. Overall, the mean velocity of flow was decreased, causing increased depth of flow for any given discharge.

Apart from demonstrating the scale of the flow resistance which may be created by aquatic vegetation, the field study provides some basis for evaluating standard resistance estimates in this context. The velocity profiles at the cross-section show how little of the flow is in contact with the actual bed and banks once vegetation is established. There is, instead, an effective wetted perimeter which will be smaller, may be smoother, but is also variable due to the flexibility of the vegetation. Resistance estimates attach a frictional significance to hydraulic radius which it may no longer have in these circumstances, and they may, therefore, be inappropriate for vegetated channels.

Simple measures of vegetation height in relation to water depth (i.e. 'relative roughness') may provide a useful resistance estimate, if the effect on velocity is fairly even within the vegetation. The profiles provide a means of checking this. There is certainly a marked contrast in the flow pattern within, and outside, the vegetation, so that they might be treated as two distinct flow situations. However, there are also considerable variations, with some high-velocity areas within the vegetation, especially through the less leafy stems at the bed; and also with different vegetative textures. The flexible, trailing stems of *Potamogeton perfoliatus* allowed more throughflow than the more dense and rigid clumps of *Oenanthe*. The effectiveness of a resistance factor based on vegetation height would therefore depend on the uniformity of the plant type, and would probably have to be adjusted for different plant forms. The 'flexural rigidity' described by Kouwen and Unny (1973) might then be taken into account.

The observed effects of aquatic vegetation lead to further speculation. The resistance to flow presented by the vegetation may stem from a number of factors. It is presumed that vegetation increases the skin resistance of a given channel form, and may also affect internal distortion resistance either by aggravating (e.g. *Oenanthe*?) or dampening (e.g. *Potamogeton*?) the turbulence of flow. There may be less direct effects, such as the stabilization of bed material and possibly bed forms, preventing adjustments to changing discharges that might otherwise occur (e.g. Simons and Richardson, 1961), and thus decreasing the efficiency of the system.

Given that the growth, if not the presence, of the vegetation is partly independent of channel form and flow, its existence throws into doubt the significance of some accepted relationships. 'Bankfull' flow, for example, will occur at a lower discharge and under very different flow conditions when vegetation is present. Bankfull flows occurring at different seasons in the same reach may not be comparable in terms of their significance to geomorphic process.

Further investigation into the effect of aquatic vegetation will meet a number of problems, particularly concerned with the semi-dependent, semi-independent characteristics that the vegetation exhibits in relation to stream flow: for example, the type and extent of vegetation may be controlled by the discharge regime (Butcher, 1933), the rate of growth and the extent of vegetation may be determined by climatic variables (also partly controlling the channel flow itself), whereas the effective height of the vegetation may, at any one moment, be determined by stream depth, discharge, and the flexural rigidity of the plants themselves. In addition, the downstream dimension of the vegetation effect will need to be considered, since a single cross-section cannot fully represent flow conditions.

Despite these problems in the investigation of the role of aquatic vegetation in natural channel form and flow, it is hoped that geomorphologists will not continue to turn a blind eye to a factor which may often have an overwhelming effect on fluvial processes.

REFERENCES

Butcher, R. W. (1933). Studies on the ecology of rivers. 1. On the distribution of macrophyte vegetation in the rivers of Britain. *J. Ecol.*, **21**, 58–91.

Chow, Ven Te (1981). *Open Channel Hydraulics*, McGraw-Hill, New York (first edition 1959).

Dawson, F. H. (1978). Seasonal effects of aquatic plant growth on the flow of water in a small stream. *Proc. EWRS 5th Symp. on Aquatic Weeds, 1978*, pp. 71–8.

Dawson, F. H. and F. G. Charlton (1988). *Bibliography on the Hydraulic Resistance or Roughness of Vegetated Watercourses*. Freshwater Biological Association Occ. Publ. No. 25, FBA, Ambleside.

Dawson, F. H. and W. N. Robinson (1985). Submerged macrophytes and the hydraulic roughness of a lowland chalkstream. *Verh. int. Ver. theor. angew. Limnol.*, **22**, 1944–8.

Kouwen, N. and T. E. Unny (1973). Flexible roughness in open channels. *American Society of Civil Engineers, Journal Hydraulics Division*, **99**, 713–27.

Powell, K. E. C. (1972). The roughness characteristics of the reach of the River Bain at Fulsby Lock. *Lincolnshire River Authority, Report of the Hydrological Section*.

Powell, K. E. C. (1978). Weed growth—a factor in channel roughness. In R. W. Herschy (ed.), *Hydrometry—Principles and Practices*, John Wiley & Sons, pp. 327–52.

Schumm, S. A. (1960). The shape of alluvial channels in relation to sediment type. *US Geol. Survey Prof. Paper*, **352-B**, 17–30.

Simons, D. B. and E. V. Richardson (1961). Forms of bed roughness in alluvial channels. *A.S.C.E. J. Hyd. Div.*, **87**, 87–105.

Westlake, D. F. and F. H. Dawson (1988). The effects of autumnal weed cuts in a lowland stream on water levels and flooding in the following spring. *Vehr. int. Ver. theor. angew. Limnol.*, **23**, 1273–7.

18 Soil Erosion and Fire in Areas of Mediterranean Type Vegetation: Results from Chaparral in Southern California, USA and Matorral in Andalucia, Southern Spain

A. G. BROWN
Geography Department,
University of Leicester

SUMMARY

Spatial variations in the magnetic susceptibility of surface soils can under certain conditions be used to assess relative erosion rates at both the basin and plot scale after fires in chaparral and matorral ecosystems. The spatial variation of surface magnetic susceptibility exhibited in degraded matorral at a site in the Guadalhorce basin, southern Spain, is related to the occurrence of dwarf fan palm (*Chamaerops humilis*) clumps. This and the existence of clumps on pedestals indicates that species of different growth-forms and population dynamics have different effects on erosion rates irrespective of cover as it is normally measured. It is suggested that variation in growth-form and associated cover fragmentation is one of the causes of the non-linear relationship between vegetation cover and erosion rates.

INTRODUCTION

This chapter forms part of an investigation by the author of the interrelationships between erosion, soil conditions and Mediterranean type vegetation. The two

Vegetation and Erosion
Edited by J. B. Thornes
©1990 John Wiley & Sons Ltd

Table 18.1   A summary of the lithological, climatic and ecological characteristics of each study area

Rattlesnake Canyon, Santa Ynez Mts., S. California, USA	Las Atalayas, Gaudalhorce Valley, Andalucia, SE Spain

Lithology

Southwesterly dipping Eocene shales and non-ferruginous sandstones with Oligocene conglomerate and Pleistocene fanglomerate outcropping in the lower third of the catchment.	Approximately horizontally bedded Miocene sandstones and conglomerate unconformably resting on near vertically bedded Jurassic limestone to the east and Oligocene marl to the northwest.

Climate

Mediterranean with episodic rainfall between November and March, summer drought, mean annual ppt, 320–480 mm with high annual variability. Mean July temp, 19° C, mean Jan, temp, 12° C. Runoff is 10–30% of mean annual ppt.	Mediterranean with most rainfall between November and April, summer drought, mean annual ppt, 500 mm. Mean monthly July temp, 26° C, mean monthly Jan, temp. 10° C. Runoff approx. 25% of mean annual ppt.

Vegetation

Hard chaparral dominated by *Ceonothus* spp., *Adenostoma fasciculatum Rhus*, *Quercus, Yucci* and *Arctostaphylos* spp. Biomass high, high litter production and multi-layer structure up to 1 m + in height. Tree-dominated riparian zone.	Low degraded matorral of tomillar and retamar type with areas in *Pinus* spp., woodland and olive cultivation. Dominant matorral species are *Thymus* spp., *Rosmarinus officinalis*, *Lygos monosperma*, *Lavendula* spp., *Daphne gnidium*, *Genista* spp., *Juniper* spp., *Stipa tenacissima* and *Chamaerops humilis*, *Tamarix* and *Nerium oleander* in the riparian zone.

Ecological history

Regarded as natural vegetation, not subject to past agriculture or grazing except by native fauna. Severe fire burnt the entire area in 1964. RI of fires approx. 50 yrs.	Regarded as an anthropogenic vegetation type derived from the understorey of Mediterranean evergreen forest. Areas have been cultivated for olives, and cereals on some slopes. Extreme grazing pressure in the past and today. Occasional light fires.

areas used in this study which are similar in geology, climate and vegetation type (Table 18.1) provide data which address two questions: (1) can mineral magnetic properties be used to assess and quantify post-fire soil erosion; and (2) is the spatial variation of soil magnetic susceptibility enhancement related to vegetation type and pattern? The second question is directly related to a traditional research theme in semi-arid geomorphology, namely the relationship between vegetation and soil erosion. Various authors have conceptualized this relationship in different ways, from a static protection factor (e.g. USLE) to the dynamic erosion–vegetation competition model (Thornes, 1985, 1988). Vegetation is the main control on the spatial distribution and absolute values of shear stress applied by raindrop impact and overland flow, as well as the infiltration capacity and shear resistance of the soil surface. Individual plants may be one of the causes of the spatially discontinuous nature of overland flow on semi-arid and arid slopes along with non-biological factors (Yair, Sharon and Larce, 1980). However, it is important to ask the question: to what extent can Mediterranean plants be regarded as substitutable (i.e. species for species, type for type) as implied in the almost universal use of cover as the major independent variable? This is important if we are to further our understanding of the role of vegetation in geomorphology through its effects on soil characteristics and spatial variations in erosion processes and rates. Traditionally geomorphologists have ascribed high erosion rates in Mediterranean type areas to the climatic regime (Langbein and Schumm, 1958) and low biomass or poor vegetation cover. However, sclerophyllous vegetation is not necessarily of low biomass, especially in areas with an annual rainfall of 400 mm or more, with values ranging from under $170\,t\,ha^{-1}$ to $350\,t\,ha^{-1}$ (Bazilivich, Rodin and Rozov, 1971; Whittaker and Likens, 1973) and it often exceeds 100 per cent cover as can be illustrated by tall chaparral, holm oak woodland, pine forest and high or scrub matorrals.

In less moisture-stressed sites biomass is not a simple function of the precipitation/evapotranspiration ratio but has evolved in response to soil conditions, grazing and fire. If allowed to accumulate (i.e. not grazed), total biomass will increase in a logistic manner until eventually reduced to practically zero by fire. Fire regenerates the vegetation by removing the accumulated woody litter, killing degenerate shrubs and allowing secondary succession to occur. It can also affect the soil directly by increasing some available nutrients, especially potassium and phosphorous (Trabaud, 1986). Indeed Mutch (1970) has suggested that these vegetation types are adapted to the fire cycle through physiological mechanisms which increase inflammability, thus in effect 'encouraging' fire. Both in structure and physiology the matorral of southern Spain resembles the chaparral of southern California despite its different recent ecological history. Grazing has largely replaced fire as a vegetation control in the Mediterranean, with biomass rarely achieving the levels associated with the natural fire cycle, although burning is used to increase grazing by stimulating grasses and new growth at the expense of degenerate woody shrubs.

Fire also has a direct effect on erosion and many studies have shown increased sediment yields in the first few years after chaparral fires (DeBano, Rice and Conran, 1979). The combination of fire with high-intensity, low-frequency storms produces what Rice, Zeimer and Hankin (1982) have termed hyperscedastic basin response. This involves not only the reduction of cover but also alterations in soil hydrological characteristics such as the production of a non-wettable layer and decreased infiltration capacity (DeBano *et al.* 1979).

There have been very few studies of post-fire erosion in the Mediterranean and results have been contradictory (Trabaud, 1986). This is undoubtedly partly because of differing periods of observation and the inherently variable post-fire response caused by the Mediterranean climate. However, calculations of the relative importance of fire over a reasonable time period (i.e. at least one fire cycle), from both chaparral and matorral ecosystems suggest the direct effect is less important than other variables such as long-term basin instability (Keller, Florsheim and Best, 1988), or uncontrolled grazing (Naveh, 1974, 1975). Fire remains important as a trigger mechanism (Keller *et al.*, 1988) and as a fundamental control on vegetation composition and dynamics.

In addition to its effects on vegetation and nutrient status, fire affects soil mineralogy including iron compounds, converting antiferrimagnetic iron oxides and hydroxides into ferrimagnetic forms such as magnetite and maghemite. This is one of the reasons for the high magnetic susceptibility associated with Mediterranean soils (Tite and Linington, 1975); the others are low-temperature oxidation of magnetite to maghemite, dehydration of lepodicrocite and natural reduction/oxidation cycles. The details of the processes of 'enhancement' (i.e. the commonly observed pedogenetic increase in magnetic susceptibility above that of unweathered parent material) will not be discussed here, but further details can be found in Mullins (1977) and Thompson and Oldfield (1986).

TOPSOIL ENHANCEMENT AND POST-FIRE EROSION

Soil pits in Rattlesnake Canyon have provided magnetic susceptibility profiles which can be related to lithology and geomorphic position (Brown, 1988). In general, the less erodible (or eroded) a soil, the greater the surface or near-surface magnetic susceptibility relative to the subsoil and lithology, i.e. the higher its topsoil magnetic susceptibility enhancement. More recent fieldwork has produced similar profiles from three lithologies in southern Spain (Figure 18.1). The main Spanish site, at Las Atalayas in Malaga Province, is also the location of investigations of spatial variations in magnetic susceptibility and vegetation. The limestone soils of the area generally have high magnetic susceptibilities probably due to the high initial hematite content of the terra rossa soils derived largely from iron-rich veins in the limestone. The marls have lower absolute values but often higher relative surface enhancement, as do the sandstones and conglomerates.

A series of sites on the sandstone/conglomerate, four of which are shown in Figure 18.2, show how enhancement changes with geomorphic position. Colluvial deposits have both high absolute and high relative (i.e. enhanced) values throughout, while on mid-slopes and interfluves enhancement is restricted to the top 1–2 cm, with colluvial hollows and lower slopes being intermediate. The surface soil (0–2 cm) at the four sites had very similar carbonate contents

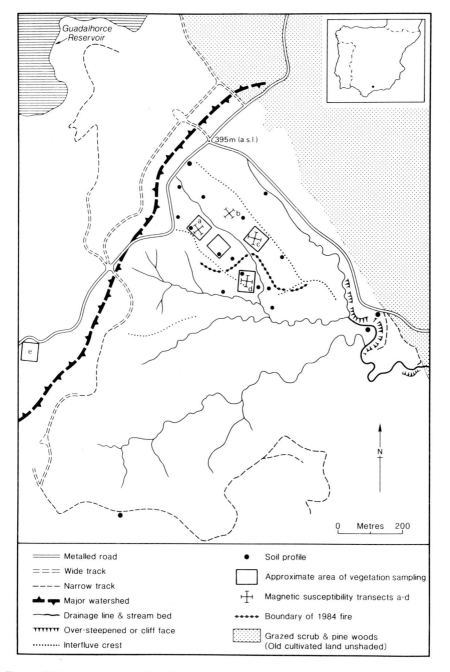

Figure 18.1 Location map of the Spanish sites including vegetation study plots, transects (a–d on Figure 18.4) and profile locations. Soil pits on the limestone at El Torcal lie 15 km to the east off the map, and Ardales badlands lie 5 km to the southwest. The area of greatest pedestal formation is at location e

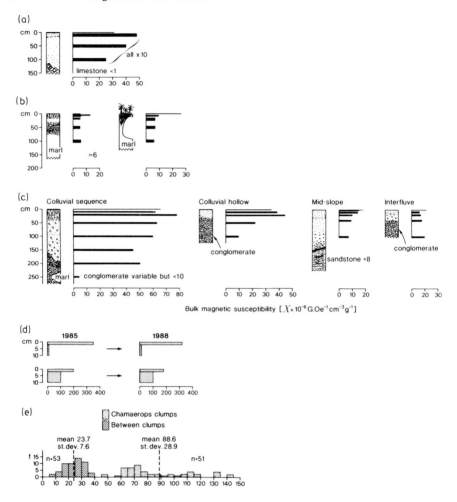

Figure 18.2 Examples of the Spanish profiles on each lithology (a–b), from different geomorphic positions (c), successive sampling of one *Chaemerops humilis* mound, top is centre and lower is edge of the clump (d) and histograms of clump and between clump samples (e). The figures by each lithology are parent material magnetic susceptibility $(\chi \times 10^{-6}$ $GOe^{-1} cm^{-3} g^{-1})$

(28.2–28.9 %), high pH (7.5–8.3) and LOI of 4.5–7.4%. While all of Rattlesnake Canyon was severely burnt in 1964, only a part of the Las Atalayas site was lightly burnt in 1984 although no samples were taken at Las Atalayas until 1985. In an attempt to measure the loss of enhanced soil following a fire, Brown (1988) has proposed a comparison of laboratory-induced enhancement (as a simulation of a wildfire) with post-fire enhancement observed in the field. Experiments by the author (Brown, 1988) have shown that there is generally a relatively low

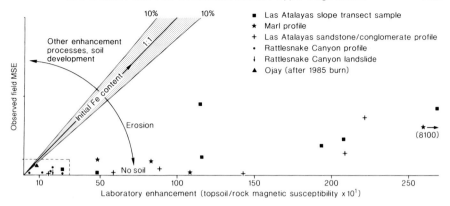

Figure 18.3   Graph of observed field surface magnetic susceptibility enhancement against laboratory ignition-induced magnetic susceptibility enhancement. On both axes enhancement is expressed non-dimensionally as soil sample magnetic susceptibility/parent material magnetic susceptibility

threshold (around 200 °C above which no further enhancement occurs and that successive burning of the same sample does not further increase magnetic susceptibility above this threshold. There are, however, considerable problems with simulating wildfires in the laboratory and the procedure used here is crude but standard, involving the ignition of 10 g of soil at 550 °C for one hour in an uncontrolled atmosphere. Surface soil taken immediately after a fire at Ojay in southern California show acceptable agreement with simulated values. The arc on Figure 18.3 between the laboratory-induced value and the observed post-fire value is therefore proportional to the loss of topsoil assuming there was full enhancement during the fire and the site is non-colluvial. This analysis suggested that in Rattlesnake Canyon the highest post-1964 erosion rates had occurred on the Juncal shale and this is in agreement with geomorphic evidence from drainage density analysis (Keller, personal communication). This measure of erosion is comparable to erosion pin methods (without the effect of the pins) rather than estimates based on sediment yield.

Under ideal conditions it may be possible to quantify surface lowering. This may be done by assuming an exponential magnetic susceptibility profile (reasonable at non-colluvial sites and on non-ferrimagnetic bedrocks) and fitting a decay constant (b). The depth of lowering can then be calculated from

$$X = X_0 e^{b(-d)}$$

where X is surface magnetic susceptibility, X_0 is the maximum possible magnetic susceptibility (i.e. laboratory-simulated fire value), $-d$ is soil depth and e is the base of natural logarithms. To facilitate calculation this can be rewritten in the form

$$-d = \left(\frac{\log (X/X_0)}{\log e} \right) / b$$

This equation can be used for any soil parameter that is exponentially related to depth, including bulk density and ^{137}Cs content. The b values for magnetic susceptibility in Rattlesnake Canyon varied between 0.09 and 0.04 depending upon the lithology, and calculations using a site with a 26° slope on the Cozy Dell shale suggest an average erosion rate of 7–8 mm yr^{-1}. Using the slope sites on the other lithologies, and area of each lithology, a crude estimate has been made of erosion in the catchment since 1964. Expressed as a yearly average rate this can be compared with other erosion estimates for the catchment which use several different methods. As can be seen from Table 18.2, the estimate is much higher than those from other sources. This is probably due to the bias towards eroding slopes, the omission of sediment storage sites and the existence of sites which have a hyperbolic rather than exponential magnetic susceptibility profile. Davis *et al.* (1988) have estimated that for a small chaparral basin in the San Rafael Mountains, California, wildfires cause between 30 and 50 per cent of the total debris deposition, the rest being caused by high-magnitude debris flows

Table 18.2 Comparison between different sediment yield/erosion estimate for Rattlesnake Canyon

Method	Source	Sediment yield (t km^{-2} yr^{-1})	Mean lowering rate (mm yr^{-1})
From the filling of Gibraltar Dam (adjacent catchment)	From data in Taylor (1983)	254	0.2
Data from other dams in the area (not incl. the catchment)	Taylor (1983)	818	0.66
Regression equation for the region	Taylor (1983)	1860	1.5
Data from channel sediment storage and events up to a 21 yr. RI.	Best and Keller (1986)	365	0.3
Using data derived from the 1969 floods for basins in the Transverse ranges.	From data in Taylor (1983)	4900	3.8
Magnetic susceptibility method using only 15°–26° slopes	this study	max. 10000 min. 3800	8 3

'this study' figures derived by applying the equation given in the text to profiles from slopes between 15° and 26° burnt in 1964, and using the best-fit b value for each lithology.

with a recurrence interval of thousands of years (Best and Keller, 1986). The recurrence interval of coincidence of a 1 in 100 year storm and a fire year for these catchments is between 1 in 3000 and 1 in 6500 years. However, the recurrence interval of the depositional event (rather than the causative event) and the recurrence interval of fire/storm coincidence are not independent as fires may contribute to long-term basin instability and act as potential trigger mechanisms (Keller *et al.*, 1988). Detailed work on instrumented burnt plots is needed to assess the full potential and error ranges of the method.

It was originally hoped that the above procedure could be used on the Spanish data; however, it cannot, probably because of the rather light burn in 1984 and the unknown date of the previous burn. This is revealed by the much greater spread of the Spanish points in Figure 18.3 than those from Rattlesnake Canyon. Additionally, the Spanish sites show a correlation between the weight loss-on-ignition and both pre- and post-ignition susceptibility (0.78 and 0.80 respectively and both significant at the 0.01 probability level). This is probably because, as previous experiments have shown, enhancement is increased with increasing organic matter content to about 10–15 per cent above which there is little or no further increase (Brown, 1988) and post-fire erosion and soil organic matter are inversely related.

POST-FIRE SPATIAL VARIATIONS IN ENHANCEMENT AND VEGETATION

The spatial variation of magnetic susceptibility was investigated using eight transects with systematic sampling after random location of the starting point but orientated normal and parallel to the slope contours; only the downslope transects are shown here (Figure 18.4). The cross-slope transects displayed similar characteristics to those downslope, which because they did not include the slope base or any areas of sediment deposition, show no cumulative downslope trend. One (a) is located in the unburnt and northeast-facing slope, two (b and c) are from the south-facing slope which was ploughed in 1985, and (d) is from the burnt area of the east-facing slope. All the transects show little autocorrelation (with the exception of c which has a suggestion of an 8 m periodicity), indicating that spatial pattern may occur under 1 m or over 15 m but not in between. Ploughing seems to have the effect of decreasing surface magnetic susceptibility and decreasing its variability, as illustrated by lower mean values and standard deviations.

The most obvious feature of the transects are the spikes which only occur under *Chamaerops humilis* (dwarf fan palm) clumps (Figure 18.5a). A comparison of 51 samples from within *Chamaerops humilis* clumps and 53 randomly located samples from between the clumps showed two discrete sample populations, and mean magnetic susceptibility to be over three times higher under

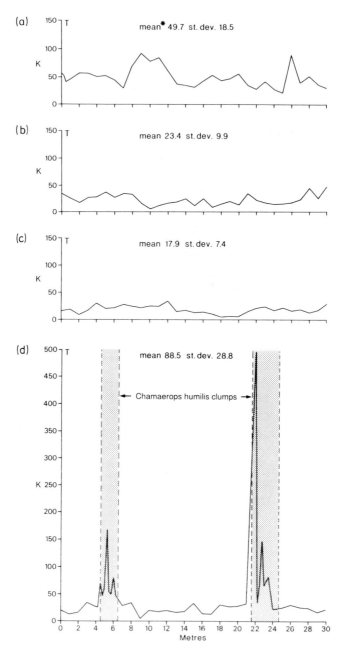

Figure 18.4   Four downslope magnetic susceptibility transects, measured using a ferrite probe. Magnetic susceptibility expressed as $K \times 10^{-6}$ Go^{-1} after calibration using single sample measurements. All descriptive statistics based on the 1 m sampling points only

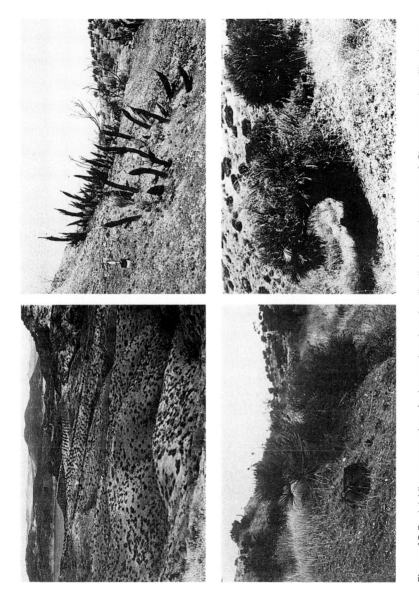

Figure 18.5 (a) View across Las Atalayas showing the distribution and pattern of *Chamaerops humilis* clumps. (b) Clump regeneration after fire 1985. (c) Clump mound on the sandstone/conglomerate. (d) Clump pedestal 0.95 m high on the marl (at location e on Figure 18.1)

the palms (Figure 18.2). One clump that was sampled in 1985 and again in 1988 showed little reduction (either at centre or edge) considering the high surface values, and ash was still present under the clumps (but nowhere else) four years after the fire. This evidence along with the local topography strongly suggests that one cause of the higher values is a lack of surface erosion under the clumps. A second possible, and related, cause may be greater litter accumulation under the clumps leading to greater enhancement than the surrounding less organic-rich soils. A minor contribution may come from the plant ash itself as ashed leaves of *Chamaerops humilis* have a higher magnetic susceptibility at 23×10^{-6} $GOe^{-1} cm^{-3} g^{-1}$ than *Thymus, Rosmarinus officianalis* or *Olea europeae* (9, 7 and $11 \times 10^{-6} GOe^{-1} cm^{-3} g^{-1}$ respectively).

Chamaerops humilis L. is the only palm native to mainland Europe and although polymorphic it is generally stemless and suckers from the base when growing in the wild (Hodge, 1982). If grazed and/or burnt, resprouting occurs preferentially around the edges of the clump causing it to expand and eventually form a ring up to 4 m in diameter. The death of the plant in the centre can provide a protected microhabitat for the germination and growth of other species such as *Lygos monosperma*. It was the first species to resprout after the fire (Figure 18.5b) along with *Asphodelus albus*, while the woody shrubs such as *Thymus* spp., *Lavendula stoechas* and *Lygos monosperma* have only been re-established in the last few years by seedling germination. The palm clumps are also often located on mounds (Figure 18.5c) or pedestals (Figure 18.5d) up to a metre high. This may be partly due to their dense root mat and growth form, which can up-warp soil and rock (best observed on laminated rocks; Barahona, personal communication), and trap windblown dust. However, the occurrence of pedestals with no up-warping of the underlying stratigraphy and mushroom shaped, because of undercutting, suggests that one cause is differential erosion. On the marl at Las Atalayas (Figure 18.5d) undercut pedestals vary from 0.5 m high near the slope top to over 1 m at the mid-slope location.

The high magnetic susceptibility of the soil on the mound and uniformly low susceptibility of the mound base and surrounding de-vegetated marl apron also suggest they are caused by differential erosion. One possible cause of this is the fan-shaped leaves which overlap and are so densely packed that intercepted throughflow cannot occur and high infiltration capacity beneath the clump prevents overland flow being generated from stemflow. The palm clumps are the only shrub covering a significant portion of some of the marl slopes in this area, and nearby on the marl at Ardales, badlands are forming. *Chamaerops humilis* is not the only Mediterranean species associated with microtopographic changes: others include the olive, various grasses (including *Stipa* spp.) and the geophytes (e.g. *Hyacinthus* spp.). Most of these species are either obligate or faculatative resprouting pyrophytes. In other semi-arid environments species of *Acacia sericomopsis* and *Acocanthera* have been observed to form mounds or pedestals, and Dunne, Dietrich and Brunegg (1978) used exposed roots of these trees and bushes to estimate erosion rates. Carrara and Carrol (1979) have used exposed tree roots and dendrochronology to estimate slope erosion rates in a semi-arid environment, but as Dunne, Dietrich and Brunengo (1978) pointed out, considerable caution is needed with this method and it cannot be used with non-woody species or where resprouting occurs.

Measurements of infiltration capacity on the sandstone/conglomerate at Las Atalaya by R. J. Rice show higher rates under the *Pinus* woodland than under the matorral scrub. This difference along with the higher and undercut pedestals on the marl, the lower mounds on the sandstone/conglomerate matorral (and old olive groves) and the lack of features indicative of surface erosion under the pine woodland, reflect relative erosion rates in the area.

The differential effect of *Chamaerops humilis* and other species (within one vegetation type) on soil conditions including magnetic susceptibility and surface erosion, must be related to variations in litter chemistry, life-form, phenology and longevity. This means it is necessary to evaluate the contribution of different plant types to the overall percentage vegetation cover; the vegetation measure most commonly used in the prediction of erosion rates. A preliminary attempt was therefore made to assess the relative cover of more and less erosion-retarding plant types and the extent to which these types control the occurrence and linkage of gaps in the vegetation.

Plant Type, Pattern and Structure

Chaparral, and matorral are multi-layered vegetation types which normally include at least four strata: a lichen mat, litter layer (which can be counted as vegetation due to the large amount of attached litter); annual herbs; woody dwarf shrubs; and a tall shrub and/or a scrub tree layer. The spatial pattern is often relatively coarse requiring large quadrats (over 2 m × 2 m) especially if palms are present. Species also vary in their cover irrespective of size due to their phenology, leaf width, rigidity, etc. Since many of these characteristics are relatively constant within life-form categories the cover was calculated differently for each category. The life-form classification used here, although based on the traditional classes of Du Rietz (1931) and Raunkiaer (1934), is modified to include more information of relevance to soil erosion (Table 18.3). In particular, the palm and cacti bushes are distinguished as a separate class due to their rather different cover, litter and growth characteristics. Life-span is also important because, if, as is normally assumed, maximum litter accumulation occurs under plants (Francis *et al.*, 1986), then the longer a plant exists in one location, the greater its effect on soil organic matter is likely to be, other factors being equal. Many of the Mediterranean shrubs and trees are long-lived, especially *Chamaerops humilis* which can live to in excess of 400 years (Hodge, 1982) and possibly very much longer. This longevity is associated with an

Table 18.3 Modified life-form classes with an indication of ground cover, soil effects and resistance to stress

Class	Subclass	Ground cover	Effect on soil conditions	Resistance to grazing/fire	Typical life-span* (yrs.)	Example(s)
Phanerophytes						
	broad leaved	L-H†	M	L	M	*Cistus* spp.
	narrow leaved	L	M-H(+)‡	H	M-L	*Lygos* spp.
Chamaephytes						
	semi-shrubby	L-H	L-M	L	S	*Helianthemum* spp.
	creeping and decumbent	L-H	L-M	L	S-M	*Ecballium* spp.
	cushion plant	L-M	L-M	L-M	M	*Genista equisetifolium*
Palm and cacti bushes		M-H	H(+)	H	L-VL	*Chamaerops humilis* *Opuntia ficus-indica* *Agave americana*
Hemicryptophytes						
	tussock grasses	L-H	H(+)	M-H	M-L	*Stipa* spp.
	non-tussock grasses	L-H	L	L	S-M	*Brachypodium* spp.
	herbs	L	L	L	1-2	*Reseda* spp.
Geophytes		L	H(+?)	M-H	M	*Hyacinthus* spp.
Lichen		L-H	?	L	L?	*Lecanora* spp.

L = low, M = moderate, H = high; *S = short, M = medium, L = long, VL = very long (*eg* 10^3 yrs +).
† may vary to species to species or due to deciduous/evergreen habit.
‡ (+) = probably has a stabilizing effect.

adaption to stress by K-type species, and in some environments longevity may be effectively controlled by the erosion rate.

Due to the size and nature of palm clumps it is impractical to measure cover in traditional ways. Two methods have been used: one was to measure inter-palm distance and diameters, and the others was to measure the area of coverage directly from photos taken as near normal to the slopes as possible. Both gave similar cover values. The cover of woody shrubs was calculated using transect coverage and of annuals, litter and bare ground/stones using small round quadrats (0.5 m diameter) and pin-frames. Surveys were undertaken in May and September 1988. Table 18.4 shows that in May 9 per cent of the slope is bare ground and that approximately 25 per cent is covered by palm clumps, which is less than covered by chamaephytes. These figures highlight a major problem with the use of cover, as the magnetic data and microtopography indicate that the palm clumps have far more effect on surface erosion rates than other vegetation types despite similar cover values. It follows that the spatial pattern of erosion-inhibiting species and their population dynamics (in space as well as in time) will affect the distribution of erosion across a slope. A test of the spatial pattern of the palm clumps using the Hopkins coefficient of aggregation (Greig-Smith, 1983) indicates that they are not randomly but regularly distributed (at 0.05 probability level); the cause of this is not known, but it is probably due to hydrological and/or allelopathic factors. Although not planted on these particular slopes, they may have been planted in the past as a source of fibres for brush and rope making or for the rhizomes which can be eaten. The linear pattern seen across some slopes may be due to planting palm as barriers (as with *Opuntia ficus-indica*) or possibly to stabilize terrace risers.

Rather than rely on total cover (or biomass) as the vegetation variable in erosion studies, some account should be taken of the differential effects of species possibly using a modified life-form classification. In addition, other

Table 18.4 Percentage areas of each vegetation type covering the Las Atalayas site in May 1988

Class/Subclass		% cover	Density (per 100 m^2)
Microphanerophytes	*Olea europea*	12*	1*
Palm and cacti bushes	*Chamaerops humilis*	25	5.7
Chamaephytes		31.5	147 (68)
Geophytes		0.7	—
Lichen and litter[†]		23	—
Stones		10.5	—
Bare ground		9	—

( ) Value for burnt area in 1988.
*Overlapped cover in old olive grove area.
[†] + 100% little cover = approx. 230 g (based on 10 quadrats).

284

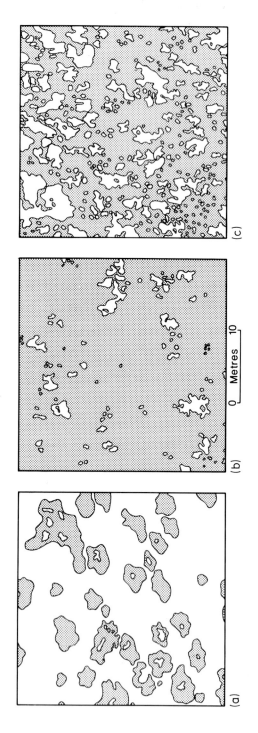

Figure 18.6 Typical cover patterns for Las Atalayas (a) taken from the large quadrat surrounding magnetic susceptibility transect d on Figure 18.1, and Rattlesnake Canyon matilija sandstone (b and c). Shaded area is dense all-year cover only. In each case top is upslope and bottom is downslope. Below are the relevant statistics:

(a) dense vegetation cover 25%, boundary 332%*, clump density 0.04 m².
(b) dense vegetation cover 94%, boundary 215%, gap density 0.10 m².
(c) dense vegetation cover 85%, boundary 630%, gap density 0.29 m².
* % of the plot perimeter.

pattern-dependent cover characteristics should be measured such as fragmentation or contagion. This necessitates the use of photography or field digitization allowing the calculation of boundary lengths as in Figure 18.6, or other measures of pattern.

CONCLUSIONS

The enhancement of magnetic susceptibility by chaparral and matorral fires may under certain conditions allow the direct estimation of post-fire sediment loss from slope soils and may be used to define spatial variations in post-fire erosion rates. Fire, although it is not in quantitative terms the major facilitator of erosion in either ecosystem studied, is of fundamental importance through its control on vegetation structure and dynamics. In the Mediteranean basin, although fire has largely been replaced by grazing it was probably more important in the past and may have played an important role in the development of the composition, structure and niche differentiation of matorral type vegetation (Neveh, 1975).

Spatial variations in the post-fire and longer-term magnetic properties of topsoil are related to the spatial pattern of vegetation. However, different types of vegetation have different effects probably related to growth-form and life-cycle factors. The common occurrence of vegetation mounds and pedestals, especially under *Chamaerops humilis*, is the most obvious reflection of the differential effects of vegetation on the spatial distribution of erosion. These differential effects are the result of the competitive success and longevity of dense clump-forming pyrophytes under heavy grazing pressure.

The inclusion of vegetation as total cover in geomorphic models or predictive equations has limited potential because of the differential effect of vegetation types and the importance of factors other than ground cover in the promotion or retardation of erosion. Apart from the methodological problems of measuring cover in chaparral and matorral ecosystems, cover does not take into account other aspects of plant eco-physiology which affect soil conditions and erosion. It is necessary to distinguish between different types of cover (using growth-forms and structure), persistence at one location (i.e. pattern) and sub-surface characteristics. Modified life-form classification includes much of this information and could be more widely used by geomorphologists. A distinction could also be made between species adaptation to erosion without influencing it (cf. passive succession) and those which do influence erosion rates (cf. active succession) either positively or negatively. In doing this it must be recognized that continuing erosion may alter plant growth rates, growth-form and reproductive strategy.

The practical measures used to combat erosion rarely rely solely on ground cover; instead they utilize specific plants which are well adapted to high erosion rates and have growth-forms reflecting this—examples include

Ammophila arenaria, and *Tamarix* spp. These species have growth and vegetative reproductive strategies (e.g. tillering) which stabilize the soil and decrease erosion and which may automatically lead to the development of a seed-bank and an increase in total vegetation cover. As biomass decreases, so pattern becomes more important in controlling the absolute erosion rate and its spatial distribution. It is the region between non-fragmented cover and low isolated clump or bush cover (i.e. approx. 50–20 per cent) that is critical as changes in dominant growth-form may lead to positive feedback which is responsible for the non-linear/exponential relationship between vegetation cover and erosion (Elwell and Stocking, 1976). Therefore the pattern, structure and dynamics of Mediterranean vegetation are key factors in controlling the rate and distribution of erosion at the plot or slope scale.

ACKNOWLEDGEMENTS

I would like to thank Dr E. A. Keller and the Chaparral Ecosystems Group for help with the chaparral work. In connection with the Spanish work I must thank Dr R. J. Rice for his help and an introduction to the area and to Dr E. Barahona for useful discussions. The work reported here was partially funded by the University of Leicester Research Board.

REFERENCES

Bazilivich, N. I., L. Y. Rodin, and N. N. Rozov (1971). Geographical aspects of productivity. *Soviet Geography*, **12**, 293–317.
Best, D. W. and E. A. Keller (1986). Sediment storage and routing in a steep boulder-bed rock-controlled channel. In J. J. DeVries (ed.) *Proceedings of the Chaparral Ecosystems Res. Conference*, California Water Resources Centre, Rept. 62, 45–56.
Brown, A. G. (1988). Soil development and geomorphic processes in a Chaparral watershed: Rattlesnake Canyon, S. California. *Catena Suppl.* **12**, 45–58.
Carrara, P. E. and T. R. Carrol (1979). The determination of erosion rates from exposed tree roots in the Piceance Basin, Colorado. *Earth Surface Processes*, **4**, 307–17.
Davis, F. W., E. A. Keller, A. Parikh, and J. L. Florsheim (1988). *Recovery of the Chaparral Riparian Zone Following Wildfire.* Unpublished paper presented at the Californian Riparian Systems Conference, 22–23 Sept., Davis, California.
DeBano, L. F., R. M. Rice and C. E. Conran (1979). *Soil Heating in Chaparral Fires: Effects on Soil Properties, Plant Nutrients, Erosion and Runoff,* USDA Forest Service, Pacific Southwest Forest and Range Experimental Station, Research Paper PSW-145.
Du Rietz, G. E. (1931). Life-forms of terrestrial flowering plants. *Acta Phytogeographica Suecica*, **3**, 1–95.
Dunne, T., W. E. Dietrich and M. J. Brunengo (1978). Recent and past erosion rates in semi-arid Kenya. *Zietschrift für Geomorphologie Suppl. Bd.*, **29**, 130–40.
Elwell, H. A. and M. A. Stocking (1976). Vegetal cover to estimate soil erosion hazard in Rhodesia. *Geoderma*, **15**, 61–70.

Francis, C. F., J. B. Thornes, A. Romero Diaz, F. Lopez Bermudez and G. C. Fisher (1986). Topographic controls of soil moisture, vegetation cover and land degradation in a moisture stressed Mediterranean environment. *Catena*, **13**, 211–25.

Greig-Smith, P. (1983). *Quantitative Plant Ecology*. Studies in Ecology, vol. 9, 3rd edn, Blackwell.

Hodge, W. H. (1982). Goethe's palm. *Principes*, **26**, 194–9.

Keller, E. A. *et al.* (1985). *Rattlesnake Canyon*, Chaparral Watershed Group, University of California, Santa Barbara, unpublished report.

Keller, E. A., J. L. Florsheim and D. W. Best (1988). Debris flows and wildfire in the chaparral: Myth and reality. *Geol. Soc. America Annual Meeting Abstract*, No. 2361.

Langbein, W. B. and S. A. Schumm (1958). Yield of sediment in relation to mean annual precipitation. *Am. Geophys. Union Trans.*, **39**, 1076–84.

Mullins, C. E. (1977). Magnetic susceptibility of the soil and its significance in soil science—a review. *J. Soil Science*, **28**, 223–46.

Mutch, R. W. (1970). Wildland fires and ecosystems—A hypothesis. *Ecology*, **51**, 1046–51.

Naveh, Z. (1974). Effects of fire in the Mediterranean region. In T. T. Kozlowski and C. E. Ahlgren (eds), *Fire and Ecosystems*, Academic Press, New York, 401–34.

Naveh, Z. (1975). The evolutionary significance of fire in the Mediterranean region. *Vegetatio*, **29**, 199–208.

Raunkiaer, C. (1934). *The Life-Forms of Plants and Statistical Plant Geography* Clarendon Press, Oxford.

Rice, R. M., R. R. Zeimer and S. C. Hankin (1982). Slope stability effects on fuel management strategies—Inferences from Monte Carlo simulations. *Procs. Symp. on Dynamics and Management of Mediterranean Type Ecosystems*, June 1981, San Diego, CA. Gen. Tech. Paper PSW-58, Washington DC, Forest Service, US Dept. Agric., 365–371.

Taylor, B. D. (1983). Sediment yields in coastal southern California. *J. Hydrol. Eng. ASCE.*, **109**, 71–85.

Thompson, R. and F. Oldfield (1986). *Environmental Magnetism*. Allen and Unwin, London.

Thornes, J. B. (1985). The ecology of erosion. *Geography*, **70**, 222–36.

Thornes, J. B. (1988). Erosional equilibria under grazing. In J. Bintliff, D. A. Davidson and E. G. Grant (eds), *Conceptual Issues in Environmental Archaeology*, Edinburgh University Press, pp. 193–210.

Tite, M. S. and R. E. Linington (1975). Effect of climate on the magnetic susceptibility of soils. *Nature, London*, **256**, 565–6.

Trabaud, L. (1986). Fire effects on soils of the Mediterranean basin region. In P. J. Joss, P. W. Lynch and O. B. Williams (eds), *Rangelands: A Resource Under Siege*, Proc. 2nd International Rangeland Conference, Australian Academy of Sciences, Canberra 1986, pp. 582–5.

Whittaker, R. H. and G. E. Likens (1973). Primary production in the biosphere and man. *Human Ecology*, **1**, 357–69.

Yair, A., D. Sharon and N. Larce (1980). Trends in runoff and erosional processes over an arid limestone hillside northern region, Israel. Hydrol. Sciences Bull., **25**, 243–56.

19 The Use of Vegetation and Land Use Parameters in Modelling Catchment Sediment Yields

D. J. MITCHELL

School of Applied Sciences, The Polytechnic, Wolverhampton

SUMMARY

Vegetation cover is widely accepted as a significant parameter in the erosion and sediment yield of drainage basins. However, the derivation of numerical parameters for modelling have been poorly developed especially in the study of large catchments. Using selected catchments from the Welsh Borderland, UK, vegetation parameters have been considered in two forms. Firstly, as seasonal factors, derived from combined sine and cosine curves, representing both soil moisture and vegetation growth–decay processes respectively. Secondly, as land use factors based on parish agricultural statistics.

The combined trigonometric function performed well as an erosion potential parameter, providing a useful technique for estimating long-term sediment yields. Land use and land management ratios from parish agricultural returns provided easily derived basin statistics which have been compared with catchment sediment yields. Interestingly, stocking ratios were found to be more significant than arable/pasture ratios in the Welsh Borderland.

The suspended sediment yields of the Farlow Catchment are probably attributed as much to high stocking ratios as the extent of arable land. In two parishes of the Farlow Catchment it was found that stocking ratios have risen from two to more than nine cattle and sheep per hectare of pasture while arable/pasture and ley grass/permanent pasture ratios have declined in recent years. Densely stocked areas with moderately steep slopes are eroded due to

Vegetation and Erosion
Edited by J. B. Thornes
©1990 John Wiley & Sons Ltd

overgrazing, poaching and river bank damage. As a result of the increases in stocking ratios erosion due to domestic animals has become an important factor in catchment studies, especially in western Britain. Erosion is also increasing in lowland arable and mixed farming areas where large numbers of farm animals are translocated to be wintered on stubble and crop residues. Rates are particularly high due to sheep grazing between October and March in the Welsh Borderland.

INTRODUCTION

Vegetation and land use are important factors in respect to the hydrology and sediment production of catchments because they are more dynamic than many other factors, with short seasonal changes as well as long-term climatic or land use management changes.

Early basin studies produced conclusive data on the influence of forest on runoff and sediment load of rivers (Bates and Henry, 1928; Dils, 1957). Subsequently other studies have considered the variation of sediment load with respect to vegetation and land use management changes (Langbein and Schumm, 1958; Douglas, 1969; Imeson, 1970; Walling, 1971; Reed, 1971; Williams and Reed, 1972; Chernyshev, 1972; Newson, 1985). In addition rapid removal of vegetation by deforestation or fire has been used to emphasize the protective role of vegetative cover (Sinclair, 1954; Imeson, 1971; Cassells, Hamilton and Saplaco, 1983).

Although the results of deforestation tend to produce the most extreme sediment yields, the effects of land use management changes are far more relevant to a much wider area. In cultivated areas, soil compaction from farm machinery has an important influence on infiltration, increased surface runoff and hence soil erosion (Steinbrenner, 1955; Chernyshev, 1972; Fullen, 1985a). Similarly compaction can also occur in densely stocked grassland areas as result of poaching which reduces infiltration, especially during the wet season (Alderfer and Robinson, 1947; Tanner and Mamaril, 1959; Federer et al., 1961; Gifford and Hawkins, 1978). In a study of rangeland in Colorado, Lusby (1970) found that runoff and sediment yield from ungrazed catchments were 30 per cent and 45 per cent less respectively than from grazed catchments. In addition, poaching by cattle in the vicinity of stream channels was found to be a significant source of available suspended sediment in the Farlow Catchment (Mitchell, 1979).

Although it is widely accepted that vegetation cover is a significant factor in the erosion and sediment yield of drainage basins, numerical parameters have been poorly developed, especially for larger catchments. Detailed vegetation and land use variables are used in mathematical models for predicting sediment yield in plot studies and small catchment surveys but the application of vegetal factors in the analysis of larger catchments becomes exceedingly vague and generalized. The use of vegetation and land use factors in empirical equations for the prediction of sediment yield have been dominated by the 'cropping management factor' and the 'conservation practice factor' from the Universal Soil Loss Equation (Wischmeier, Smith and Uhland, 1958) with variable specialized parameters for the remainder of the equations.

The progression of models based on plot studies to catchment models is difficult and complex. Some models such as the USLE, derived for plot studies, have been modified for wider applications including variable source areas and catchment studies in selected regions (Onstad and Foster, 1975; Onstad, Piest and Saxton, 1976). An extension from plot studies to small catchments seems a reasonable progression but the application of the above techniques to larger catchments becomes more difficult.

The application of these two parameters from USLE to drainage basins in the United Kingdom does not seem very practicable, especially with insufficient empirical data from plot studies of contrasting crop types and management techniques, therefore some other methods may be more appropriate. Examining the concept of competition between erosion and vegetation growth, Thornes (1985) reviewed the contemporary approaches to modelling available and the idea of ecological models to evaluate the existing complex relationships. Development of mathematical models is dependent on the derivation of numerical parameters so that verification and application can be assessed. Using catchments of various sizes in the Welsh Borderland, numerical parameters have been calculated in two ways: firstly as seasonal factors and secondly as land use factors based on Ministry of Agriculture Fisheries and Food parish agricultural statistics.

STUDY AREA: CATCHMENTS IN THE WELSH BORDERLAND

Small unit areas can be investigated in detail over short time scales whereas in-depth investigations of large catchments are more difficult; therefore the intensity of field investigations varies inversely with area. With a full awareness of the interpretative problems of comparing small and large catchments, it is intended to make comparison between a detailed investigation of the Farlow Catchment (20 km²) (Mitchell, 1979) and the more generalized survey of the Wye Catchment (4010 km²) (Figure 19.1). Using results from 1974 data, a comparison has been made between the two catchments (Table 19.1) so that when cross-referencing is used, the similarities and differences can be appreciated.

The Wye Catchment

Vegetation and land use changes within the Wye Catchment have been extremely

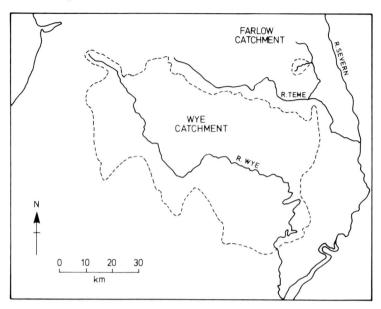

Figure 19.1　Location of the Farlow and Wye catchments

Table 19.1　Comparison between the Farlow and Wye catchments in 1974

Variable		Farlow catchment	Wye catchment
Area (km²)		20.138	4010.000
Relief Highest	(m)	503.000	762.000
Lowest	(m)	137.000	9.000
Range	(m)	376.000	743.000
Relief ratio		0.078	0.007
Geology %	Upper Palaeozoic	15.000	45.000
	Old Red Sandstone	85.000	55.000
Land use %	Rough grazing	13.600	14.700
	Pasture	55.800	57.000
	Arable	23.100	14.800
	Woodland & orchard	6.600	12.500
	Others	1.700	1.000
Hydrometry	Precipitation (mm)	829.000	1482.000
	Evapotranspiration (mm)	421.000(AE)	392.000(PE)
	Runoff (mm)	540.000	667.000
	Runoff coefficient (%)	65.000	45.000
Suspended sediment yield (t km⁻²)		44.280	51.700
Solute yield (t km⁻²)		107.380	76.000

AE: Actual evapotranspiration
PE: Potential evapotranspiration

important factors in determining the variation of erosion during the Holocene, as shown by rates of sedimentation (Bartley, 1960; Jones, Benson-Evans and Chambers, 1985). An examination of contemporary effects of vegetation cover on sediment yield cannot easily be used to reconstruct influences of earlier climatic changes but they are of greater use in estimating the influence of man's impact on the basin sediment system. Examples of present land use changes and land management can be used to suggest possible effects of changes in the vegetal cover, but it is important to consider that technological changes in equipment and techniques can present unrepresentative results.

The progressive decline in forest cover from more than 90 per cent during the Atlantic Period to less than 10 per cent at present has been instrumental in the erosional history because of the influence of trees on the water balance of the catchments and the protective role of the canopy cover and root systems. Re-afforestation and more recently monitored de-forestation of extensive areas in central Wales have provided a unique opportunity to assess the relative effects of trees and grassland on water yield and catchment erosion. The establishment of research catchments by the Institute of Hydrology to investigate the effects of afforestation took advantage of the contrasting land use management of the grassland Upper Wye and afforested Upper Severn catchments. Results from the Plynlimon catchment (Newson, 1979, 1985) compared well with similar studies in the USA (Bates and Henry, 1928), indicating distinct reductions in runoff from the forest catchment due to interception and evapotranspiration and a reduction in peak flow and rise times (Table 19.2). In normal circumstances decreased runoff would lead to a reduction in erosion but for the consequence of the preparation of land for afforestation in central Wales. Using results from the Cyff (Upper Wye) and Tanllwyth (Upper Severn) catchments of Plynlimon, Painter et al. (1974) suggested that forest ditching has an immediate effect on sedimentation, giving a 1000-fold increase in total sediment yield compared with the pre-ditching situation. As a consequence of afforestation, erosion of

Table 19.2   Effect of afforestation on runoff characteristics, 1970–77

Land use	Wye	Severn
Grassland	98%	40%
Forest	2%	60%
Precipitation mean (mm) (P)	2348	2213
Runoff (mm) (Q)	1944	1364
Runoff coefficient	83%	62%
$P - Q$ mm	405	849
Av. rise time (hr)	2.0	2.3
Peak flow	100%	66%
Av. suspended sediment (mg l^{-1})	38	69
Bedload (m^3 yr^{-1})	2.5	8.4

From Newson (1979, 1985).

suspended sediments and bedload on the Upper Severn is greater than in the Upper Wye (Table 19.2) (Newson, 1979, 1980, 1985).

Besides the disruption in the afforested plantations of the upper catchments, increased land drainage and ploughing for reseeding grassland has caused a dramatic change in the land use (Howe, Slaymaker and Harding, 1967; Slaymaker, 1972). Using results from 1911 to 1964, Howe *et al.* (1967) showed that flood heights increased at Hereford on the River Wye and Shrewsbury on the River Severn. A Shrewsbury flood height of 5.1 m was expected once in 25 years during the period 1911–40, but in the subsequent period 1940–64 the same height had an occurrence of 1 in 4 years. This increased occurrence of flooding can be associated with the extension of drainage density due to forest ditches and land drainage in the upper catchments, permitting more rapid runoff of precipitation from upland areas. These results from headwater basins in contrast with larger catchments serve to indicate the problem of scale with the increase in complexity of catchments (Arnett, 1978). In an examination of long-term changes in discharge of the River Severn with changes in land use, Thornes (1983) concluded that 'although we can detect changes relatively easily, explaining them is significantly more difficult'. By considering the distance from the channel of various land use changes taking place there is the need to examine more carefully the hydrological sensitivity of land use changes.

In lowland catchments field drainage leads to similar conditions as found in the uplands. In a theoretical model prepared by Gregory (1971) it is implied that an extension of artificial drainage systems should lead to a reduction in drainage density and hence channel erosion, but where field drains replace slow infiltrating soils, such as those of the Bromyard series, runoff is more rapid, leading to higher discharge and greater erosion of the remaining stream channels (Mitchell, 1979). In addition localized outfall from field drains in the Farlow Catchment leads to extensive erosion by back cutting in a highly entrenched channel network. With the problem of increased runoff and risk of flooding, routine management of river banks coupled with extensive tree clearance has become a major function of river authorities. Erosion has become a serious problem in many middle and lowland sections of rivers due to increased flow rates. In the early 1950s the former Lower Wye Drainage Board carried out an extensive tree clearance scheme along the Monnow to increase the flow of the river, reducing the risk of flooding (Fryer, 1960). Increased velocity of the river has caused an even greater increase in bank erosion. In parts of the Middle Monnow Valley the extent of the erosion can be plotted by lines of alder *Alnus* marking the original channel with unstable banks more than 36 m from the original channel.

Intensification of farming techniques has developed in the piedmont and lowland areas of the Wye Catchment. As a result of continuous cultivation on sandy soils in certain areas, soil erosion has become a more significant factor due to the compaction of soils with low organic content, as described in

Shropshire by Reed (1979, 1983) and Fullen (1985a,b). High sediment yields occur from catchments with steep slopes composed of soils of high erosion potential. Although not as obvious as in semi-arid areas of Australia or America, stocking ratios and animal husbandry can be influential factors in the acceleration of erosion. Thomas (1964) surveyed an area of 3.64 km^2 of Plynlimon and estimated 0.18 km^2 (5 per cent) was affected by sheet erosion induced by sheep, with a total of 0.70 km^2 (19 per cent) likely to be eroded if sheep grazing remained unrestricted. Lowland areas with a high arable/pasture ratio tend to have lower stocking ratios but in most cases domestic animals are kept on steeper slopes and marshy lands near river courses. Bare soil on steep slopes and bank instability due to animals, especially cattle, sheep and rabbits, were found to be significant factors in channel and slope erosion of the Farlow Catchment (Mitchell, 1979).

The Farlow Catchment

Detailed vegetation and land use of the Farlow Basin was mapped on a scale of $1:2500$. In respect to the Farlow Basin, land use has an important influence on the suspended sediment yield. In autumn and winter arable land is vulnerable to increased erosion compared with grass and woodland, while in summer, when the whole catchment is vegetated, sediment production is more uniform. These differences are more noticeable in respect to the distribution of arable land within the catchment which lead to large variations in sediment rating curves of sub-catchments. In places sheet erosion and rill development occur on arable land as a result of poor farming practice, similar to conditions described by Chernyshev (1972). As a consequence of soil erosion, sediment rating curves, especially of winter data, tend to be much steeper in arable areas compared with pastoral catchments (Table 19.3) (Mitchell, 1975; Mitchell and Gerrard,

Table 19.3 Suspended sediment rating curve constants and coefficients compared with land use of the Farlow Catchment, Shropshire

Subcatchment	Major land use	Suspended sediment rating curves, constants and coefficients					
		Winter			Summer		
		a	b	r	a	b	r
Ingardine Brook	Arable/ley grass	1.245	1.234	0.801	2.592	1.012	0.814
Wheathill Brook	Arable/ley grass	0.206	1.244	0.817	2.353	0.631	0.613
Silving Brook	Pasture/moorland	9.972	0.590	0.547	3.024	0.862	0.702
Cleeton Brook	Moorland	3.459	0.757	0.663	7.124	0.601	0.732
Shirley Brook	Moorland	0.528	0.548	0.553	4.041	0.132	0.178

Suspended sediment concentration $= a\,Q^b$.
$r =$ Correlation coefficients.
From Mitchell (1975).

1987). Cultivation leads to an increase in sediment yield especially in winter but the nature of stream bank vegetation is an equally important factor influencing the yield from the Farlow Catchment, comparing well with the findings of Hadley and Schumm (1961) in South Dakota and of Zimmerman, Goodlett and Comer (1967) in Vermont.

Along deeply incised stretches of the Farlow Brook and its tributaries instability of riparian vegetation has resulted in tree falls and accumulation of extensive channel debris. The effects of this debris on sediment supply vary considerably. In places wind-thrown trees have diverted streams, causing rapid erosion, while in other areas debris dams form natural 'sediment traps' reducing the load of the river. The interrelation between the vegetation components operating in the Farlow Catchment is summarized in Figure 19.2, indicating the possible effects of vegetation and land use on basin yield.

SEASONAL PARAMETERS

Three different methods have been largely used to numerically evaluate seasonal effects. The first is the identification of standard periods have been used in the estimation of sediment yield (Dragoun, 1962; Jones, 1966; Mitchell, 1979) and the calculation of sediment rating curves (Mitchell, 1975, 1979; Walling, 1977; Loughran, 1977). The second method incorporates air temperature as a seasonal factor (Flaxman, 1972), providing a readily available parameter, especially if distinct temperature regimes exist. The third method, which is the most suitable for application to computers, is the use of day factors (Benson, 1962; Keller, 1970; Walling, 1974). In sediment yield models based on Chestuee Creek, Tennessee Valley Authority, Benson (1962) numerically represented the months by arbitrarily setting January as 30 degrees adding 30 degrees for each subsequent month. Using stepwise multiple regression, Keller (1970) analysed the water quality of five small catchments in the north-east Pre-Alps of Switzerland. As well as water temperature he found that a seasonal factor

$$S = \sin \left[\frac{(\text{day of year}, 4\pi)}{365} + 5.38 \right]$$

was a significant parameter. In a detailed statistical study of a small catchment near Exeter, prior to urbanization, Walling (1974) used both cosine day and sine day factors to provide measures of seasonal effects relating to calendar year and water year respectively.

In British catchments two seasonal features are normally present: firstly changes in soil moisture storage and secondly growth and decay of vegetation. In order to model these two features on a daily basis the distribution of numerical values on trigonometrical curves using $2\pi d/365$ (radians) to calculate the angle

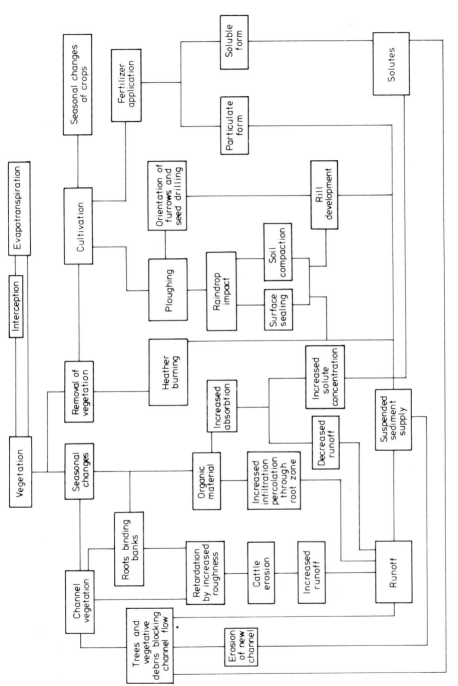

Figure 19.2 Vegetation components operating in the Farlow catchment (Mitchell, 1979)

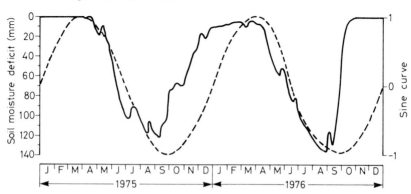

Figure 19.3 Seasonal changes in soil moisture deficit for the Wye catchment compared with the sine curve

in comparison with the number of days (*d*) from the beginning of the year is used. Soil moisture is best represented by the sine curve (SIN) with the maximum soil moisture occurring during the winter and the minimum at the end of the summer so that the winter period October to March has an increasing soil moisture curve, and the summer, April to September, a continuously decreasing value. Using average weekly values for the Wye Catchment for 1975 and 1976, soil moisture deficit (SMD) compares well with the sine curve (Figure 19.3). Although these two years were particularly noted for their drought conditions, they provide relatively smooth soil moisture depletion curves with measured values throughout the joint winter period. In spite of a relatively good comparison between the sine curve and increasing SMD, a distinctive displacement of the two curves results during September and October. Using this relationship, the greatest runoff and surface erosion will theoretically occur during late winter when maximum saturation of the soil and aquifers occur. The sine curve can also be equated with the annual cycle of the base flow although, depending on the water storativity and transmissivity of rocks, a greater temporal displacement in the curve exists.

The protective effect of vegetation is best represented by the cosine curve (COS), which is the converse of the vegetation growth and decay curve. Using data obtained in southern England by Anslow and Green (1967), the protective effect of seasonal dry matter production of tall fescue *Festuca arundinacea* has been compared with the cosine curve (Figure 19.4). Although the reversed representation of the increase in dry matter during spring and early summer relates closely to the cosine curve, the natural decline in productivity in late summer and autumn does not influence the protective role of the vegetal cover unless the grass is cut or grazed. Using the cosine curve the maximum value occurs at the beginning of the year when the protective role of vegetation has

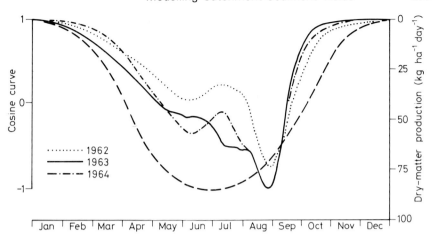

Figure 19.4 Seasonal changes in dry-matter production of tall fescue *Festuca arundinacea* (adapted from Anslow and Green, 1967), compared with the cosine curve

a minimal influence over catchment yield. The lack of vegetation reduces interception and evapotranspiration, resulting in high runoff conditions on soils without a protective covering and roots to bind the soil. In contrast the minimum impact occurs at the beginning of July when evaporation losses and vegetation cover reduces and modifies runoff and erosion. Similarly the uptake and release of solutes by vegetation will also respond to the cosine curve although a temporal displacement may occur to account for the growth and decay processes.

To take account of both these conditions a combined curve (COSSIN) which is the locus of the mid-point between the sine and cosine curves displaced by unity has been generated using

$$\text{Dayfactor} = \frac{1}{2}\left[\left(\cos\frac{2\pi d}{365} + 1\right) + \left(\sin\frac{2\pi d}{365} + 1\right)\right]$$

which will give a maximum value of 1.71 on day 45 (14th February) and a minimum of 0.29 on day 225 (12 August). This locus compares well with the curve generated from the gradients of monthly suspended sediment—discharge rating curves derived from point samples at Redbrook, the lowest gauging station on the River Wye (Figure 19.5). These three seasonal parameters COS, SIN and COSSIN were used with 33 hydrometric variables to construct a series of multiple regression models for the prediction of daily suspended sediment yield. In order to construct and calibrate these statistical models, data from the Wye Catchment above Redbrook (4010 km²) and nine sub-catchments were used for a seven-year period (1974–80). The COSSIN parameter had the highest coefficient ($r = 0.616$) in a Pearson correlation matrix with suspended sediment estimated by multi-rating curve method for the Wye at Redbrook, improving relationships with the

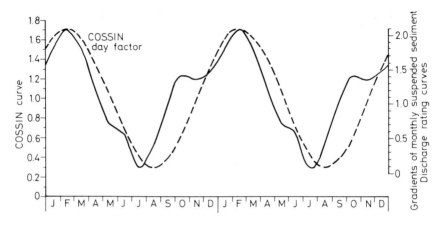

Figure 19.5 A comparison between gradients of monthly suspended sediment–discharge rating curves for the River Wye at Redbrook and the COSSIN curve

cosine curve ($r = 0.581$) and leaving the sine curve parameter, reflecting soil moisture and groundwater, noticeably insignificant ($r = 0.291$). Although seasonal factors for the sub-catchments showed variable format, COSSIN indicated the best response for the combined stepwise multiple regression model with an exponent of 0.3179. With a maximum value of 1.7071 on 14 February for the COSSIN curve, the numerical weighting of the factor increases suspended sediment yield by 3.500 times compared with 1.2360 times for the lowest value on 12 August. The same seasonal parameters were converted into monthly parameters by recording the value of the mid-point of each month. Using nine hydrometric variables and three seasonal factors the best stepwise multiple regression equation for predicting monthly suspended sediment yield gave an exponent of 0.5889 for the COSSIN parameter, providing a multiplier of 10 during February compared with only 1.5 in August (Table 19.4). As these seasonal factors are only dependent on the input of the day number from 1 January, they can be used for long-term estimates of sediment yield as long as the associated conditions remain in phase with the current climatic regime. Using monthly precipitation, runoff and the COSSIN seasonal factor for six catchments with hydrometric data for up to 48 years, estimates of long-term suspended sediment yield have been made (Table 19.5).

LAND USE DATA

Although seasonal curves can be used as percentage cover or erosion potential parameters, more detailed land use data are needed to assess the true impact of land use and land management. National surveys have provided useful

Table 19.4 Monthly multiple regression model for the prediction of suspended sediment yields of catchments of the Wye Basin

$$SSA = 1.0016 \frac{P^{0.7801} \ Q_{AV}^{0.7845} \ Q_{MAX}^{0.1728} \ Q_{INTER}^{0.049} \ 10^{0.1311COS} \ 10^{0.5889COSSIN}}{PPE^{0.0246} \ Q_{MIN}^{0.5362}}$$

Multiple correlation coefficient 0.895

SSA	= Suspended sediment yield (kg km^{-1})
P	= Precipitation (mm)
PPE	= Effective precipitation (precipitation – potential evapotranspiration, mm)
QAV	= Average monthly discharge (m^3 s^{-1})
QMAX	= Maximum monthly discharge (m^3 s^{-1})
QMIN	= Minimum monthly discharge (m^3 s^{-1})
QINTER	= Interflow monthly total (mm)
COS	= Monthly cosine seasonal parameter
COSSIN	= Monthly cossin seasonal parameter

Table 19.5 Estimated long-term suspended sediment yield using precipitation, runoff and COSSIN seasonal parameter for selected Wye catchments

Catchment	Period (years)	Total yield (t km^{-2})	Monthly mean (t km^{-2})	1974–80 estimated as % of observed
Wye at Ddol Farm	43	630	1.22	89
Wye at Erwood	43	933	1.81	82
Wye at Belmont	48	1254	2.18	78
Lugg at Lugwardine	41	873	1.77	87
Monnow at Grosmont	32	898	2.34	85
Wye at Redbrook	48	1311	2.28	86

information for land use assessment, usually associated with some specialized investigation. Using field records based on a 1:63 360 map, Davies (1936, 1940) produced a grassland map of Wales on a scale of 1:253 440. More recently the Soil Survey has followed this early survey with a map showing the soil suitability for grassland at a scale of 1:2 000 000. The Soil Survey of England and Wales has also prepared a detailed national map of a bioclimatic classification based on thermal, moisture and site exposure categories which influence plant growth (Bendelow and Hartnup, 1980). The scale of these surveys tends to limit their use to generalized observations covering the whole catchment. At larger scales the most comprehensive published survey is the first Land Utilisation Survey of Britain with land use mapped on a scale of 1:63 360 from 1931 to 1938. Associated with this survey, county reports as a part of *The Land of Britain* (Stamp, 1936–46) were published for the five counties covering the Wye Catchment. Although the Second Land Utilisation Survey (Coleman, 1970)

Table 19.6 The annual variation of the agricultural statistics for the parish of Clyro, 1974–1979

	1974		1975		1976		1977		1978		1979		1974–79 average
	ha	%	ha	%	ha	%	ha	%	ha	%	ha	%	
Wheat	113		153		155		108		184		219		155
Barley	281		224		216		207		208		274		235
Oats	98		80		111		79		55		25		75
Mixed corn	8		10		1		1		2		2		4
Potatoes main	11		10		9		18		16		12		13
Sugar beet	0		0		0		41		44		0		14
Other crops	79		69		47		45		42		43		54
Bare fallow	14		21		6		23		7		4		13
TOTAL ARABLE	604	21.6	576	20.7	545	19.6	521	18.5	559	18.4	579	20.8	564 19.9
Temporary grass	511		578		554		524		534		418		520
Permanent grass	1 519		1 450		1 513		1 605		1 575		1 601		1 544
Rough grazing	99		124		98		94		94		98		83
TOTAL PASTURE	2 129	76.4	2 028	73.0	2 067	74.3	2 129	75.4	2 109	73.3	2 019	72.8	2 080 74.2
Woodland	36		34		50		50		46		43		43
Orchard and small fruits	3		2		4		4		1		1		3
Other land	15		15		21		27		68		35		30
TOTAL AREA	2 784		2 777		2 781		2 822		2 876		2 774		2 802

Arable–pasture ratio	0.283	0.284	0.264	0.245	0.265	0.287	0.271
Total cattle and calves	3 385	3 287	3 366	3 431	3 433	3 536	3 406
Total sheep and lambs	16 564	15 906	15 775	14 625	15 109	14 497	15 413
TOTAL CATTLE & SHEEP	19 949	19 193	19 141	18 056	18 542	18 033	18 819
Cattle/sheep ratio	0.204	0.206	0.213	0.235	0.227	0.244	0.222
Cattle/ha pasture	1.590	1.621	1.628	1.612	1.628	1.751	1.638
Sheep/ha pasture	7.780	7.843	7.632	6.869	7.164	7.180	7.411
Cattle and sheep/ ha pasture	9.370	9.464	9.260	8.481	8.792	8.931	9.050

printed on a 1 : 25 000 scale would supersede the earlier survey in precision and relevance to present-day land use, unfortunately the survey has only been partly published.

Besides these national surveys a number of localized investigations into land use management and changes have been prepared as parts of special investigations. As a basis of the catchment studies by the Institute of Hydrology the 'natural' vegetation and artificial influences have been mapped by the Environmental Information Service, the Institute of Terrestrial Ecology and Newson in the Upper Wye above Cefn Brwyn Gauging station (Newson, 1976). As an investigation in the land use changes, especially rough pasture, improved farmland and woodland between 1948 and 1983 in Mid-Wales uplands, Parry and Sinclair (1985) examined a large proportion of the upland catchments west of the River Wye with two detailed sample areas in the Upper Elan Catchment and the Ifron Catchment. In this 35-year period they found the main overall change in land use was the increase in coniferous plantation (3 per cent to 25.5 per cent) at the expense of rough pasture (78 per cent to 55.4 per cent), but except for the south-west portion of the Irfon Catchment and isolated afforestation schemes, the majority of these changes have taken place outside the Wye Catchment.

Although these surveys provide spatial representation of land use and land management, areal variables are required to generate mathematical models, and land use data have been obtained from the Ministry of Agriculture, Fisheries and Food agricultural parish returns. Despite some inaccuracies (Clark, Knowles and Phillips, 1983) associated with holdings located in more than one parish, the agricultural returns provide a detailed set of land use and stocking data. Also the availability of these data, although in variable format, since 1866 provides an historical data set for long-term erosion assessment. The 318 parishes of the Wye Basin were sub-divided according to the catchment areas. Where parishes overlapped across the Wye watershed and inter-catchment boundaries the data were then allocated proportionately to each area, assuming that the land use was homogeneously distributed throughout each parish. The land use parameters for this investigation have been obtained from the census returns for 1977 to represent the 1974–80 period. Although slight variations occur from year to year, generally farmers tend to practice crop rotation so that basic farm types and crop areas remain fairly constant in the short term, as illustrated by the agricultural statistics for Clyro for 1974 to 1979 (Table 19.6). In some of the upland catchments where large areas of non-farmland exist, adjustments are required to account for forestry, common land and other uses.

Eight land use parameters were derived from the parish data for the ten catchments of the Wye and compared with average annual suspended sediment yield for 1974–80 (Table 19.7). The generally low proportion of arable land in the Wye Catchment and hence the low arable/pasture ratio reduces the numerical association with sediment yield as shown in the correlation matrix

Table 19.7 Suspended sediment yields and land use parameters for the Wye catchments 1974–80

Variable	Ddol Farm Wye	Climery Irfon	Disserth Ithon	Erwood Wye	Belmont Wye	Butts Bridge Lugg	Lug-Wardine Lugg	Yarkhill Frome	Grosmont Monnow	Redbrook Wye
Dependent										
SS (t km^{-2})	17.0650	34.5870	38.9750	28.4950	35.2540	35.6830	26.7910	12.0230	36.1430	32.7320
ORG (t km^{-2})	2.1040	2.7680	1.2340	7.4100	6.5460	0.5100	0.7230	0.1140	0.7840	6.1110
INORG (t km^{-2})	14.9610	31.8200	37.7410	21.0860	28.7080	35.1730	26.0680	11.9090	35.3590	26.6220
Land use										
Arable	0.0355	0.0196	0.0460	0.0358	0.0728	0.2283	0.2163	0.2930	0.1502	0.1471
Pasture	0.9278	0.6605	0.7763	0.8211	0.7966	0.6979	0.5707	0.5754	0.6057	0.7025
Woodland	0.1352	0.3020	0.0980	0.1343	0.1231	0.0267	0.0349	0.0649	0.0187	0.1240
Arable/Pasture R	0.0380	0.0300	0.0590	0.0440	0.0910	0.3270	0.3790	0.5090	0.2480	0.2090
Cattle/ha. Pasture	0.3880	0.4330	0.7350	0.5950	0.8260	1.3350	1.4720	1.8420	1.2210	1.1830
Sheep/ha. Pasture	6.8260	11.9250	9.2670	9.3220	8.7070	7.7680	7.0480	4.1550	7.0790	7.2670
Cattle & Sheep/ha	7.2140	12.3580	10.0020	9.9170	9.5330	9.1030	8.5200	5.9970	8.3000	8.3900
Cattle/Sheep R	0.0568	0.0363	0.0793	0.0639	0.0948	0.1719	0.2089	4.4433	0.1725	0.1628

Table 19.8 Correlation matrix for suspended sediment and land use variables of the Wye catchments

Independent variables	Dependent variables		
	Log SS	Log organic SS	Log inorganic SS
Arable	−0.386	−0.771	−0.237
Pasture	0.036	0.610	−0.126
Woodland	0.098	0.532	−0.008
Arable/pasture ratio	−0.438	−0.801	−0.281
Cattle/ha pasture	−0.313	−0.722	−0.174
Sheep/ha pasture	0.694	0.621	0.609
Cattle and sheep/ha pasture	0.719	0.537	0.677
Cattle/sheep ratio	−0.616	−0.786	−0.486

Table 19.9 The suspended sediment yield of the Farlow Catchment 1972–75 (Mitchell, 1979)

Period	Suspended sediment yield (t km^{-2})		
	Organic	Inorganic	Total
1972–73			
Winter	20.54	63.66	84.20
Summer	55.25	77.53	132.78
Total	75.79	141.19	216.98
1973–74			
Winter	6.20	31.79	37.99
Summer	0.23	0.45	0.68
Total	6.43	32.24	38.67
1974–75			
Winter	4.53	19.92	24.45
Average annual	34.70	77.33	112.04

(Table 19.8). Most of the catchments except for the Lugg, Frome and the Wye below Hereford, are dominated by pasture, including rough grazing. If the spatial influence of the arable/pasture ratio is considered (Figure 19.6), the distribution of the high ratios can be equated with the high sediment yields in localized areas of the Lugg and Lower Wye valleys. As the Wye Basin is more noted as a stock-farming region, the stocking ratios provide a more significant relationship with sediment yield than arable/pasture ratios, as indicated by the higher positive linear correlation coefficient of $r = 0.719$ with total cattle and sheep per hectare of pasture. Furthermore, the logarithmic relationship between the two variables shows that the increase in suspended sediment yield is proportional to the stocking ratio with an exponent of 1.57 (Figure 19.7). The spatial distribution of the highest stocking ratios (Figure 19.8) emphasizes the importance of the

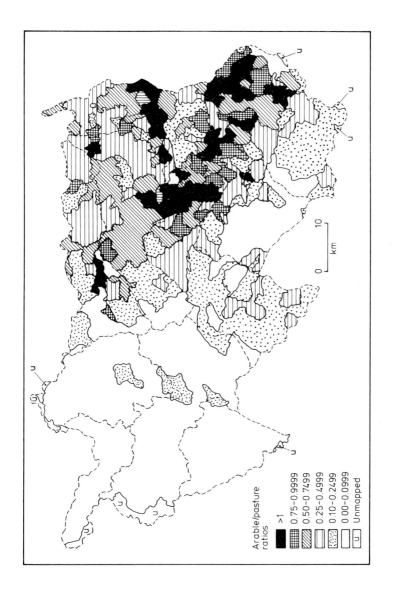

Figure 19.6 Arable/pasture ratios of the Wye catchment based on the MAFF parish agricultural statistics for 1977

Arable/pasture ratios

>1
0.75–0.9999
0.50–0.7499
0.25–0.4999
0.10–0.2499
0.00–0.0999
Unmapped

10

km

0

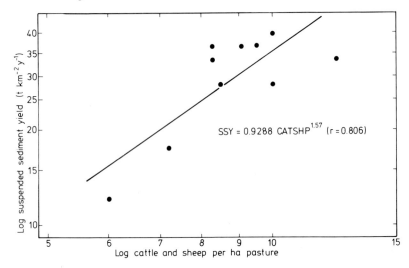

Figure 19.7 The relationship between cattle and sheep per hectare of pasture and average sediment yield for the Wye catchments

intermediate areas between the harsh uplands and arable farming lowlands, forming a large part of the Irfon, Ithon, Monnow and Upper Lugg catchments. These densely stocked catchments with moderately steep slopes are subjected to erosion due to close cropping of grass, poaching and river bank damage, resulting in high sediment yields. Furthermore, if these areas of high stocking ratios possess soils of high poaching risk, erosion due to animals will be even more significant.

The grassland suitability classes based on the work of Harrod (1979) and Harrod and Thomasson (1980) assessed the balance between potential grass yield and risk of poaching (Figure 19.9). Restrictions similar to the trafficability by machinery apply to poaching by animals; therefore soils are classed on an increasing risk scale of 1 to 5 (Rudeforth and Hartnup, 1985). If the poaching risk map (Figure 19.10) is examined in conjunction with stocking ratios (Figure 19.8) the results help to explain the high sediment yield in particular catchments which hitherto appeared anomalous (Figure 19.11).

The suspended sediment yields of the Farlow Catchment (Table 19.9) are probably attributed as much to high stocking ratios as to the extent of arable land. Using parish data for Wheathill and Silvington, which cover a part of the Farlow Catchment (Figure 19.12), the arable/pasture ratio, ley grass/permanent pasture ratio and stocking ratio (cattle and sheep/ha pasture) have been plotted from 1900 to 1981. Arable/pasture ratios remain low until the rapid increase during the 'ploughing up' scheme during the Second World War, after which ratios have remained fairly constant except for a marked

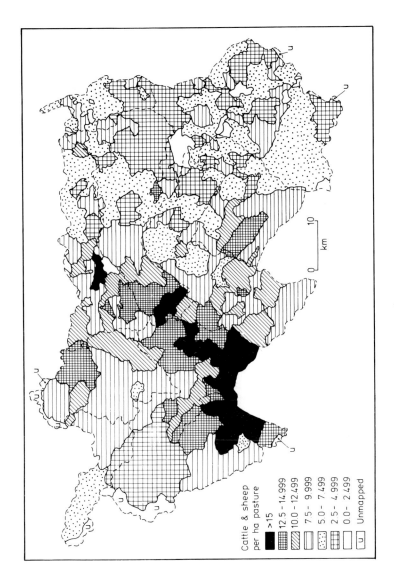

Figure 19.8 Stocking ratios of the Wye catchment. Cattle and sheep per hectare of pasture (including rough grazing) based on the MAFF parish agricultural statistics for 1977

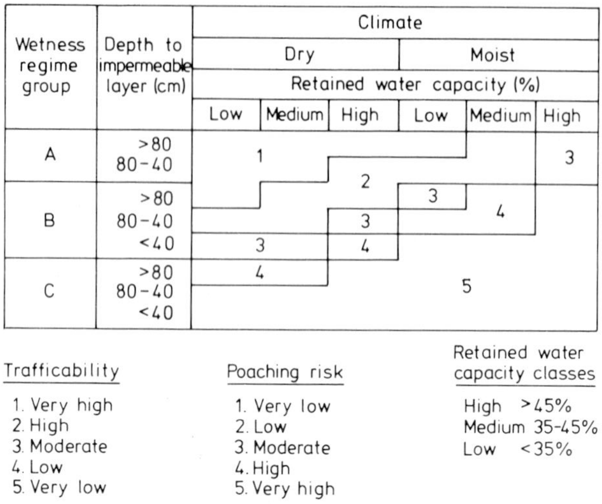

Wetness regime group	Depth to impermeable layer (cm)	Climate Dry			Climate Moist		
		Retained water capacity (%)					
		Low	Medium	High	Low	Medium	High
A	>80 / 80-40	1					3
			2				
B	>80				3	4	
	80-40			3			
	<40	3	4				
C	>80	4			5		
	80-40						
	<40						

Trafficability	Poaching risk	Retained water capacity classes
1. Very high	1. Very low	High >45%
2. High	2. Low	Medium 35-45%
3. Moderate	3. Moderate	Low <35%
4. Low	4. High	
5. Very low	5. Very high	

Figure 19.9   The influence of climate and soil moisture conditions on trafficability and poaching risk (from Harrod, 1979; Harrod and Thomasson, 1980)

reduction in the early 1960s. Grassland was dominated by permanent pasture until after the Second World War when the use of short leys increased, associated with arable crop rotation. Furthermore, with the demand for increased productivity, ley grass/permanent pasture ratios increased rapidly especially between 1955 and 1970 at the expense of arable crops, demonstrating a five-year cycle of reseeding. Since 1973 ley grass/permanent pasture ratios have declined to almost pre-war values, associated with the cessation of land improvement grants and the completion of farm improvement plans. The most significant feature in the changing land use of the two parishes of the Farlow Catchment is the large increase in stocking ratios since the war, rising from two to more than nine cattle and sheep/hectare of grassland. With a decline in arable/pasture and ley grass/permanent pasture ratios, catchment sediment yields are increasingly controlled by stocking ratios. Erosion due to animals is becoming a much more significant factor, especially in western Britain, but even in intermediate arable and mixed farming areas large numbers of farm animals, especially sheep, are translocated during winter months, that is during the greatest 'erosion period'. This transhumance, particularly practised in the Welsh Borderland, leads to increased erosion on even gradual slopes, especially when large flocks of sheep are wintered on stubble and the residue of other crops.

CONCLUSION

Vegetation and land use are clearly important factors controlling sediment yield, but the use of numerical parameters in statistical models has been poorly

Figure 19.10 Poaching risk on grassland in the Wye Basin based on data from the Soil Survey of England and Wales

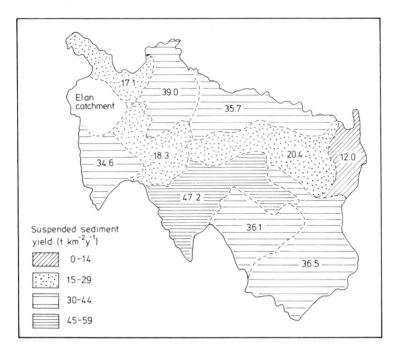

Figure 19.11   Average suspended sediment yields for the Wye catchments, 1974–80

developed, especially for larger catchments. Therefore in order to extend the use of vegetation and land use parameters in sediment yield models two methods have been proposed and applied to basin studies:

1. To take account of both seasonal changes in soil moisture and vegetation, a combined curve which is the locus of the mid-points of the sine and cosine curves, displaced by unity to give positive results, has been used. In subsequent statistical analysis this COSSIN seasonal parameter performed well and, assuming that the present seasonal cycle has remained relatively constant, the parameter has been invaluable in estimating long-term sediment yields.
2. To avoid cumbersome surveys of large catchments, land use and land management parameters have been derived from the parish agricultural returns of the MAFF. Despite some inaccuracies and some assumptions, the agricultural returns provide a detailed set of land use and stocking data. Although high arable/pasture ratios are associated with high sediment yield, stocking ratios were found to be more significant in the study catchments of the Welsh Borderland. Densely stocked areas with moderately steep slopes are subjected to erosion due to close cropping of grass, poaching and river

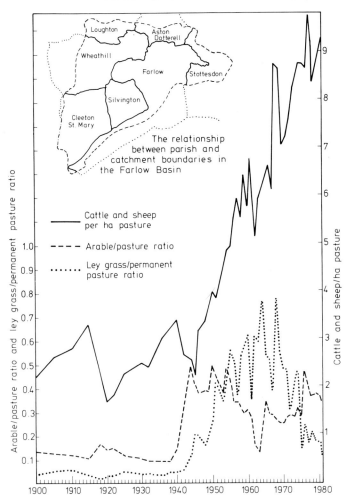

Figure 19.12   Land use changes of Wheathill and Silvington parishes, 1900–81, from MAFF parish agricultural statistics

bank damage. Furthermore, the historical database of the agricultural statistics indicates the dramatic increase in stocking ratios compared with other land use indices.

These two approaches for deriving vegetation and land use parameters for drainage basin studies have been applied to selected catchments in the Welsh Borderland but a wider application to a greater number of catchments with contrasting vegetal cover would provide a larger database, enabling further sediment yield models to be constructed.

ACKNOWLEDGEMENTS

The author is grateful to R. Hartnup of the Soil Survey and Land Research Centre, the Wye Division of the Welsh Water Authority, the Ministry of Agriculture, Fisheries and Food and the Institute of Hydrology for providing data; The Polytechnic, Wolverhampton for financial assistance; Dr T. Hocking, Dr M. A. Fullen for his comments, and Mr G. Cole for cartographic assistance.

REFERENCES

Alderfer, R. B. and R. R. Robinson (1947). Runoff from pastures in relation to grazing intensity and soil compaction. *J. Am. Soc. Agron.*, **39**, 948–58.

Anslow, R. C. and J. U. Green (1967). The seasonal growth of pasture grasses. *J. of Ag. Sc., Cambridge*, **68**, 109–27.

Arnett, R. R. (1978). Regional disparities in the denudation rate of organic sediments. *Zeit. Geomorph.*, Supp. 29, 169–79.

Bartley, D. D. (1960). Rhosgoch Common, Radnorshire; Stratigraphy and pollen analysis, *New Phytol.*, **59**, 238–263.

Bates, C. C. and A. J. Henry (1928). Forest and stream flow experiment at Wagon Wheel Gap in Colorado, *US Monthly Weather Review*, Supp. No. 36, 1–79.

Bendelow, V. C. and R. Hartnup (1980). *Climatic Classification of England and Wales*. Technical Monograph 15, Soil Survey of England and Wales, Harpenden.

Benson, M. A. (1962). Factors influencing the occurrence of floods in a humid region of diverse terrain. *US Geol. Survey Water Supply Paper 1580B*.

Cassells, D., L. Hamilton and S. R. Saplaco (1983). Understanding the role of forests in watershed protection. In R. A. Carpenter (ed.), *Natural Systems for Development*, Macmillan, London.

Chernyschev, Y. P. (1972). Hydrologic characteristics of soil erosion in the Central Chernozem Region. *Soviet Hydrology selected papers*, No. 4, 325–31.

Clark, G., D. J. Knowles and H. L. Phillips (1983). The accuracy of the Agricultural Census. *Geography*, **68**(2), 115–26.

Coleman, A. A. (1970). The conservation of wildscape: A quest for facts. *Geog. J.*, **136**, 199–205.

Davies, W. (1936). The grassland of Wales—A survey. In D. G. Stapledon (ed.), *A Survey of the Agricultural and Waste Lands of Wales*, Faber & Faber, London.

Davies, W. (1940). *The Grassland Map of England & Wales*, Vol. 48, Pt. 2, pp. 112–21.

Dils, R. E. (1957). *The Coweeta hydrologic laboratory Asheville N.C.*, South Eastern Forest Expt. Sta., p. 40.

Douglas, I. (1969). Sediment yields from forested and agricultural lands, *Proc. Symp. on 'The role of Water in Agriculture'*, University of Wales (Aberystwyth) Memorandum No. 12, E1–E22.

Dragoun, F. J. (1962). Rainfall energy as related to sediment yield. *J. Geophys. Res.*, **67**, 1495–501.

Federer, C. A., C. B. Tenpas, D. R. Schmidt and C. B. Tanner (1961). Pasture soil compaction by animal traffic. *Agron J.*, **53**, 53–4.

Flaxman, E. M. (1972). Predicting sediment yield in Western United States. *Proc. ASCE. J. Hydraulic Div.*, **98**(HY12), 2073–85.

Fryer, N. T. (1960). Farmland lost through river erosion. *Country Life*, 28 July 176–8.

Fullen, M. A. (1985a). Soil compaction, hydrological processes and soil erosion on loamy sands in East Shropshire, England. *Soil and Tillage Research*, **29**(6), 17–29.

Fullen, M. A. (1985b). Erosion of arable soils in Britain. *Intern. J. Environmental Studies*, **26**, 55–69.

Gifford, G. F. and R. H. Hawkins (1978). Hydrologic impact of grazing on infiltration: A critical review. *Water Resources Research*, **14**, 305–13.

Gregory, K. J. (1971). Drainage density changes in South West England. In K. J. Gregory and W. L. D. Ravenhill (eds), *Exeter Essays in Geography*, Exeter, pp. 33–53.

Hadley, R. F. and S. A. Schumm (1961). Sediment sources and drainage basin characteristics in the Upper Cheyenne River Basin. *US Geol. Sur. Water Supply Paper*, **1531**. 137–98.

Harrod, T. R. (1979). Soil suitability for grassland. In *Soil Survey Applications*, Soil Surv. Tech. Monogr. No. 13.

Harrod, T. R. and A. J. Thomasson (1980). *Grassland Suitability Map of England & Wales* 1:2M, Ordnance Survey, Southampton.

Howe, G. M., H. O. Slaymaker and D. M. Harding (1967). Some aspects of the flood hydrology of the upper catchment of the Severn and Wye. *Trans. I.B.G.*, **41**, 35–58.

Imeson, A. C. (1970). *Erosion in three East Yorkshire catchments and variations in dissolved suspended and bed load.* Unpublished PhD thesis, University of Hull.

Imeson, A. C. (1971). Heather burning and soil erosion on the North Yorkshire Moors. *J. App. Ecol.*, **8**, 537–42.

Jones, B. L. (1966). Effects of agricultural conservation practices on the Hydrology of Corey Creek Basin, Pennsylvania 1954–60. *US Geol. Surv. Water Supply Paper* 1532-C, 55.

Jones, R., K. Benson-Evans and F. M. Chambers (1985). Human influence upon sedimentation in Llangorse Lake, Wales. *Earth Surface Processes and Landforms*, **10**, 227–36.

Keller, H. M. (1970). Factors affecting water quality of small mountain catchments, *Proc. Wellington Symp., IAHS*, Publ. 97, Vol. 2, pp. 162–9.

Langbein, W. B. and S. A. Schumm (1958). Yield of sediment in relation to mean annual precipitation. *Trans. AGU*, **39**(6), 1076–89.

Loughran, R. J. (1977). Sediment transport from a rural catchment in New South Wales. *J. Hydrol.*, **34**, 357–75.

Lusby, G. C. (1970). Hydrologic and biotic effects of grazing versus non-grazing near Grand Junction, Colorado. *US Geol. Survey Prof. Paper*, **700B**, 232–6.

Mitchell, D. J. (1975). *The relationship between basin characteristics and sediment rating curves.* Unpublished paper given to the B.G.R.G. Basin Sediment Systems Meeting on 23 November 1975.

Mitchell, D. J. and A. J. Gerrard (1987). Morphological responses and sediment patterns. In K. J. Gregory, J. Lewin and J. B. Thornes (eds), *Palaeohydrology in Practice: A River Basin Analysis*, Chapter 9, Wiley, Chichester.

Newson, M. D. (1976). The physiography, deposits and vegetation of the Plynlimon catchments, *Institute of Hydrology Rept. No. 30*, NERC.

Newson, M. D. (1979). The result of ten years experimental study on Plynlimon, mid-Wales, and their imporance for the water industry. *J. Inst. of Water Engrs.*, **33**, 321–33.

Newson, M. D. (1980). The erosion of drainage ditches and its effect on bedload yields in mid-Wales, Reconnaissance case studies, *Earth Surface Processes*, **5**, 275–90.

Newson, M. D. (1985). Forestry and water on the uplands of Britain—the background of hydrological research and options for harmonious land use. *J. Forestry*, **79**, 113–20.

Onstad, C. A. and G. R. Foster (1975). Erosion modelling on a watershed. *Trans. ASAE*, **18**, 288–92.

Onstad, C. A., R. F. Piest and K. E. Saxton (1976). Watershed erosion model variation for southwest Iowa. *Proc. Third Fed. Inter-Agency Sed. Conf*, pp. 1.22–1.34.

Painter, R. B., K. Blyth, J. C. Mosedale and M. Kelly (1974). The effect of afforestation on erosion processes and sediment yield. *I.A.H.S.*, **113**, 62–7.

Parry, M. and G. Sinclair (1985). *Mid-Wales Upland Study*, Countryside Commission.

Reed, A. H. (1979). Accelerated erosion of arable soils in the United Kingdom by rainfall and runoff. *Outlook on Agriculture*, **10**, 41–8.

Reed, A. H. (1983). The erosion risk of compaction. *Soil and Water*, **11**, 29, 31, 33.

Reed, L. A. (1971). Hydrology and sedimentation of Corey Creek and Elk Run Basins, North-Central, Pennsylvania. *US Geol. Surv. Water Supply Paper*.

Rudeforth C. and R. Hartnup (1985). Grassland in Wales related to soils and climate. *BSSS* York, April 1985.

Sinclair, J. D. (1954). Erosion in the San Gabriel Mountains of California. *Trans. Am. Geophys. Un.*, **35**, 264–8.

Slaymaker, H. O. (1972). Patterns of present subaerial erosion and landforms in Mid-Wales, *Trans. IBG*, **55**, 47–68.

Stamp, L. D. (ed.) (1936–46). *The Land of Britain*, The report of the Land Utilisation Survey of Britain.

Steinbrenner, E. C. (1955). The effect of repeated tractor trips on the physical properties of forest soils. *Northwest Sci.*, **29**, 155–9.

Tanner, C. B. and C. P. Mamaril (1959). Pasture soil compaction by animal traffic. *Agron. J.*, **51**, 329–31.

Thomas, T. M. (1964). Sheet erosion induced by sheep in the Plynlimon area, Mid-Wales. *Brit. Geomrph. Res. Group Occ. Pub. 2*, pp. 11–14.

Thornes, J. B. (1983). Discharge: Empirical observations and statistical models of change. In K. J. Gregory (ed.), *Background to Palaeohydrology*, Wiley, Chichester, pp. 51–67.

Thornes, J. B. (1985). The ecology of erosion. *Geography* **70**(3), 222–35.

Walling, D. E. (1971). *Instrumented catchments in south-east Devon: Some relationships between drainage basin characteristics and catchment responses.* Unpublished PhD thesis, University of Exeter.

Walling, D. E. (1974). Suspended sediment and solute yields from a small catchment prior to urbanisation. *Trans. IBG Special Pub. 6*, pp. 169–92.

Walling, D. E. (1977). Limitations of the rating curve technique for estimating suspended sediment loads, with particular reference to British Rivers. *I.A.H.S.* Pub, **122**, 14–38.

Williams, K. F. and L. A. Reed (1972). Appraisal of stream sedimentation in the Susquehanna River Basin. *US Geol. Surv. Water Supply Paper*.

Wischmeier, W. H., D. D. Smith and R. E. Uhland (1958). Evaluation of factors in the soil loss equation. *Ag. Eng.*, **39**(8), 458–62, 474.

Zimmerman, C., J. C. Goodlett and G. H. Comer (1967). The influence of vegetation on channel form of small streams. *Symp. on River Morphology Intern. Ass. Sci. Hyd.*, p. 255.

20 Vegetation Cover Density Variations and Infiltration Patterns on Piped Alkali Sodic Soils: Implications for the Modelling of Overland Flow in Semi-arid Areas

HAZEL FAULKNER

Department of Geography, Polytechnic of North London

SUMMARY

An overland flow model for simulating event hydrographs was attempted for a semi-arid watershed on the western slope of Colorado. As part of the calibration of the hillslope generation stage of the model, an infiltration survey was undertaken on the watershed, which has complex vegetation density variations and piped alkali sodic parent materials. The sample survey of 24 sites was conducted by means of a double-ring ponding infiltrometer, and was designed to include a variety of estimated site cover density values, as well as slope angle, aspect and soil conditions. Results support the links suggested in many places in the literature between aspect and vegetation cover density and to a certain extent between cover type (but not density) and soil conditions. Some support for the usual assumption that vegetation cover density is directly linked to infiltration capacity rates can be gained from the results, in particular all bare sites had very low values of final infiltrability, f_c. Links between vegetation cover, soil piping and infiltration patterns are considered, and a comparison between infiltration rates and rainfall intensities of significant watershed events allows the runoff implications to be additionally explored. Support for the precalibration

Vegetation and Erosion
Edited by J. B. Thornes

method is finally obtained by comparing results of the simulation of a real event with field discharges monitored during the event.

INTRODUCTION

The influence exerted by vegetation over the intensity and frequency of overland flow production and surface wash erosion in semi-arid areas is well documented (Bryan and Campbell, 1986). In a general way, the nature of the cover influences surface litter extent and type, ultimately affecting the weathering matrix and the soil macropore arrangements, and several authors (e.g. Musgrave and Holtan, 1964) have demonstrated that infiltration rates are dramatically improved for soils with a significant humic horizon. Because of the additional protective role of vegetation in reducing effective rainfall intensity at the ground surface, runoff is less likely to be generated when cover is dense. In catchments with cover, these effects may vary seasonally with differing patterns of plant growth and evapotranspiration loss, and also spatially, because of slope angle and aspect effects (Hack and Goodlett, 1960). Consequently, the possibility that Hortonian overland flow never occurs in humid temperate areas has been suggested (Whipkey, 1969), and the view that Hortonian overland flow production is spatially uniform in semi-arid areas is now in doubt, too (Bryan and Campbell, 1986).

A variety of 'vegetation-based' models are available for assessing runoff production outside semi-arid areas (e.g. Gurnell and Gregory, 1987), but no general semi-arid model tackling spatial variability exists. Horton is still relevant, but the applicability of his infiltration theory in particular has been criticized for sites with a locally dense cover (Scoging, in Bryan and Yair, 1982), and on crusted sites (Romkens et al. 1986). Nevertheless, many other authors find that the Hortonian infiltration models holds up well for simulating *site* production of runoff in semi-arid areas (Al-Azawi, 1985; Davidoff and Selim, 1986), and his overland flow model has the great advantage of simplicity when compared with other more complex models (e.g. that of Richards, reviewed in Th. van Genuchten, 1987).

This chapter describes an attempt to incorporate some of the sources of spatial variability in infiltration rates into an areally distributed overland flow model for a small watershed in the Alkali basin in western Colorado which experiences high-intensity, short-duration summer storms of a variety of sizes and recurrence intervals. As a starting point, very simple assumptions were made concerning the relationship between runoff and vegetation, and these model assumptions were later modified ('recalibrated') by means of an infiltration survey of differing covers and soil types in an essentially Hortonian way. To a large extent the chapter focuses on the field calibration of the model rather than the model itself, since some of the model results have been published elsewhere (Faulkner, 1987).

After a description of the 'generation' phase of the model, and of the study area, infiltration curves under differing covers, soil and slope conditions are considered in the context of the intensity of significant watershed events, allowing the discussion to consider the runoff implications of these curves, and to suggest a method of precalibrating model runs.

THE HILLSLOPE GENERATION MODEL

To develop a catchment-scale areally distributed watershed model it was necessary to devise firstly an appropriate algorithm for the hillslope phase which will result in site hillslope hydrographs through runoff duration, T; and then secondly a suitable numerical scheme for routing these flows through the network to produce an output hydrograph. The latter stage in model construction is not considered in this chapter.

For the hillslope generation model, the watershed was divided into 103 sub-catchments, representing over 200 lateral contributing sideslopes. Using sites 44, 45 and 46 as examples, the approach adopted can be described using Figure 20.1. Firstly, the contributing sideslopes between adjacent network sites were identified (Figure 20.1a), then the runoff volume generated on each sideslope during a rainfall event with duration t_r and total rainfall of r cm was calculated by assuming initially that only bare areas generate overland flow and that on these sites runoff is 100 per cent. Figures 20.1(b) and (c) show that bare area can be calculated by overlaying a vegetation density map, (b), over the sub-catchment map (a). Over 500 contributing areas were separately calculated in this way, and between each site the incremental runoff volume generated, VOL in m, is simply

$$VOL = r\Sigma AB \qquad (1)$$

which is the area underneath the incremental hillslope hydrograph (Figure 20.1d). It is noted that this is a hydrograph summing *all* lateral contributions between network sites.

Secondly, the hillslope hydrograph base (runoff duration T) was calculated as the sum of rainfall duration t_r, and the time taken to drain the slope after cessation of rainfall, t_d:

$$T = t_r + t_d \qquad (2)$$

For contributing sideslopes with a maximum slope length of l metres (measured from Figure 20.1a), and assuming a mean hillslope velocity of 0.0457 m/sec (Emmett's (1978) mean overland flow velocity measured across a wide range of environments), then

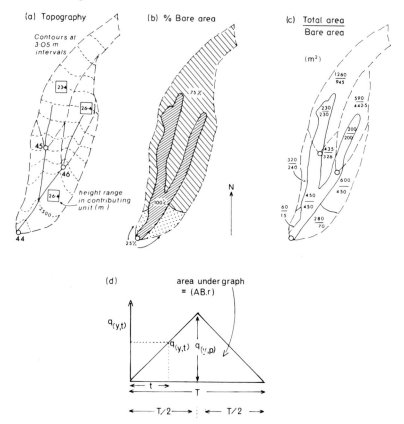

Figure 20.1 (a) to (c) Stages in the calculation of runoff volume between watershed sites; (d) the lateral hydrographs between sites. (For notation see text.)

$$t_d = l/v \qquad (3)$$

The final assumption was that of a simple triangular lateral input function (used by Ragan, 1966). Since we can calculate VOL and T, the lateral hillslope flow at any time, t, during the runoff event at site y can be calculated from AB and l at each site (Figure 20.1d). This gives us $q_{(y, t)}$, the lateral input in the network routing scheme for all y and all t, working on a data file of 103 site values of AB and l. Additionally, changing r and t_r allow events of a variety of rainfall totals and duration to be simulated.

Aspects of the Model to be Tested

The main assumptions in the model can be summarized as follows:

1. 100 per cent infiltration on covered sites, no infiltration on bare sites (unmodified VOL).
2. Instantaneous runoff at the start of rainfall (unmodified T).
3. Emmett's mean v appropriate to all events across the catchment.
4. A simple triangular shape approximates the hydrograph in the hillslope phase.

The fieldwork presented here explores the validity of the first two assumptions by means of an infiltration survey. This suggests a method for pre-calibrating model runs for events of particular rainfall totals and durations.

THE STUDY AREA

The study area (Figure 20.2) is a small (0.4 km²) sub-catchment of the US Forest Service 'Alkali Creek Soil and Water Project' basin in western Colorado. Alkali Creek flows into West Divide Creek, which joins the Colorado 20 km north, near Silt. This larger basin drains almost 300 km² and forms part of the variable relief plateau referred to locally as the western slope (inset on Figure 20.2). The main sequence exposed is the Wasatch formation, which consists of shales and claystones (Figure 20.3) interbedded with horizontal lenses of variable-textured arkostic, calcareous sandstones which appear locally well indurated. The Wasatch is notorious for erodible soils with high exchangeable sodium percentages (ESPs) which commonly form on this unit.

In the study area, elevation ranges from 2372 to 2560 metres a.s.l. and slopes in the area are generally steep throughout. Channel gradients vary from 0.01 to 0.25, but slope angle can rise to 0.5 in the headwater areas. Figure 20.4 shows a series of minor topographic undulations formed by the Wasatch sandstone ledges, and soils are more loamy when admixtures from the sandy units are involved. Consequently these sandstone units affect not only topography, but also soils and vegetation type. Three main soil variations occur on the clay shales and sandstone units of the area (Figure 20.3), occupying the headwaters, the shale slopes, and the main channel alluvium areas.

The headwater areas are vegetation-free in most locations; and most of the soils are Solonetz. This soil type has been defined by Kelley (1951) as a clay-rich, alkali–sodic or saline–alkali soil in which the relocation of colloids by leaching produces a dense subsoil horizon prone to swelling. They have a columnar structure, the columns being generally 3 to 5 cm in diameter and 10 to 15 cm deep in the study area (Figure 20.5a and b). Subsurface accumulation of soluble salts by leaching generally causes dispersion of the clays, and if a suitable hydraulic gradient and outfall site are available this leads to piping (Baillie *et al.*, 1986). In the study area, high ESPs were noted by Heede (1971), and he found that values of ESP over 12 rendered local soils liable to dispersion, flocculation, and piping. High ESPs have been shown to be responsible for

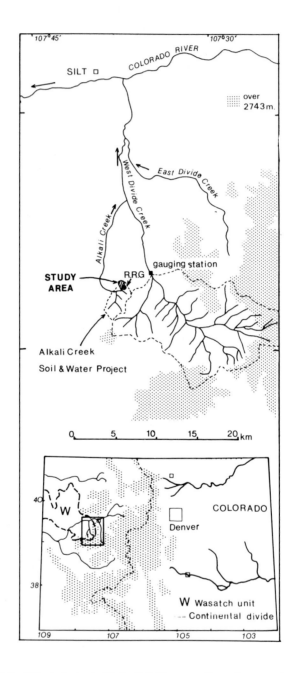

Figure 20.2 The study area. (*Note*: R.R.G. = recording rain gauge location.)

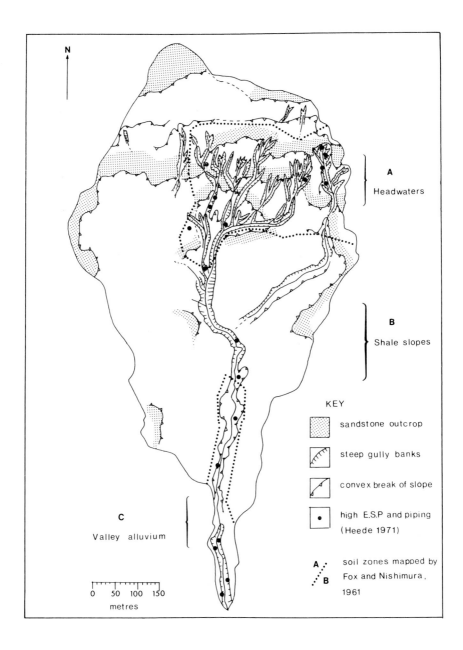

Figure 20.3 Topography, soil characteristics and location of piped sites in the study area

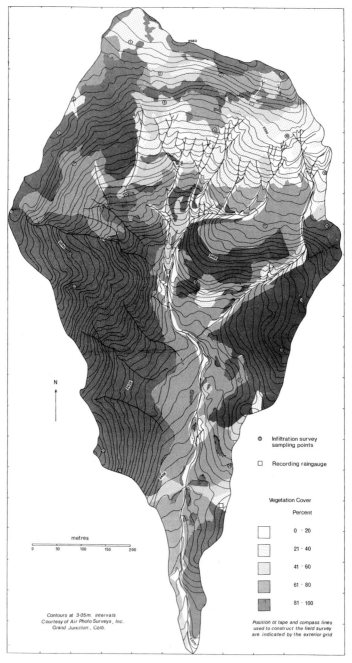

Figure 20.4 Vegetation density mapped on a cell basis, and the location of randomly selected infiltration sites

Figure 20.5(a)   Surficial hexagonal cracks on bare solonetz soils in the study area

overall infiltration reduction (Kazman, Shainberg and Gal, 1983), but it may be assumed that when pipes develop due to the vertical cracking common on dispersive clays, these pipes may be providing 'preferred pathways' for infiltrating water (Hillel, 1987) (Figure 20.5c). Thus high ESP soils might be expected to result in a high level of spatial variability in terms of final infiltrability.

Apart from gully bed and banks, most of the watershed outside the headwaters possesses a good vegetation cover. The Solentz soils which are present on these shale slope areas are therefore protected from high-intensity rainfall. The soils show no sign of piping here (Fox and Nishimura, 1957); rather the soil profile develops a thin but distinctive A horizon which seems to be associated with a reduction in ESP values. Soil depth in this sense appears to vary with aspect.

On the main channel alluvium, soils are better, drainage is improved, and some soils support western wheat or western bluegrass. Despite the cover and the improved nature of these alluvial materials, they can still pipe, and in these deposits Heede (1971) described an impressive pipe complex which extended

Figure 20.5(b)   Columnar structure, crust and vertical cracks of local soil

back into the main gully banks 5.5 metres which was 'associated with a dense vegetation cover' although 'no pipe inlets were found at a distance greater than 25 ft. (7.6 m) from the present edge of the gully'.

The vegetation cover within the watershed was mapped on a grid basis both by type and by percentage cover, this latter being assessed on the ground using the air survey as an aid. The map formed the basis of the areally distributed overland flow model (Figure 20.1) but additionally served as a basis for sampling randomly for infiltration. Using north–south compass bearings and a tape held horizontally, a series of cells were laid out across the watershed which had sides

Figure 20.5(c) The effect of 1 cm rain on the infiltration profile of soils under Fescue and Bottlebrush Squirreltail grasses in the study area

6.1 metres long. The position of the tape and compass lines used to construct the percentage vegetation cover map are indicated on Figure 20.4 on the external grid, and produced a total of 142 cells.

In each cell, percentage vegetation cover was estimated and type noted. At sites where quaking aspen and scrub oak were present, cover was always 100 per cent because of the closed tree canopy. Elsewhere, where tree stands were separated and mixed with low-level shrub, such as silver sage-brush, or serviceberry in association, cover ranged from 60 to 100 per cent. Where tree species were absent and brush was predominant, cover was rarely 100 per cent, being mostly sage. On the concavities at the base of the long slopes and in the alluvium in the lower channel, western wheat was found, along with other hydrophytic plants such as Thurber's fescue, Idaho fescue and bottlebrush squirreltail. These provided quite a dense cover, and were occasionally mapped as 100 per cent density. Apart from this anomaly, there was generally a strong relationship between percentage cover and cover type.

Most south and south-west facing sites above 2470 metres are devoid of cover, whereas all north, north-east and east-facing slopes throughout have a cover of 100 per cent. Below 2470 metres, south-facing and west-facing slopes have generally less cover than sites at the same elevation which face east or south-east. All flat sites on the lower part of the watershed support the dense grass associations, but these are absent above 2470 metres, and on flat sites are replaced by sagebrush, or the occasional isolated oak tree.

To draw together some of these observations, the variables influencing cover type and density are as follows:

1. *Aspect*. This is the main control on vegetation cover type and density in the watershed. The link between aspect, vegetation and surface was noted by Hack and Goodlett (1960), who found that 'the drainage network is more developed on the south-west than in the north-east side of the (Appalachian) mountains', and by Hadley (1962), who noted 'sheet erosion is 50 to 75 percent greater on slopes that have southerly and south-westerly exposures than slopes that have other exposures'.
2. *Elevation*. This appears to be the second most important control, since above 2470 metres western wheat is always replaced by silver sagebrush on flat sites, and the brush associations found on south and south-west facing slopes below this elevation are absent at the higher elevations. It is though that lower elavations have a more favourable soil moisture environment than higher up the watershed.
3. *Slope angle* appears to be the third most significant variable because it affects the extent of erosion on bare sites above 2470 metres and controls whether or not sagebrush or the aspen–oak association predominates in the lower elevations.

THE INFILTRATION SURVEY

Since the area is semi-arid with a summer characterized by high-intensity storms, and from the description of the vegetation and soils above, it is anticipated that the summer storms are likely to produce overland flow primarily on bare sites on the south-west facing slopes. At least initially the assumption that covered sites will not produce overland flow during most watershed events appears supported by field observations, although infiltration down pipes may confuse this picture. The infiltration survey was conducted with these possibilities in mind.

Cells in the vegetation map matrix were sampled randomly across the watershed as a basis for the infiltration survey. The highest point in each grid square was arbitrarily chosen as the point at which to sample each square, and the specific location of each of these points in relation to the cells of the vegetation survey is indicated on Figure 20.4.

At each of the 24 sample sites, the vegetation density as mapped in the complete cell on Figure 20.4 was noted, as well as the local site vegetation type, the maximum slope gradient through the site and the site aspect, employing the classes south and south-west (A) west west and east (B) and north and north-east (C).

The canopy density as viewed directly above site, and the ground cover conditions (litter or bare ground) were separately noted, as well as infiltrometer site conditions. These data are given in Table. 20.1.

Methods

The infiltration survey was conducted on the 24 field sites using a home-made double-ring infiltrometer constructed as described by Evanko (1950). The inner ring had a diameter of 6 in. (15.02 cm) and an outer ring diameter of 8 in. (20.32 cm), being sizes in common usage. The choice of double-ring ponding infiltrometer for these tests requires some justification. Although a type F or type FA rainfall simulator is preferred by others (Te Chow, 1964; Thornes, personal communication), Hills (1970) points out that 'the spray rig is a rather elaborate and expensive piece of equipment and it demands a rather large water supply'. It was clearly impossible in the steep, inaccessible terrain of Alkali Creek to consider using a simulator rig. Nevertheless, there is always doubt about representativeness when methods which do not simulate raindrop impact are employed. For instance, Te Chow (1964) has shown that in a comparison between double-ring ponding infiltrometers and a type F rainfall simulator on bare crusted sites, initial infiltration rates may be eight times greater when using the ponding method rather than the sprinkler. However, differences were considerably less on covered sites where slaking effects can be ruled out, and he did also show that final infiltrability, f_c (Thornes, 1976) converged to similar values, in contrast to initial f rates. Free, Browning and Musgrave (1940) found correlations of 0.68 between values of final infiltrability, f_c, obtained by the cylinder method and those obtained by sprinkler, and Johnson (1963) points out that because a larger sample size can be obtained when a simple, versatile method is used, statistical representativeness in some ways improves. Nevertheless, the results and conclusions drawn here must be seen against these methodological problems.

At each site water was fed directly into both rings, the inner ring having been previously calibrated using a fine calibration scale attached to the inner ring. The outer and inner rings were initially filled, and subsequent readings taken against the scale, using the meniscus position. The advantage of the method over the ponded water reservoir method preferred by Hills (1970) is that no disturbance occurs to the water level position, and although hydraulic head varies as time proceeds, Bouwer (1963) suggested a method whereby corrections could be made for the remaining head in both rings after 1 hour. At each site the infiltrometer was pushed into the ground to a depth of 2 cm and readings against

Table 20.1 Infiltration survey: site conditions (Sites listed in order of decreasing cover density)

Site no.	Aspect class	Veg. cover (%)	Cover Type	Local infiltrometer site conditions	Soil type (P* indicates presence of pipes)	Parent material	Site slope (m/m)	Infilt. capacity (mm/hr)
22	C	100	Aspen and oakbrush	Closed canopy: ground litter	Solonetz—but with good organic horizon, 10cm	Sst. Lens	0.44	54
18	B	100	Aspen and oakbrush	Closed canopy: ground litter	Solonetz—organic horizon, 5 cm	Clayshales over sst. lens	0.20	48
19	B	100	Oakbrush	Closed canopy: ground litter	Solonetz—but with good organic horizon 10 cm	Sst. lens	0.20	42
13	C	100	Oakbrush	Closed canopy: ground litter	Solonetz—but with good organic canopy 10 cm	Sst. lens	0.07	23
8	C	100	Aspen and oakbrush	Closed canopy: ground litter	Solonetz—organic horizon 5 cm	Clayshales	0.26	18
23	C	100	Aspen and oakbrush	Closed canopy: ground litter	Solonetz—organic horizon 5 cm	Clayshales	0.48	18
4	B	100	Oakbrush	Closed canopy: ground litter	Solonetz—poor organic horizon	Clayshales	0.28	1
20	A	100	Oakbrush	Closed canopy: ground litter	Solonetz—but with good horizon 10 cm	Clayshales	0.32	24
16	A	85	Oakbrush	Partially closed canopy: ground litter	Solonetz—organic horizon 5 cm	Clayshales	0.25	18
24	B	85	Oakbrush	Fairly open canopy: litter & bare patches	Solonetz	Clayshales	0.20	9
9	B	81	Isolated patch of oakbrush	Some open patches: little litter	Degenerate Solonetz: open cracks, no organic horizon (P*)	Clayshales	0.27	12
17	C	80	Oakbrush	Partially closed canopy: litter	Solonetz—organic horizon 5 cm	Clayshales	0.48	18

Site no.	Aspect class	Veg. cover (%)	Cover Type	Local infiltrometer site conditions	Soil type (P* indicates presence of pipes)	Parent material	Site slope (m/m)	Infilt. capacity (mm/hr)
21	A	80	Oakbrush & western wheat	Partially closed canopy: bare patches	Solonetz—organic horizon 5 cm	Clayshales	0.22	12
11	C	80	Aspen and oakbrush	Partially closed canopy: bare patches	Solonetz—no organic horizon	Clayshales	0.34	0
15	B	75	Oakbrush	Partially closed canopy: litter	Solonetz—but with organic horizon, 5 cm	Sst. lens	0.33	32
3	A	60	Sagebrush	Dense network of roots	Solonetz—no organic horizon	Clayshales	0.35	24
6	B	58	Sagebrush	Local bare site	Degenerate Solonetz: cracks columnar structure	Clayshales	0.28	9
14	B	45	Sagebrush	Bare: pan of small gully	Laminations over Solonetz	Clayshales	0.17	4
2	B	40	Oakbrush and sagebrush	Scattered trees: locally bare site	Degenerate Solonetz: cracks	Clayshales	0.27	6
5	C	39	Aspen and oakbrush	Scattered Trees: locally bare site	Degenerate Solonetz: cracks	Clayshales	0.26	6
1	B	38	Sagebrush	Bare site	Degenerate Solonetz: cracks	Clayshales	0.19	1
10	A	25	Sagebrush	Bare site	Degenerate Solonetz: cracks (P*)	Clayshales	0.22	5
12	A	0	Sagebrush	Bare site	Degenerate Solonetz: cracks (P*)	Clayshales	0.27	0
7	A	0	Sagebrush	Bare site with rills	Degenerate Solonetz: cracks	Clayshales	0.30	0

The numbers on the curves refer to the survey sites shown on Figure 20.4

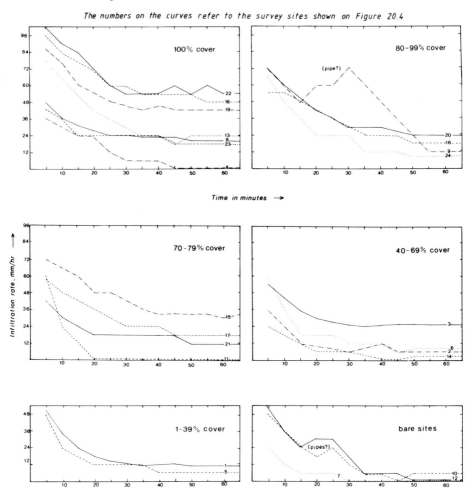

the scale were taken every 5 minutes and converted to hourly equivalents. The final infiltrability f_c was adjusted by the Bouwer method where necessary when considerable head remained in both rings after 1 hour.

Results

Several observations can be made from the infiltration curves (grouped percentage cover on Figure 20.6), from the tabulated site conditions Table 20.1) and from the f_c and cover histograms displayed by aspect, slope angle and site condition respectively on Figures 20.7, 20.9 and 20.10.

Firstly, Figure 20.7 clearly shows that aspect, via its influence on vegetation cover densities, affects the infiltration patterns on the watershed. South-facing

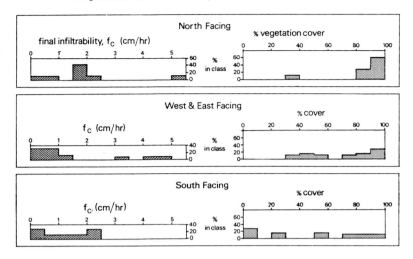

Figure 20.7 The relationship between f_c, percentage cover and aspect

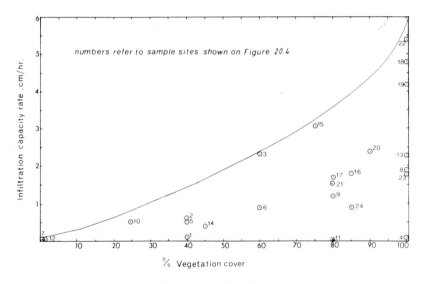

Figure 20.8 The relationship between f_c and percentage vegetation cover

sites appear to have a greater chance of overland flow than north-facing sites. Secondly, whether or not initial infiltration or final infiltration is considered, all infiltration rates as monitored in the field are low. Under oak woodland on clay loams, Hills (1970) quotes infiltration capacity rates of between 15 and 320 cm/hr, whereas at Alkali Creek the maximum rate in any 5 minute interval was the initial rate of 10 cm/hr at site 22 under a cover of 100 per cent oakbrush.

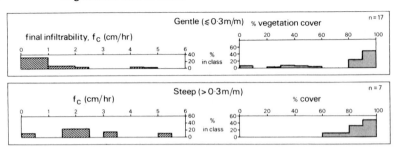

Figure 20.9   The relationship between f_c, percentage cover and slope angle

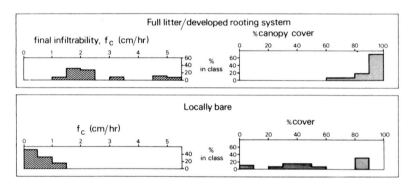

Figure 20.10   The relationship between f_c, percentage cover and site conditions

Values of final infiltrability on bare sites on Table 20.1 compare well, however, with bare site f_c values cited by Kincaid, Osborn and Gardner (1966), and on grazed range and poor pasture land by Musgrave and Holtan (1964). However, the dense-cover sites have low final infiltrability values compared with full tree cover values cited by Whipkey (1969). Clearly soil factors become increasingly influential as a source of variability below an established cover, whereas bare sites always have low values unless tilled (Musgrave and Holtan, 1964). This idea is substantiated by an inspection of Figure 20.8 which relates f_c to percentage cover for the Alkali Creek survey data, showing the wide range of values of f_c at 80–100 per cent cover.

So thirdly, an attempt was made to interpret this variability at covered sites in more detail. Although the nature of the data precludes a statistical treatment, Figure 20.9 demonstrates that slope angle may be important, since infiltration appears to be better on the gentler sites. These are mostly in the lower part of the watershed, however, where 100 per cent cover is likely to be a grass, rather than oak sagebrush. This not only implies that cover type may be as important as percentage cover, but also that slope angle is not the real casual variable here.

Local site conditions were finally considered. Figure 20.10 and Table 20.1 suggest that percentage cover is related to ground conditions (except where piping modifies f_c) and that there are clear links between ground cover, the development of an organic horizon, and consequently higher f_c values. So Figure 20.10 lends some support to the use of bare areas alone (AB values) in the calculation of runoff volumes, as discussed in Figure 20.1.

The only remaining note of caution associated with the use of bare areas only as a variable in flow generation models at Alkali Creek is the nature of the piped Solonetz soils on some of the bare badland areas. At site 9, for example, where higher ESPs were noted locally by Heede (1971), piping rather than an organic horizon caused a higher f_c despite the good cover, and Figure 20.6 shows how pipes can modify the shape of infiltration curves. In the badland areas, for example, when cover dropped below 50 per cent, none of the degenerate Solonetz soils demonstrated an organic horizon, and all were cracked on the surface, and at sites 10 and 12 the presence of pipes was strongly inferred from large, occasionally eroded, surface cracks. An exception in this category is site 14, which although bare, was located on the fan at the lower end of the small discontinuous tributary gully. Here the soil displayed some lamination associated with deposition at that location.

The infiltration survey supports the concentration of attention mainly on the bare areas for use in the overland flow model, but there is obviously some loss to infiltration even on these sites. To assess the importance of this loss, infiltration data were compared with the rainfall characteristics of the watershed.

CLIMATIC FACTORS: RAINFALL INTENSITIES OF LOCAL STORMS

The average January temperature at Collbran (1889 m, 24 km south of the study area) is $-5.6\,°C$ whilst the July average is $20.4\,°C$. Mean annual precipitation in the study area is 471.8 mm (1961–74), and this low total falls partly in the winter months as snow which accumulates slowly, to form a snowpack which persists all spring. Isothermal conditions occur in the snowpack in the early spring causing a sudden, dramatic snowmelt flood lasting five to six weeks. In the summer months, by contrast, high-density, short-duration storms caused by convective cloudburst activity are common and these events produce overland flow on unprotected surfaces from which runoff and erosion are severe. The detailed characteristics of these events, when viewed in conjunction with the infiltration curves, were necessary for the recalibration of the overland flow model. Consequently, the local recording raingauge records are analysed here in some detail.

Six years of recording raingauge records are available for the study area covering the period 1968–74. From this rather limited trace, the summer storms were listed in terms of their intensity, duration, and total event precipitation (this last being the product of the first two). Use was made of these data to identify a range of events with known recurrence probabilities to simulate both those 'common' events, and infrequent or 'freak' events, taking the largest/longest event on record as a comparison. The recurrence probabilities of the event totals for the 95 events on record were established using the Log Pearson III Event Frequency analysis method, and then (since the overland flow model uses event total r, and event duration t_r, as variables for input) event totals were linked to particular durations by regression.

The probability of the total event precipitation being equalled or exceeded is shown in Figure 20.11. The events with probabilities ranging from 0.5 to 0.99 are identified and plotted, giving 11 points plotted across the range of possibilities which are to be used in subsequent discussion (Tables 20.2 and 20.3). Figure 20.12(a) links the total event precipitation (the product of intensity and duration) to the event duration. Although the significance levels of this regression are spurious for obvious reasons, it is possible to use the relationship to link particular event totals to particular event durations with some level of confidence. The duration of the 11 events identified on Figure 20.11 were

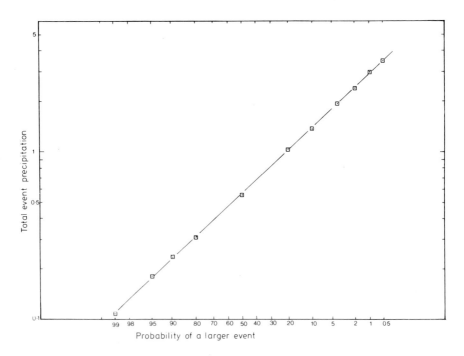

Figure 20.11   The local R.R.G. trace analysed: recurrence probabilities of event totals

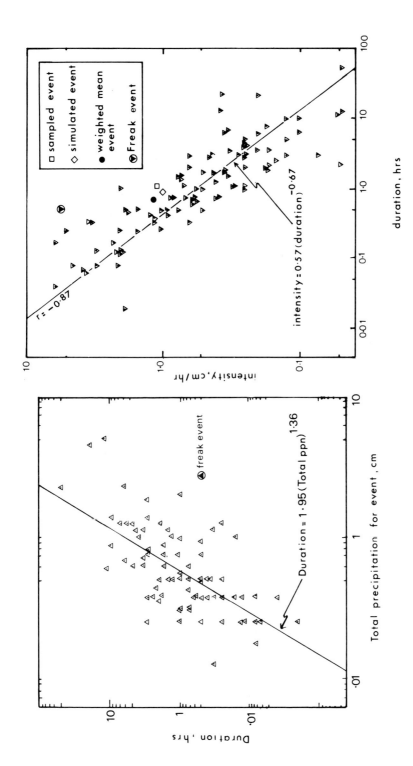

Figure 20.12 (a) Event duration related to event total. (b) Intensity related to event duration

Table 20.2 Characteristics of selected summer storms, and the selection of events for simulation runs

Probability of larger value of total ppn for event (%)	Intensity (mm/hr)	Duration (hrs)	% event runoff (from Table 20.3) (B)	Total ppn for event (mm) (A)	mm. of ppn becoming runoff (mm) (A×B)	Duration less f_c time (hrs)	Frequency at class midpoint (%)	
99	26.26	0.100	0.5	1.132	8.0057	—	2	
95	17.51	0.187	10.1	1.79	.1808	—	3	
90	13.15	0.287	11.0	2.45	.2640	—	15	
80	10.47	0.394	16.0	3.09	.494	0.0637	30	* Run 7
50	5.27	0.888	28.3	5.62	1.471	0.558	25	** Run 8
20	3.65	1.945	28.3	10.00	2.833	1.615	11	* Run 9
10	2.70	3.044	35.0	13.90	4.860	2.714	7	
4	2.00	4.758	36.7	19.31	7.09	4.4277	2	
2	1.61	6.619	31.0	24.62	7.63	4.289	1	
1	1.37	8.358	28.8	29.21	8.41	8.028	2	
0.5	1.16	10.35	23.5	35.11	8.25	10.405	2	

Run 10 = The 'freak' event which occurred on August 24 1971, in which 27.94 mm fell in 0.5 hrs at an intensity of 55.88 mm/hr. The percentage event runoff for this event (Table 20.3) was calculated as 81%, so that 25.15 mm became runoff, and that the duration less time to f_c will be 0.166 hr (10 min)

Table 20.3 The calculation of percentage event runoff, using known event intensities and durations, and using the field-monitored infiltration data shown on Figure 20.4. The method used is illustrated on Figure 20.12 (locally bare sites only are used)

Intensity (mm/hr)	Duration (hr)	% probability of larger value of total ppn for event	% of event listed becoming runoff on sites								Average %
			1	2	5	6	7	10	12	14	
26.26	0.100	99	0	0	0	0	2	0	0	2	0.5
17.51	0.187	95	0	15	12	3	31	0	0	20	10.1
13.15	0.287	90	0	15	5	0	35	0	0	22	11.0
10.47	0.394	80	0	24	7	5	55	7	0	33	16.0
5.27	0.888	50	0	18	15	0	80	40	35	37	28.3
3.65	1.945	20	0	0	0	0	90	14	70	52	28.3
2.70	3.044	10	0	0	0	0	95	7	65	60	35.0
2.00	4.758	4	0	0	0	0	95	5	75	65	36.7
1.61	6.619	2	0	0	0	0	98	0	85	61	31.0
1.37	8.358	1	0	0	0	0	99	0	90	40	28.8
1.16	10.735	0.5	0	0	0	0	100	0	95	0	23.5
*55.88	0.5	—	75	86	93	78	100	77	80	85	81.0

* The freak event which occured on August 24, 1971

calculated in this way, therefore, and these data are listed as the second and third columns of Table 20.2.

Plotting the precipitation intensity against duration for the 95 events in the six years of record (Figure 20.12b) reveals a good relationship between these properties, significant at the 5 per cent level. It is perhaps not surprising that the longer events, occurring mainly as prolonged, low-intensity showers in October, have the lowest intensities. The high-intensity events, which occur in July and August, are only of short duration. Once again the relationship predicts the event intensity from event duration and the data can be added to Tables 20.2 and 20.3.

RUNOFF: TOWARDS A CALIBRATION METHOD

Knowing the intensity and duration of the 11 events of interest allows the runoff percentage for each to be calculated for all bare sites when superimposed on

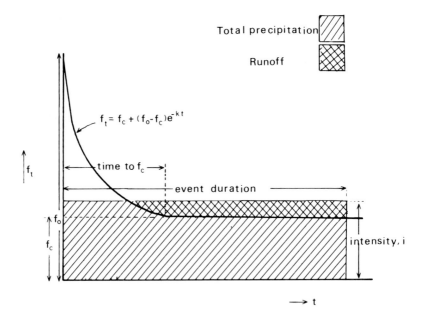

Runoff time = (event duration − time to f_c) + slope length / v

Runoff = (ppn) (% of rectangle above the curve)

Figure 20.13 The method whereby percentage runoff for particular events is calculated

field infiltration curves, and Figure 20.13 demonstrates the essentially Hortonian aspects of this recalibration procedure. In this, the actual runoff from an event with a constant infiltration rate can be calculated as the area under the intensity plot which remains after subtraction of the infiltrating volume. Using this method, we can infer for instance that during an event with an intensity of 10 mm/hr falling over an hour, after 15 minutes site 7 would produce runoff at a rate of 5 mm/hr, rising to 10 mm/hr at the end of the event. Sites 2, 4 and 11 would be the next to produce runoff, again first at a fairly low rate, rising after an hour to rates equal to the difference between 10 mm and the final infiltrability of the site. Of the locally bare sites (1, 2, 5, 6, 7, 10, 12 and 14), all would generate Hortonian flow at a level which can therefore be assessed by comparing the event duration and intensity with the infiltration curves for these sites. Of the sites with local cover on Table 20.1, only sites 1, 4 and 24 could be expected to produce any rapid-response runoff during the simulated event, the rainfall being lost either as eventual evaporation from an interception store, or by entering the channel as throughflow later in the event, or by being lost to plant use or a low water table. Under a well-developed canopy, ground levels of rainfall intensity must, however, be lower than the above canopy, so that even on sites 1, 2 and 24 there may be no runoff during an event the size of that under consideration. On the well-covered sites some increase in the spatial extent of overland flow may occur, especially when events such as that of 24 August 1971 are considered. The largest event on the raingauge record, this storm produced 26.94 mm of rain in half an hour, falling therefore at an intensity of 55.88 mm/hr (plotted as a 'freak' event on Figure 20.12). Mostly, however, it is clear that covered sites produce no runoff.

Model Recalibration

The infiltration curve observations, whilst largely supporting the use of AB in equation (1), indicate two sources of error in the hillslope generation calculations: the use of the unmodified rainfall amount r as a basis for volume calculations, and the assumption of instantaneous runoff on the bare sites. Taking the first of these problems, and considering only bare or locally bare sites for reasons considered above, the procedure described by Figure 20.13 applied to the bare or locally bare sites on the infiltration survey during a 10 mm event, as an example, gives for sites 1, 2, 5, 6, 7, 10, 12 and 14 the following runoff percentages: 10, 15, 13, 8, 57, 25, 26 and 48 respectively. The average of these values is 25 per cent, indicating that the precipitation which becomes runoff is not 10 mm, but is more likely to be 2.5 mm on average.

A further source of error in the simulation is the assumption that the runoff duration is equal to rainfall time, plus the time taken to drain the slope. For all the bare or locally bare sites in the infiltration survey, runoff duration is in fact equal to the rainfall time less the time to infiltration capacity, plus the time taken to drain the slope. The average time to infiltration capacity on these curves is 20 minutes. As a result, the equation presented earlier is more suitably calculated as:

$$T = (t_r - 1200) + l/v \tag{3}$$

where t_r is in seconds.

The planned simulations must therefore be preceded by, first, a comparison of the event intensity and duration information with the bare site infiltration curves as illustrated on Figure 20.6, and then the time to f_c assessed, and used to alter the hillslope hydrograph timebase, as suggested by equation (3). On this basis any sort of event can be simulated provided this precalibration method is followed.

The simulations do not consider the fate of the infiltrated water. Clearly on much of the watershed with over 100 per cent cover, some of the infiltrating water is likely to reach the channel via pipeflow during the event, or much later as delayed flow. Neither of these sources of flow has been considered here, and it is only possible to note that such contributions are likely to extend the recession limb of real hydrographs when compared to the simulated ones. It is assumed that these effects are not a significant control on the hydrographs, however, especially during larger events.

The Runoff of Significant Watershed Events

Each of the 11 events for which the probabilities of a greater value have been identified on Figure 20.11 and tabulated on Table 20.2, can be regarded as representing the midpoint of a class of events. For instance, the 95 per cent probability event, in which 1.79 mm of total precipitation occurs, can be regarded as the midpoint of a class running from $(1.132 + 1.79)/2 = 1.461$ mm to $(1.79 + 2.45)/2 = 2.12$ mm, and so on. Using classes identified in this way, the percentage frequency of events in each class can be calculated for all the 95 events on record. The percentage frequencies for the classes whose midpoints are the fifth column on Table 20.2 are listed as the last column. These data reveal that the 90 per cent probability event, the 80 per cent probability event, the 50 per cent probability event and the 20 per cent probability event classes together represent 82 per cent of all events analysed, the others representing very improbable cases. Taking the three smallest events in terms of precipitation on the list, it is clear from this that little runoff will occur during these events, since despite very high intensities, their short duration ensures little or no runoff when compared with the infiltration curves. All precipitation for these events can therefore be considered to infiltrate on most locally bare sites and so these are insignificant, both in terms of frequency and runoff.

Turning to the events with the highest precipitation totals (which have between 10 and 0.5 per cent probabilities of a larger precipitation total), the method used here implies that these events produce larger totals because of extraordinarily long durations, falling at low intensities for up to 10 hours. Representing altogether only 13 per cent of all events on the watershed, these clearly represent the low-intensity, long-duration events which are occasionally experienced in the late summer months in the study area. It can be shown that the very low intensities mean that only on sites where f_c approaches zero will runoff occur for such events. When this does occur, the long duration will produce such a long timebase for the hillslope hydrographs that the simulation results predict a slow trickle of runoff over a long runoff period. Once again, these are not likely to be major runoff events. It must be concluded from this that the remaining three asterisked events on Table 20.2, which represent common occurrences on the watershed, are the only common events likely to produce hydrographs with a significant peak, and even here runoff volumes appear low.

As a contrast to these three events, the record of events for Alkali Creek was inspected for the largest event which had both a very high intensity and high duration. This event, which occurred on 24 August 1971, plots well away from the regression lines on Figure 20.12, where it has been indicated as a black triangle because it has an unusually high intensity (55.8 mm/hr) and a relatively long duration (0.5 hr). In statistical terms this event represents a 'freak' occurrence. Because of its high intensity and duration, the percentage runoff for this event (as calculated using the method shown on Figure 20.13) is 81 per cent so that of the 27.94 mm which fell, 25.15 mm became runoff (Table 20.3). The duration of the event, less time to f_c, becomes 10 minutes (0.166 hrs).

Inspecting the results of these calculations for all watershed events (Table 20.3),

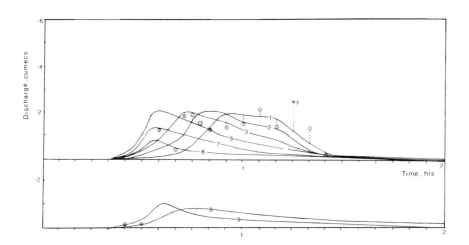

Figure 20.14   On 4 June 1981, 13.1 mm of rain fell in 1.25 hours. The diagram shows the computer run for this event after pre-calibration. Data points are field-recorded discharges at various watershed sites (1 to 9). Correlation $r = 0.78$ for 15 data points

we reach the important conclusion that all common events produce only moderate runoff, and that there are as many events in a year which have attenuated or non-flashy runoff patterns as there are those which have a flashier regime. Very occasionally an event will occur on the watershed which is so dramatic in its impact as to completely override the effect of the more common events, and it is clearly only these events which are geomorphologically significant.

Obviously these conclusions cannot be fully explored without presenting the results of the simulations using the full runoff model (both generation and routing phases). The full description, calibration and application of the runoff model are beyond the scope of the present chapter. It is encouraging to report, however, that a suitably pre-calibrated run of the full model produced hydrographs which matched up quite well to discharge data collected at significant network sites during a field-monitored event (Figure 20.14).

CONCLUSIONS

As one stage in the development of a Hortonian overland flow model, simple assumptions were made concerning runoff volume from bare sites and concerning the duration of runoff. This chapter has presented the pre-calibration of this model which was concerned with the testing of infiltration assumptions. The following conclusions can be drawn.

All bare sites will generate runoff during most (but not all) of the watershed events on record. None of the covered sites is likely to produce runoff during 'normal' events but occasional freak storms are likely to extend the contributing area. Low infiltration rates are found beneath covered sites, however, especially when the cover is grass and the sites are flat and indurated. Occasionally very high infiltration rates can be found which can be linked to patterns of alkali sodicity and consequent piping in the parent materials. Aspect exerts a very dominant effect on vegetation response and infiltration patterns on the watershed as a whole.

Additionally, a simple infiltration survey is effective in developing a pre-calibration method for runoff models which have areally distributed runoff sources. Infiltration data, used in conjunction with a vegetation cover map, a slope angle map, and raingauge records can be used to simulate Hortonian runoff on semi-arid watersheds with partial vegetation cover.

Such calculations show that 'normal' summer storm events are largely insignificant in that runoff production from them is low, and the most significant watershed events must have a recurrence interval beyond the six years of R.R.G. records analysed here.

ACKNOWLEDGEMENTS

I would like to thank the US Forest Service, Rifle Ranger Officer for data and field support, ILEA for a grant to support field work and Dr A. M. Harvey for his useful criticisms of an earlier draft.

REFERENCES

Al-Azawi, S. A. (1985). Experimental evaluation of infiltration models. *J. Hydrology* (NZ), **24**(2), 77–88.

Baillie, I. B., P. H. Faulkner *et al.* (1986). Problems of protection against piping and surface erosion in C. Tunisia. *Env. Cons.*, **13**(1), 27–33.

Bouwer, H. (1983). Theoretical effects of unequal water levels in the infiltration rate determined with buffered cylinder infiltrometers. *J. of Hydrology*, **1**, 29–34.

Bryan, R. B. and I. A. Campbell (1986). Runoff and sediment discharge in a semi-arid drainage basin. *Zeit für Geomorph.*, Sp.bd., **58**, 121–43.

Bryan, R. B. and A. Yair (1982). *Badland Geomorphology and Piping*, Geobooks, Norwich.

Davidoff, B. and H. M. Selim (1986). Goodness of fit for eight water infiltration models. *Soil Soc. Am. J.*, **50**(3), 759–764.

Emmett, W. W. (1978). Overland flow. In M. J. Kirkby (ed.), *Hillslope Hydrology*, ch. 5, John Wiley, pp. 145–70.

Evanko, A. B. (1950). A tin can infiltrometer with improvised baffle. *Northern Rocky Mtn. Forest and Range Expt. Stn. Missoula, Mont. Research Note.* 76.

Faulkner, P. H. (1987). Gully erosion in response to both snowmelt and flash flood erosion, Wn. Colorado. In V. Gardiner (ed.), *International Geomorphology 1986*, Part I, John Wiley, pp. 947–69.

Fox, C. J. and J. Y. Nishimura (1957). Soil Survey Report of the Alkali Creek Watershed, *USDA Forest Service*. US Printing Office, M-5123.

Free, G. R., G. M. Browning and G. W. Musgrave (1940). Relative infiltration and related physical characteristics of various soils. *USDA Tech. Bull.*, 729.

Gurnell, A. M. and R. J. Gregory (1987). Vegetation characteristics and the prediction of runoff. *Hydrological Processes*, **1**(2), 125–142.

Hack, J. T. and J. C. Goodlett (1960). Geomorphology and forest ecology of a mountain region in the Central Appalachians. *US Geol. Survey Prof. Paper* No. 343.

Hadley, R. F. (1962). Some effects of microclimate on slope morphology and drainage basin development. In *USGS Research in 1961*. US Geological Survey, 13–22.

Heede, B. H. (1971). Characteristics and processes of soil piping in gullies. *USDA Forest Service Res. Paper* RM-68.

Hillel, D. (1987). Unstable flow in layered soils: A review. *Hydrological Processes*, **1**(2), 143–7.

Hills, R. L. (1970). The determination of the infiltration capacity of field soils using the cylinder infiltrometer. *BGRG Tech. Bull.*, **3**.

Johnson, A. I. (1963). A field method for measurement of infiltration. *US Geol. Survey Water Supply Paper*, 1544-F.

Kelley, C. (1951). Chemistry of saline sodic soils. In F. Bear (ed.), *Soil Chemistry*, 1963, Ch. 3.

Kazman, Z., I. Shainberg and M. Gal (1983). Effect of low levels of exchangeable sodium and applied phosphogypsum on the infiltration rate of various soils. *Soil Science*, **135**(3), 184–92.

Kincaid, D. R., H. B. Osborn and J. H. Gardner (1966). Use of unit-source watersheds for hydrologic investigations in the semi-arid south west. *Wat. Res. Res.*, **2**, 381–92.

Musgrave, G. W. and H. N. Holtan (1964). Ch. 12 in Chow, Ven Te (ed.), *Handbook of Applied Hydrology*, McGraw-Hill.

Ragan, R. (1966). Laboratory evaluation of numerical flood routing techniques for channels subject to lateral inflows. *Wat. Res. Res.*, **2**(1), 111–21.

Romkens, M. J. M., R. L. Baumhardt, M. B. Parlange, R. D. Whistler, Y. Parlange and S. N. Prasad (1986). Rain-induced surface seals, their effect on ponding and infiltration. *Annales Geophysical*, Series B, **4B**(4), 417–24.

Rouse, W. R. (1970). Relations between radiant energy supply and evapotranspiration from sloping terrrain; An example. *Canadian Geographer*, **14**, 37.

Springer, E. P. and T. W. Cundry (1987). Field-scale evaluation of infiltration parameters from soil texture for hydrolic analysis. *Water Res. Res.*, **23**(2), 325–34.

Te Chow, Ven (ed.) (1964). *Handbook of Applied Hydrology*, McGraw-Hill.

Th. van Genuchten, M. (1987). Progress in unsaturated flow and transport modelling. *Rev. of Geophysics*, **25**(2), 135–40.

Thornes, J. B. (1976). Semi-arid erosional systems, *Geographical Papers,* No. 7, London School of Economics.

Whipkey, R. Z. (1969). Storm runoff from forested catchments by subsurface routes. *Int. Ass. Sci. Hydrol. Sym. from Leningrad*, **85**, 773–9.

Zingg, A. W. (1940). Degree and length of landslope as it affects soil loss in runoff. *Agric. Eng.*, **21**, 59–64.

21 Response of Four Different Mediterranean Vegetation Types to Runoff and Erosion

MARIA SALA
Department of Physical Geography, University of Barcelona

and

ADOLFO CALVO
Department of Geography, University of Valencia

SUMMARY

Runoff, sediment and litter removal have been measured in four different vegetation types of the Catalan Ranges (NE Iberian coast), comprising a beechwood (*Fagus sylvatica*) and an evergreen-oak (*Quercus ilex*) woodland in the Montseny mountain, and a shrubland (*Erico-Thymelaeetum tinctoriae*) and a woodland (*Quercetum ilicis galloprovinciale*) in the Prades mountains.

Differences within each site are marked and they occur both along the slopes and between plots. The distribution, size and morphology of the vegetation, together with litter accumulation in concavities, seem clearly responsible for most of these differences.

In the two woodlands in the Montseny area differences found are thought to be related to the different phenology of leaf evolution, and its implications for soil and surface conditions. A higher rate of litter formation in the deciduous beechwood contributes to an increase in the degree of water absorption and thus favours runoff and fine sediment entrainment. A lower rate of litter and soil formation makes the evergreen oak slopes more susceptible to debris displacement by gravity processes.

Sediment removal in the evergreen oak woodland compared with the shrubland in the Prades mountains also shows the importance of vegetation control in slope runoff and sediment removal. A lower amount of litter accumulation, soil formation and canopy covering

Vegetation and Erosion
Edited by J. B. Thornes

makes the shrubland slopes less protected, and thus with a higher amount of sediment available for erosion. On the other hand, shrubs are more effective in retaining slope debris. Rainfall simulation has shown very low rates of runoff both in the woodland and shrubland and has reinforced the theory of strong spatial differences produced by vegetation type and distribution.

INTRODUCTION

Research on rainfall/runoff/erosion relationships in two Mediterranean woodland types which differ in substratum and in vegetation, started in 1982, showed that differences in each of the study areas could be related to the presence and distribution of vegetation (Sala, 1988). Further investigation was carried out comparing the response of a woodland and a shrubland to rainfall impact (Llorens, 1987). Initial measuring techniques consisted of a series of troughs installed at different positions along the slopes and according to vegetation, collecting runoff produced by natural rainfall. At present data are obtained using a small portable rainfall simulator. The results have been analysed at the slope and plot scales.

DESCRIPTION OF SITES

The Montseny mountain is located 60 km north of Barcelona, 20 km from the coastline (Figure 21.1). Average annual rainfall ranges from 1200 mm at the summit (1700 m) to 700 mm at the footslopes (250 m). Depending on rainfall and lithology, vegetation communities vary from evergreen oak and cork oak to chestnut and beech woodlands. Two study sites are considered in this chapter, Santa Fe and La Castanya.

At Santa Fe (1170 m a.s.l.) the underlying lithology is granitic (adamellite) and has been deeply weathered to more than 15 m of grus (Cervera, 1986). Removal of fine material has produced a tor and boulder landscape, with tors in the interfluves and grus and soil-covered slopes. Soils are inceptisols and entisols, with a sandy loam structure and a pH of 4.5–6.5 (Terrades et al., 1984). The average organic material content varies between 1.0 and 5.3 per cent and the sandy B horizon developed in the grus is often saturated, although water seldom reaches 30 cm from the surface during the wet period and lies at about 150 cm during the dry season (Terrades et al., 1984). The vegetation consists almost exclusively of *Fagus sylvatica* (Fagetum). It is interesting to note that beech trees have roots which keep near the surface, expanding horizontally and sometimes vertically upwards. Thus they are very abundant in the A horizon, which explains the absence of other plants in this woodland and the high water absorption. Another peculiarity of beech is their regular canopy, ascending branches and smooth trunks, all of which tend to produce high rates of stemflow,

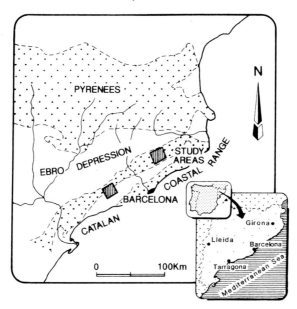

Figure 21.1 Location of study sites

found by Rodà (1983) to be 10–14 per cent of the incidental rainfall. Leaves fall in October and November, mainly during the second half of October (80 per cent), at a rate of 0.3 kg m^{-2} yr^{-1} (Terrades *et al.*, 1984). Litter distribution on the slopes is variable with bare patches alternating with well covered areas, especially on the footslopes where the thickness can reach more than 1 metre. Lateral transport, mainly by wind, and surface irregularities redistribute the litter over the slopes with litter relation to trees (Terrades *et al.*, 1984).

At La Castanya (800 m a.s.l.) the geology consists of slates formed by quartz, sericite, albite and chlorite covered by a mantle of rock debris in a silt and clay matrix up to 1 m thick in places. This unevenly distributed and variably sorted mixture of coarse and fine sediments makes the hydrological responses of the slopes to rainfall and sediment removal extremely uneven. Widespread rock outcrops, especially near the summits, contribute to the variability of the surface with slope angles between 22 and 35°. Soils are colluvial rankers 0.4 to 1.5 m deep (Avila and Rodà, 1988). Vegetation cover is formed by an evergreen oak forest and shrubs are scarce. Stemflow accounts for 8–10 per cent of the incident rainfall. Leaves fall all the year round and litter cover is homogeneous and thin, although it is not evenly distributed on the slopes. Work by Rodà (1983) shows that in evergreen oak woodlands throughfall is highly variable. It does not occur at low rainfall, and is positively correlated with distance from the trunk.

The Prades mountain is located 80 km south of Barcelona and, as in the case of Montseny mountain, is part of the Pre-Littoral Coastal Range and 20 km from the coast (Figure 21.1). Average annual rainfall ranges from 750 mm to 550 mm. Due to its southern position and lower altitude (1200 m a.s.l.) this is a drier environment than Montseny mountain. Vegetation is either an impoverished evergreen oak woodland or a shrubland varying in species depending on the underlying rock. The study sites are located in the slopes of the l'Avic drainage basin.

In the l'Avic site (750 m a.s.l.) the underlying lithology is similar to that of La Castanya, consisting of schists and slates with a regolith cover formed by a poorly sorted deposit with a percentage of particles larger than 2 mm between 53 per cent in the upper slope and 75 per cent in the middle slope, and 57 per cent at the lower part (Llorens, 1987). The regolith depth measured by seismic methods (Escarré et al., 1986) varies from 0.6 m near the divides to 1.5 m at the footslope. Either on bed rock or on the slope deposits the soil is a xerochrept with a profile with a B horizon 20–25 cm deep, an A1 horizon of 5 cm, an A_o of 1–2 cm, and an A_{oo} 2–5 cm in depth, depending on the type of vegetation cover and slope. The percentage of organic material varies between 16 per cent in the upper part of the slopes and 25 per cent in the lower part. (Escarré et al., 1986). Slope gradients range from 31° in the upper part to 25° in the middle part. Vegetation on the south-facing slopes includes two main associations: an evergreen oak woodland (*Quercetum ilicis gallo provinciale*), and a shrubland (*Cisto-sarothamnetum catalaunici*). The woodland in l'Avic has a higher percentage of shrubs than that of La Castanya and consequently a wider range of responses to interception, throughflow and stemflow, which vary between species. Stemflow has been found highest for *Arbutus* (twice as much as for the green oak). Biomass is lower than in the La Castanya woodland, with 115 t ha^{-1} (160 t ha^{-1} in La Castanya), and the lowest of all published data on green oaks. The green oak itself is the highest producer, with a total of 100 t ha^{-1} of which 5.8 t ha^{-1} comes from leaves. The yearly production of organic material is variable (200–443 g m^{-2} yr^{-1}) of which 122–187 g m^{-2} yr^{-1} are leaves (Escarré et al., 1986).

EXPERIMENTAL DESIGN

In Santa Fe and La Castanya (Montseny mountain) four sites, three along a slope and one in an intermediate position, were instrumented with 12 Gerlach troughs, three in each site (Figure 21.2a and b). Measurements were made at monthly intervals or after storms during 1982, then seasonally in the next three years (Sala, 1988). In l'Avic (Prades mountain) 12 Gerlach troughs were installed on a slope and distributed in groups of three in four sites: sites A and C in the upper and middle part of a shrubland area, and sites B and D in the same

SANTA FE EXPERIMENTAL SITES

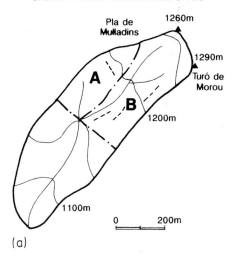

(a)

LA CASTANYA EXPERIMENTAL SITES

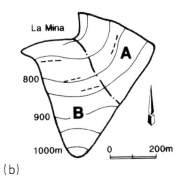

(b)

L'AVIC EXPERIMENTAL SITES

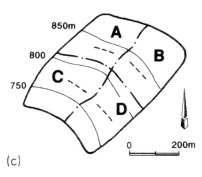

(c)

Figure 21.2   Santa Fe (a), La Castanya (b) and L'Avic (c) sites

positions of a neighbouring woodland area (Figure 21.2c). Measurements, although discontinuous and seasonal, were made at the same time in the two vegetation types for the period 1983–5 (Llorens, 1987). Slope lengths are 100 m approximately.

In the woodlands the troughs in each site were located systematically at tree bases, near rock outcrops, and in litter-filled concavities. In the shrubland similar positions were selected but in this case areas between plants were devoid of litter.

The equipment used in rainfall simulation was described in Calvo et al. (1988). The rain was applied at a constant rate during 20–35 minutes at an intensity of 60 mm/h in a circular ring 55 cm in diameter. Positions in relation to vegetation were selected similar to the trough locations.

RESULTS

Rainfall and runoff are always significantly correlated, although better in the beechwood of Santa Fe and in the shrubland of L'Avic than in the evergreen woodlands of La Castanya and L'Avic (Figure 21.3). The low proportion of water collected as overland flow (13 per cent of the rainfall at La Castanya, 10 per cent at Santa Fe, 1.7 per cent in the L'Avic woodland, and 0.7 per cent in the shrubland) indicates that infiltration is high in all environments. A great deal of water must be taken by tree roots in the upper soil horizons of Santa Fe (Cervera, 1986) or directed to the bedrock layer through the stony soils in La Castanya, because subsurface flow is low in these slopes (Bonilla, personal communication). At l'Avic soils are similar to those in La Castanya.

Average annual values for sediment removal are moderate as is common on vegetated slopes, and do not differ in order of magnitude in the two studied environments of the Montseny mountain ($152 \, \text{g m}^{-1} \text{yr}^{-1}$ in Santa Fe and $203 \, \text{g m}^{-1} \text{yr}^{-1}$ in La Castanya), but the granulometric composition of the eroded material is markedly different because a high proportion (80 per cent) of coarse material is removed from the slopes of La Castanya (Sala, 1988; see Table 21.1). In L'Avic sediment production is lower than in the Montseny, but is 1.9 higher on the shrublands than on the woodlands (Llorens, 1987). The highest percentage of coarse material (80 per cent) is trapped in the shrubland site (Llorens, 1987; Table 21.1). Sediment transport is poorly correlated with runoff (Figure 21.4), being lowest in the shrubland, highest in the beech forest and similar in both evergreen woodlands.

Along the slopes, runoff and erosion appear to be more closely related to surface properties and/or vegetation type than to position on the slope. More water and sediment are trapped in the upper plots, located at the base of a rock outcrop, while differences between the middle and lower ones are not remarkable (Sala, 1988). Both in La Castanya and in Santa Fe differences in runoff are clearly related to the rock/soil ratio, as used by Yair and Lavee (1985).

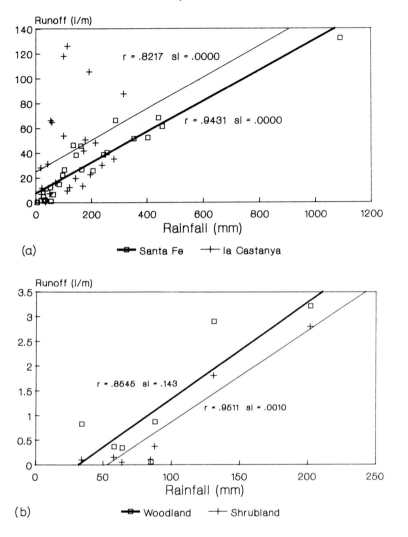

Figure 21.3 Rainfall/runoff relationships in the Montseny sites (a) and in the Prades sites (b)

The role of vegetation in the production of overland flow and erosion was more clearly observed from data for individual troughs. The most extreme and discontinuous events, both in amounts of water and of sediment collected, were measured in the troughs located under the influence of trees. Although in all cases runoff is well correlated with rainfall, the troughs under trees (SF1, LC2) plot out as a separate group, with two families of data that can be related to different responses to rainfall and the season (Figure 21.5). The greater amount of interception means that runoff commences at higher rainfall amounts under

Table 21.1 Percentage of fines in the eroded material

Site	< 2 mm ϕ %
Beech woodland (Santa Fe	72
Evergreen oak woodland (La Castanya)	19
Evergreen oak woodland (L'Avic)	41
Shrubland (L'Avic)	80

From Llorens (1987) and Sala (1988)

trees and that stemflow increases runoff at high-rainfall events. In the beechwood higher values generally occur in the summer (higher stemflow) and in winter (lower interception) while in the evergreen oak woodland (La Castanya) higher responses are in the autumn and spring, i.e. closest relationship with rainfall events.

The troughs located between trees, in concavities usually filled with litter, produce a runoff strongly correlated to rainfall (SF2, LC1, Figure 21.6a). Finally, areas where the influence of vegetation is minimized have the strongest runoff rates (for example, bounded trough LC3) or better correlation with rainfall (troughs draining a percentage of bare rock like SF3), as shown in Figure 21.6b. Variability is highest for runoff/erosion relationships even for troughs with a very good rainfall/runoff correlation (Figure 21.7), although always best correlated and significantly different from zero in Santa Fe.

Because a study of the individual trough behaviour in l'Avic was not possible with the limited and discontinuous data available, a series of rainfall simulation experiments were carried out on the shrubland and woodland areas of the l'Avic sites. Six rainfall experiments were undertaken, three in the shrubland area and three in the woodland area, all located in the middle of the slope. Table 21.2 contains the main features of the plots and the results of the rainfall simulations. Runoff in the woodland was very irregular (Figure 21.8a), probably due to the very small amount of water and sediment produced (ten times smaller on average than in the shrubland plots), and as a consequence some errors may have been recorded. Plots W1 and W3, both covered by litter (80–100 per cent), have mean runoff rates of only 0.08 and 0.05 mm/h respectively, while plot W2, with 60 per cent litter and 40 per cent debris on the surface, produced a mean runoff rate of 1.9 mm/h.

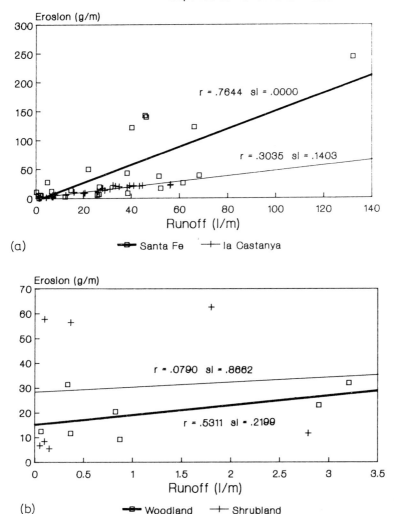

Figure 21.4   Runoff/erosion relationships in the Montseny sites (a) and in the Prades sites (b)

In the shrubland area (Figure 21.8b) plot S1, located on a bare surface between two shrubs and with 60 per cent cover of debris, produced the highest runoff rates, with fairly constant rates of 13 mm/h. Plot S2 had also a high and steady response (9 mm/h), although rainfall reached the bare ground (20 per cent debris and no litter) through a 50 cm tall shrub of *Cistus albidus*. Plot S3 produced very little runoff, about 0.2 mm/h maximum; in this case rainfall was produced over a 30 cm high *Cistus salviifolius* covering 90 per cent of the plot, in addition to 30 per cent coverage of litter.

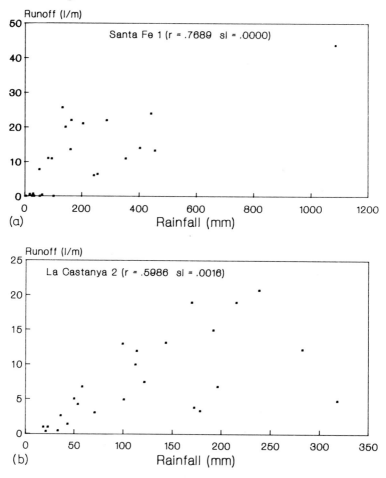

Figure 21.5   Rainfall/runoff relationships in Santa Fe (a) and La Castanya (b) in troughs close to trees

DISCUSSION

Comparison between rainfall, runoff and sediment dynamics in a woodland and a shrubland indicates in all cases a higher erosion rate in the latter. Runoff, although always correlating well with rainfall, differs widely from plot to plot within each environment. This is partly due to spatial variations in surficial material properties and is also strongly influenced by vegetation, i.e. the importance of stemflow, interception and percentage of litter coverage.

The differences between the two woodland communities can be related to the response of particular species to interception and stemflow, and to litter

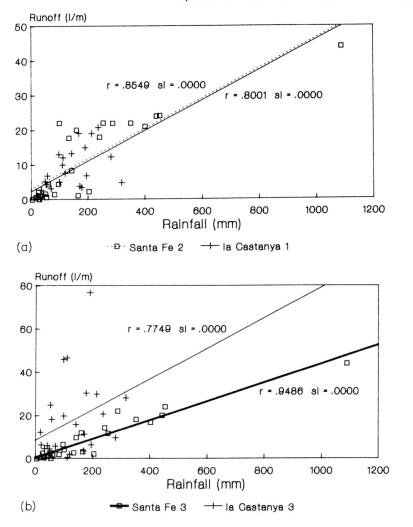

Figure 21.6   Rainfall/runoff relationships in the Santa Fe and La Castanya in troughs located in concavities (a) and where the influence of vegetation is minimized (b)

production and distribution. The correlation between runoff and rainfall is stronger where vegetation plays a secondary role (for example, where the trough is in the open or when the percentage of rock outcrop is high), and is more variable under the influence of a diversified vegetation community.

Higher runoff and soil loss rates occur on the bare surfaces of a shrubland while lower rates occur under trees in a woodland. Plant cover increases the irregularities of runoff and erosion on the slopes and thus makes the system less integrated, but at the same time this slows down erosion processes. In

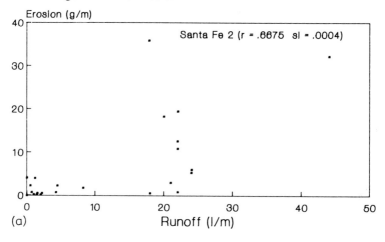

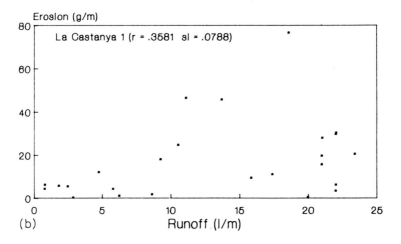

Figure 21.7 Runoff/erosion relationships in Santa Fe (a) and La Castanya (b)

certain deciduous species stemflow produces high pulses of runoff and erosion while at high percentages of litter cover runoff response is more adjusted because of storage. Small shrubs are probably the best agents in retaining debris sliding on slopes.

The different responses to rainfall, both in the generation of overland flow and erosion, can be explained by basic soil and vegetation characteristics and their conditions immediately before, during and after the rainfall events. Work by Sevink (1988) in the Montseny areas has shown large differences in chemical and physical properties of organic soil profiles, such as nutrient status, water-holding capacity and water repellence, and he considers that hillslope runoff

Table 21.2 Rainfall simulation results in l'Avic landscapes

	Shrubland			Woodland			Mean values	
	S1	S2	S3	W1	W2	W3	Shrub	Wood
Slope	26	25	30	25	29	26	27	27
Aspect	218	210	205	202	202	202	211	202
% Stones	60	20	50	5	40	20	43	22
% Tree	0	0	0	0	0	70	0	23
% Bush	0	50	90	0	0	0	47	0
% Grass	5	0	30	0	0	0	12	0
% Litter	0	0	30	100	60	80	10	80
% Soil moisture	22	22	25	66	25	24	23	38
Runoff start (min, sec)	3'45"	3'35"	4'00"	2'00"	1'03"	3'45"	3'47"	2'16"
Total runoff (mm)	3.01	2.62	0.05	0.04	0.59	0.28	1.89	0.30
Mean runoff rate (mm/h)	10.38	9.29	0.17	0.08	1.90	0.05	6.61	0.68
Total soil loss (g/m^3)	0.83	0.83	0.04	0.00	0.08	0.04	0.57	0.04
Sediment conc. (g/l)	0.30	0.30	0.90	0.00	0.10	1.50	0.50	0.53

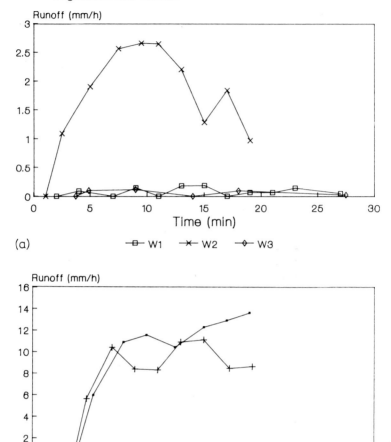

Figure 21.8 Results obtained in the Prades sites during rainfall simulation experiments in the woodland (a) and in the shrubland (b)

is probably strongly influenced by the seasonal variation in the organic soil profiles. We have found that the highest runoff percentages related to rainfall were in summer for the Santa Fe sites and in autumn for La Castanya, while minimum values were registered during spring in Santa Fe and during winter in La Castanya. The rhythm of seasons is steadier and better correlated with other variables in the beech woodland of Santa Fe.

The debris slopes of La Castanya and l'Avic seem to behave similarly to slopes in a dry Mediterranean climate, where effectiveness of rainfall is greatest in spring and autumn (Thornes, 1980) due to high-intensity events. In Santa Fe the system functions more like a humid environment (Dunne, 1978; Anderson and Burt, 1978) because of higher water retention of its thicker regolith and soils throughout the seasons, provoking a more regular overland flow response to rainfall and consequent effectiveness in the washing of material.

It is interesting to note that in the evergreen-oak sites particles bigger than 2 mm ϕ (often bigger than 10 mm ϕ) are commonly trapped, probably because woodland vegetation structure (low percentage of shrubs) and soils (more water retention) provide an appropriate sliding surface for debris to move. In the shrubland, most of the material collected in the troughs is fine, which points to the importance of the retention of bigger particles by small bushes.

In all environments individual plants produce differences due to their role in interception and stemflow production. In general the results indicate greater variability with increasing vegetation interference (trees, shrubs or litter accumulation).

The problem arises when sampling for large areas with irregular and heterogeneous vegetation. Future research should perhaps be focused on the behaviour of the different species individually.

ACKNOWLEDGEMENTS

Research in Montseny mountain was initiated with a grant from Fundacion Juan March (1981) and later funding has been provided by CIRIT (1982, 1983). We are also grateful for facilities provided by the field centres of El Vilar de la Castanya (Servei de Protecció de la Natura, Direcció General del Medi Rural, Generalitat de Catalunya), of Santa Fe del Montseny (Servei de Parcs Naturals, Diputació de Barcelona) and Poblet (Servei del Medi Natural, Diputació de Tarragona).

REFERENCES

Anderson, M. G. and T. P. Burt (1978). Experimental investigation concerning the topographical control of soil water movement on hillslopes. *Zeitschrift für Geomorphologie Supp.*, **29**, 52–63.

Avila, A. and F. Rodà (1988). Export of dissolved elements in an evergreen-oak forested watershed in the Montseny Mountain (NE Spain). In A. C. Imeson and M. Sala (eds), *Geomorphic Processes in Environments with Strong Seasonal Contrasts*, Vol. I, *Hillslope Processes, Catena Supplement*, **12**, 1–11.

Calvo, A., J. Gisbert, E. Palau and M. Romero (1988). Un simulador de lluvia portátil de fácil construcción. In M. Sala and F. Gallart (eds), *Métodos y técnicas para la medición en el campo de procesos geomorfológicos*. *Sociedad Española de Geomorfologia, Monografia* No. 1, 6–15.

Cervera, M. (1986). Spatial variation of surface wash and erosion in the slopes of Santa Fe. In M. Sala, F. Gallart and N. Clotet (eds), Geomorphic processes in environments with strong seasonal contrasts, *COMTAG Symposium Excursion Guide*, Barcelona, pp. 69–72.

De Ploey, J. (1981). The ambivalent effects of some factors of erosion. *Mm. Inst. Geol. Univ. Louvain*, **XXXI**, 171–81.

Dunne, T. (1978). Field studies of hillslope flow processes. In M. J. Kirkby (ed.), *Hillslope Hydrology*, Wiley, pp. 227–90.

Escarré, A. *et al.*, (1986). Balance hidrico, meteorización y erosión en una pequeña cuenca de encinar mediterráneo (Proyecto LUCDEME). *ICONA, Monografias*, **47**, 57–115.

Llorens, P. (1987). Processos de transport de sediment als vessants en relació a diferents unitats de paisatge en una àrea de muntanya mediterrània. In M. Sala and J. Martin-Vide (eds), *Mesura i experimentació en Geografia Fisica, Notes de Geografia Fisica*, Barcelona, pp. 63–72.

Rodà, F. (1983). *Biogeoquimica de les aigües de pluja i de drenatge en alguns ecosistemes forestals del Montseny*. Unpublished PhD, Departament d'Ecologia Universitat Autónoma de Barcelona.

Sala, M. (1988). Slope runoff and sediment production in two Mediterranean mountain environments. In A. C. Imeson and M. Sala (eds), *Geomorphic Processes in Environments with Strong Seasonal Contrasts*, Vol. I *Hillslope Processes, Catena Supplement* **12**, 13–29.

Scoging, H. M. and J. B. Thornes (1979). Infiltration characteristics in a semi-arid environment. *IASH publ.* 128, 159–68.

Sevink, J. (1988). Soil organic horizons of Mediterranean forest soils in NE-Catalonia (Spain): Their characteristics and significance for hillslope runoff and effects of management and fire. In A. C. Imeson and M. Sala (eds), *Geomorphic Processes in Environments with Strong Seasonal Contrasts*, Vol. I *Hillslope Processes, Catena Supplement*, **12**, 31–43.

Terrades, J. *et al.* (1984). *Introdució a l'ecologia del faig al Montseny*. Diputació de Barcelona, Barcelona.

Thornes, J. B. (1976). *Semi-arid erosional systems: Case studies from Spain*, Geogr. Papers, No. 7, London School of Economics.

Thornes, J. (1980). Erosional processes of running water and their spatial and temporal controls: a theoretical viewpoint. In M. Kirkby and R. C. P. Morgan (eds), *Soil Erosion*, Wiley, 129–82.

Yair, A. and H. Laves (1985). Runoff generation in arid and semi-arid zones. In A. Anderson and T. Burt (eds), *Hydrological Forecasting*, Wiley, 183–220.

22 Runoff Hydrographs from Three Mediterranean Vegetation Cover Types

C. F. FRANCIS and J. B. THORNES
University of Bristol

SUMMARY

Rainfall simulations on plots 5 by 2 m under degraded matorral, shrub matorral and high matorral with pines produce decreasing runoff and soil losses in that order. These results support and provide parameters for an erosion competition model. The main difference in magnitude is between the degraded and shrub matorral, suggesting that shrub matorral, even with a modest percentage cover can provide erosion protection similar to that of trees. The rates for high-intensity storms are comparable with those obtained from much larger natural plots exposed to similar rainfall.

INTRODUCTION

In developing models to investigate the competitive behaviour between vegetation cover and soil erosion, Thornes (1988) assumed the relationship between cover and runoff and soil loss to be exponential. This is in accordance with the results of Elwell and Stocking (1976), Dunne, Dietrich and Brunengo (1978) and Lee and Skogerboe (1985) for a variety of environments. In order to establish and parameterize this relationship for covers characteristic of semi-arid Spain, where the erosion problem is at times acute, a medium-term investigation of the ecology of matorral and its relationship to erosion has been undertaken.

It is widely assumed that in south-east Spain desertification is a common and active process and that this is manifested by intense erosion due to lack of cover following deforestation. A corollary of this belief is that it is assumed that replacement of matorral by tree species is the obvious and sometimes the only

Vegetation and Erosion
Edited by J. B. Thornes
©1990 John Wiley & Sons Ltd

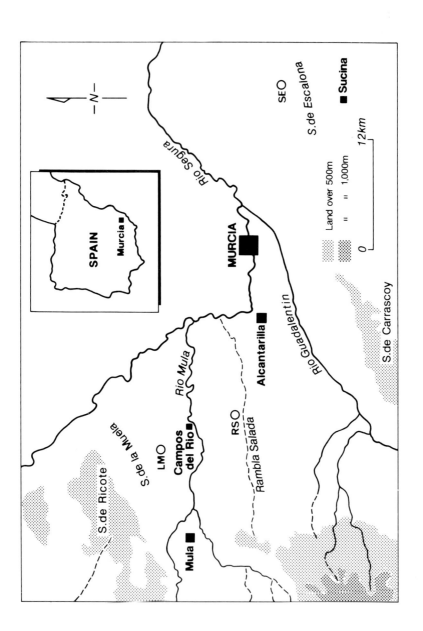

Figure 22.1 Site location

solution. It is generally asserted that trees, by providing the most dense possible cover, invariably offer a much higher level of protection from erosion. It is also part of our objective therefore in this project to attempt to establish the relative roles of trees, matorral and degraded matorral in providing protection from erosion. In this chapter a series of rainfall simulation experiments to measure erosion are described to throw light on the erosion problem in a regional context where cover is frequently at or about the critical 30 per cent level (Esteve Chueca, 1972; Dargie, 1987) and where changing agricultural policy may significantly increase the rate of abandonment of land from arable to other uses.

Finally we wish to provide some evidence which will go towards explaining the locally high variations in estimation of sediment yield in this semi-arid environment as a basis for predicting regional erosion rates from satellite and other remotely sensed imagery.

SITE LOCATION AND DESCRIPTION

The sites selected lie in the lowlands of the Province of Murcia within the semi-arid region of south-east Spain (Figure 22.1). The mean annual rainfall over much of the area is less than 300 mm (299.3 mm in Murcia itself) with pronounced seasonal and annual variations. The rainfall maximum occurs in October (mean 51.6 mm) with a secondary peak in April (36.9 mm) and the autumn storms are often intense and prolonged in duration and are associated with high runoff, erosion and flooding. The summer months are relatively dry with only 9.2 per cent of the annual rainfall falling between June and August. Taken together with the high temperatures (mean maximum and minimum values at Murcia in August are 32.4 °C and 21.0 °C), these ensure extreme summer drought. Periods of more prolonged drought occur; for example, between 1978 and 1984 the mean annual precipitation was 172.4 mm (excluding 1980 when 398.9 mm of rain fell). More detailed accounts of the climate are presented by López Bermúdez (1973, 1985) and Capel Molina (1984).

Much of this region falls within the thermomediterranean bioclimatic level where the climax vegetation is the association *Chamaeropo humilis–Rhamneto lycioides sigmetum*, a sclerophyllous matorral, which is only found in remnants due to intense anthropogenic activity (Alcaraz Ariza and Peinado Lorca, 1987). This association has largely been replaced by various types of degraded shrubland (matorrales and tomillares) and steppe pastures, descriptions of which have been published by Esteve Chueca (1972), Dargie (1987) and Alcaraz Ariza and Peinado Lorca (1987).

The three sites chosen in this study are characterized by different types of vegetation. Two of these, at La Muela and Rambla Salada, lying to the north-west of Murcia, are shrubland areas whilst the third at Sierra Escalona to the

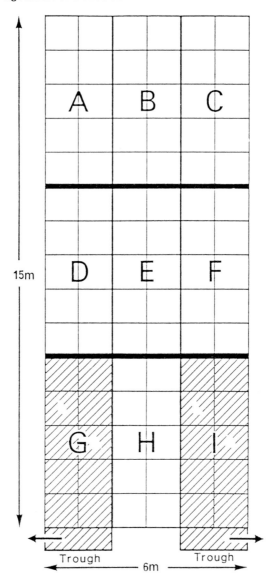

Figure 22.2 Sampling design for experimental plot

south-east of Murcia, is a pine forest with a shrub understorey. All the sites are north-facing slopes with calcareous rich soils.

At La Muela the vegetation is a tomillar shrub (so named because of the large amount of tomillo or *Thymus* species). The main cover species include *Rosmarinus officinalis*, *Thymus hyemalis*, *Helianthemum almeriense*, *Sideritis leucantha*,

Rhamnus lycioides, and *Salsola genistoides*. The site is located north of the village of Campos del Rio on a cuesta of calcareous sandstone overlying Miocene marls. The soils are calcareous regosols with a very stony upper horizon and covering due to the break-up and transport of the sandstone capping rock. Published papers on a very similar site describe the ecological conditions (Francis *et al.*, 1986), the vegetation (Fisher *et al.*, 1987), and the soil erosion (Romero Díaz *et al.*, 1988).

The second site in the Rambla Salada is on gently sloping ground draining into a deeply entrenched ephemeral channel. The calcareous regosols grade rapidly into Andalucian marls. Occasional thin pebble layers in the bedrock are exposed to form locally stony surfaces. The vegetation community is a matorral shrubland where the main cover species are *Anthyllis cytisoides*, *Thymus hyemalis*, *T. zygis*, *Artemisia campestris*, *A. herba alba*, *A. barrelieri*, *Helianthemum almeriense*, *Cistus albidus*, and *Thymelaea hirsuta*.

The last site lies on the steep cuesta slope of the Sierra Escalona and is underlain by late Tertiary sandstones and marls. The vegetation community consists of a woodland of *Pinus halepensis* between 2 to 3 m high, with occasional individuals reaching 5 to 6 m high. The bush understorey is primarily composed of *Rosmarinus officinalis*, but with isolated examples of *Juniperus oxycedrus*, *Genista valentina*, and *Pistacia lentiscus*. Smaller woody shrubs include *Thymus membranaceus*, *Helianthemum cinereum*, *Fumana ericoides*, and *Fumana hispidula*. A herbaceous layer includes *Centurium erythraea* and *Polygala rupestris*.

SAMPLING DESIGN AND METHODOLOGY

Each location of three plots, 6 m wide by 15 m long, were subdivided into nine 10 m² rectangles 2 m by 5 m. These rectangles were further subdivided into 1 m² quadrats (Figure 22.2) and these formed the basis of the sampling design.

Vegetation Characteristics

The amount of vegetation cover at La Muela and Rambla Salada and understorey cover at Sierra Escalona were measured by point count analysis and the presence of vegetation (naming the species), bare ground, stone, or litter was recorded and presented as a percentage. For each rectangle the mean vegetation cover was found by averaging over the ten quadrats. The surveys were undertaken in September/October 1987 on La Muela, January/February 1988 on Rambla Salada and November/December 1987 on Sierra Escalona. The pine tree cover was estimated separately. The radius of the canopy of individual trees was measured in four compass directions, and then the location and canopy area were mapped. The amount of cover provided by the trees was then measured for a subsample using a gridded mirror and data were presented as a percentage of total cover. The amount of vegetation cover on the rectangles at Sierra Escalona was then estimated by averaging the percentage vegetation cover of the shrub understorey and that of the pine trees in proportion to the areas covered by shrubs and pine trees. Under the pine trees no attempt was made to merge the percentage shrub cover provided by the understorey with that provided by the trees. On pine areas therefore the percentage cover will be an underestimate.

The above-ground biomass was harvested on selected rectangles covering a total area of $210 \, m^2$, with eight rectangles on La Muela and Rambla Salada and five rectangles on Sierra Escalona, and recorded as woody shrubs, standing dead or herbaceous plants/grasses. Each group was weighed in the field to an accuracy of 25 g, and subsamples were taken from each. The subsamples were reweighed to 0.1 g, dried at 80 °C for 48 hours, and reweighed and the ratio of dry to wet weights used to correct the field values to dry weight. The biomass was measured in July 1988 during the simulation experiments.

On Sierra Escalona only the understorey above-ground biomass could be harvested because it was not possible to cut down the pine trees. The biomass of the pine trees was estimated using dimension analysis whereby the above-ground biomass is correlated with the tree diameter at breast height. Such correlations have been established for various tree species (Whittaker and Niering, 1975; Whittaker and Marks, 1975), but data for *Pinus halepensis* have not been found. Table 22.1(a) shows the estimated biomass of individual trees with diameters of 5, 10, and 15 cm at breast height for five groups of trees. These regressions give similar estimates of biomass for the smaller sized trees, but differ widely for the larger trees. In the study the diameters at breast height of those trees occurring wholly within the rectangle ranged between 1 and 16 cm, with five of the seven trees having diameters of 2 cm or less. The equations for *Picea rubens* and *Pinus cembroides* were used to estimate the biomass of trees contained entirely within the $10 \, m^2$ rectangle and overhanging branches. The trees with overhanging branches had diameters at breast height ranging between 2 and 12 cm.

Simulation Experiments

The simulation experiments were performed in rectangles G and I at the downslope end of the plots (Figure 22.2). At La Muela and Rambla Salada four Gerlach troughs were installed, using plots 1 and 2 at La Muela and 1 and 3 at Rambla Salada, whilst at Sierra Escalona six troughs were installed, two in each plot. The troughs were emplaced in December 1986 to allow the soil and vegetation time to settle afterwards. The first trial experiments were conducted in August 1987, but the main period of simulations was July 1988.

Table 22.1 (a) Comparison of regression equations to estimate the dry weight of individual trees (kg) once data have been transformed to base 10; (b) regressions to estimate branch wood and bark dry weight (kg) for trees with varying diameter

Tree type			Diameter at breast height		
			5 cm	10 cm	15 cm
(a)					
Shrubs and trees[1]	A	2.4667	8.6	36.7	85.9
	B	2.0980			
Deciduous[1]	A	2.2380	8.5	45.7	122.1
	B	2.4223			
Picea rubens[1]	A	2.3151	6.9	31.5	76.3
	B	2.1830			
Pinus cembroides[2]	A	2.7593	8.3	26.0	51.0
	B	1.6563			
Quercus hypoleucoides[2]	A	2.6775	8.3	28.2	57.9
	B	1.7728			
(b)					
Shrubs and trees[1]	A	1.8518	2.00	8.45	19.59
	B	2.0748			
Deciduous trees[1]	A	1.0823	0.97	6.45	19.51
	B	2.7276			
Picea rubens[1]	A	0.9115	0.49	2.85	7.98
	B	2.5428			
Pinus cembroides[2]	A	2.2107	2.47	7.98	15.85
	B	1.6915			
Quercus hypoleucoides[2]	A	1.9187	1.87	7.14	15.66
	B	1.9353			

[1]From Whittaker and Marks (1975); [2]from Whittaker and Niering (1975).
The estimates for shrubs and trees are based on samples from the Brookhaven National Laboratory, New York and Oak Ridge, Tennessee; those for deciduous trees and *P. rubens* are from Hubbard Brook, New Hampshire; and those for *Pinus cembroides* and *Quercus hypoleucoides* are from samples in the Santa Catalina Mountains, Arizona.

The Gerlach troughs were based on a design described by Dunne, Dietrich and Brunengo (1980) and were 2 m long, with a steeply raking floor which narrowed at the downslope end. This prevented sedimentation within the trough and facilitated the periodic sampling of the runoff in beakers during the experiments. Before each experiment the Gerlach trough and lip were cleaned carefully to remove dust and organic matter. Water samples were collected during each experiment to measure the discharge, and some of these were filtered in the field to collect the organic and mineral debris in the flow. These two components will be separated later. The runoff rates were used to estimate the cumulative volume of runoff during the entire run.

Two types of simulation experiment were undertaken involving slightly different sampling designs. For the high-intensity sprays two SPRAYCO 1110191423 nozzles were used, mounted on an 'A' shaped frame with the two nozzles 2 m apart and 2 m above the ground. Trial runs showed that the sprayed area was more or less confined within the 2×5 m rectangle. On Sierra Escalona the frame was raised so that the nozzles were 4 m above the ground and sprayed above the pine trees.

The low-intensity sprays used two TOPLAS nozzles which were mounted facing upwards on pipes 1 m or 0.5 m high. (The spray was forced up and then fell down rather than falling straight down.) Water from the low-intensity sprays usually extended beyond the plot area, in particular landing on the concrete and metal apron of the Gerlach trough. This water contributed to runoff in the trough, and for low flows results in a significant source of error.

Water was brought to the site in four 200 litre tanks for each experiment and either two or four tanks of water were pumped onto the site. During the change-over period between barrels the spray usually stopped for a short period. This often resulted in a short burst of increased runoff and sediment losses and sampling was avoided in this period.

During each experiment 20 or 24 small beakers were arranged over the rectangle in rows of four to measure the application rate of water to the ground. Due to interception losses (especially on Sierra Escalona from the taller pine trees) the application rate refers to the quantity of water reaching the ground surface, and not the amount being pumped out of the nozzles. In trial runs this system was used to map spatial variations in the spray itself on unvegetated flat ground. For the experiments the application rate was averaged over the site. The raindrop sizes and distributions were measured using the stain paper technique (Brandt, 1986).

A wind break, 6 m long and 2 m high, was constructed using angle-iron posts and heavy-duty polythene and placed alongside the runoff plots. Despite this precaution, strong winds were problematic on several occasions.

Four types of experimental runs were undertaken. At La Muela and Rambla Salada a low-intensity spray for dry conditions (dry) was followed immediately by a low-intensity spray for wet conditions (wet) with a period of about half an hour between each. At Sierra Escalona, field evidence indicated that low-intensity sprays were highly unlikely to produce any runoff because of the much higher interception losses, not only from the pine trees, but also from the thick litter of pine needles. The high-intensity spraying was undertaken on all plots at La Muela and Rambla Salada, and on plots 1i, 1g (dry and wet run), 2i, 3i, and 3g at Sierra Escalona. Plot 2g has no pine trees growing on it. Following the high-intensity sprays on La Muela and Rambla Salada, three of the four rectangles were cropped to measure the biomass, and another low-intensity spray was conducted (bare) when the soils were still damp from the high-intensity experiments.

RESULTS

Vegetation Cover and Biomass

The mean biomass measured on La Muela was $0.443\,\mathrm{kg\,m^{-2}}$, and on Rambla Salada was $0.687\,\mathrm{kg\,m^{-2}}$. On Sierra Escalona the mean biomass for the bush understorey was $0.406\,\mathrm{kg\,m^{-2}}$ which rose to $2.544\,\mathrm{kg\,m^{-2}}$ using the regression for *P. cembroides* or $3.220\,\mathrm{kg\,m^{-2}}$ using the model for *P. rubens*. The estimate of biomass at La Muela is close to a reported value of $0.35\,\mathrm{kg\,m^{-2}}$ at a nearby location (Fisher *et al.*, 1987). The estimates for the shrubland at first appear low when compared with reported values from other Mediterranean ecosystems, being closer to biomass estimates for desert scrublands. For example, Whittaker and Niering (1975) cite biomass values between 0.263 and $1.310\,\mathrm{kg\,m^{-2}}$ for desert grasslands and shrublands in Arizona, compared with 1.12 and $1.88\,\mathrm{kg\,m^{-2}}$ for wetter scrubland and open woodland, and values between 11.4 and $79.0\,\mathrm{kg\,m^{-2}}$ for various types of mesic forests. Biomass data for Californian and Chilean mediterranean shrublands vary between 2 to $9\,\mathrm{kg\,m^{-2}}$ (Dodge, 1975; Kummerow, Montenegro and Krause, 1981; Gray, 1983). However, Godron *et al.* (1981) indicate a relationship between mean canopy height and biomass. They suggest that a community of *Thymus vulgaris* with a mean height of about 24 cm would have a biomass of about $0.4\,\mathrm{kg\,m^{-2}}$, whereas a *Rosmarinus officinalis* community standing about 65 cm high would have a biomass in the order of $1.5\,\mathrm{kg\,m^{-2}}$. Using this relationship a community of *P. halepensis* between 2 to 3 m high would have a biomass between 5 and $8\,\mathrm{kg\,m^{-2}}$. However, this should be taken as a maximum figure as Godron *et al.* indicate that *P. halepensis* has a relatively low above-ground biomass for its height. Indeed the one value given for this species is well below the regression line as a stand 7.5 m high had a biomass of about $8\,\mathrm{kg\,m^{-2}}$. These latter values accord well with the results presented here. The lower biomass values (and inferentially lower mean heights of the shrub community) may be indicative of the degraded state of many shrublands in south-east Spain due to overgrazing and other anthropogenic pressures in comparison with less disturbed state of Mediterranean shrublands elsewhere.

Comparison of the percentage cover on all 27 rectangles at La Muela and Rambla Salada give mean and standard deviations of 34.93 per cent and 6.60 and of 48.54 per cent and 10.69 respectively. Though the samples are not strictly independent within each group, a difference of means test indicates a significant difference between the two sites at the 0.001 level. Studies have shown the percentage vegetation cover and biomass to be correlated by a semi-log function. Here a semi-log regression of biomass (x) against percentage vegetation cover (y) gives a correlation coefficient (r) of 0.705 using the Rambla Salada and La Muela data. When the data for Sierra Escalona are added, the correlation falls to 0.618. Both of these correlations are significant by the 0.001 level.

Characteristics of the Rainfall Simulator

The median volume drop diameters for the low and high intensities were 0.9 and 2.36 mm respectively. The value for the high intensity compares well with values calculated using three separately derived equations (Table 22.2); however, the low-intensity D50 is well below other estimates for 25 mm hr^{-1} storms. The drop size distributions, shown in Figure 22.3, indicate that the distribution for the low-intensity simulations is markedly peaked and more closely resembles distributions for rainfall intensities of 0.1 to 1.3 mm hr^{-1} than 13.0 to 25.5 mm hr^{-1}. The high-intensity simulation has a flatter and wider drop size distribution than empirical values for natural rain.

A summary of the application rates is given in Table 22.3. The low-intensity simulation experiments had application rates which ranged from 18.3 to

Table 22.2 Median volume drop diameters (mm) calculated using various regressions

	25 mm hr^{-1}	100 mm hr^{-1}	
Laws and Parsons (1943) D50 = 1.238 I$^{0.182}$	2.22	2.86	Washington and Europe
Marshall and Palmer (1943) D50 = 0.92 I$^{0.21}$	1.81	2.42	Coalescence and frontal rains
Brandt (1986) D50 = 1.416 I$^{0.123}$	2.10	2.49	Brazil
Simulation experiment	0.90	2.36	

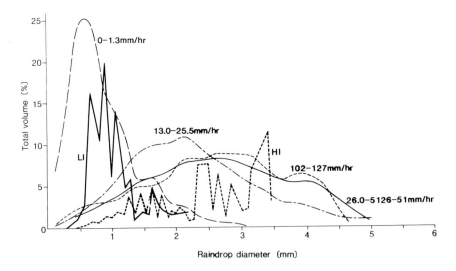

Figure 22.3 Drop size distributions for natural rainstorms with different intensities (after Carter *et al.*, 1974) and for the simulation experiments

31.5 mm hr^{-1} with a mean value of 25.8 mm hr^{-1} and standard deviation of 3.42. A one-hour storm of this magnitude has a return period of about 5 years in Murcia, so that with respect to the intensities this is a realistic low-intensity storm for the region. However, the small median drop sizes suggest that the kinetic energies may not be correct. The kinetic energies have not been calculated because of the difficulty of estimating the velocity of the drops under pressure. A one-hour rainfall with a 100-year return period is estimated to have about 48.7 mm in Murcia (Elias Castillo and Ruiz Beltran, 1979). Although this estimate is crude (the database is only 15 years), it is evident that the high-intensity application rates used in the experiments, ranging between 65.6 and 123.8 mm hr^{-1}, are extreme values. This largely illustrates the difficulty of obtaining practical simulators which replicate natural rainfall while covering a meaningful area. Nevertheless, for the purposes of the study, the high-intensity sprays provide a useful comparison for the low-intensity events and an analogy for extreme events.

The high-intensity sprays have a mean value of 100.7 mm hr^{-1} (standard deviation 17.41) and are more variable than the low-intensity sprays. This

Table 22.3 Application rates (mm hr^{-1}) during the simulation experiments at La Muela, Rambla Salada and Sierra Escalona

		La Muela		Rambla Salada		Sierra Escalona	
High intensity							
	1.	1g	99.3	1g	116.7	1g(d)	65.6
	2.	1i	101.8	1i	101.8	1g(w)	94.1
	3.	2g	87.9	3g	102.0	1i	70.1
	4.	2i	119.3	3i	110.7	2i	98.6
						3g	118.5
						3i	123.8
Low intensity							
dry	1.	1g	22.7	1g	24.1		
	2.	1i	21.8	1i	27.6		
	3.	2g	27.8	3g	25.1		
	4.	2i	25.5	3i	18.3		
wet	1.	1g	25.9	1g	26.2		
	2.	1i	22.4	1i	26.8		
					(25.4)		
	3.	2g	31.3	3g	29.2		
	4.	2i	25.1	3i	21.5		
bare	1.	1i	26.5	1g	24.3		
	2.	2g	30.7	3g	31.5		
	3.	2i	22.8	3i	30.6		
Mean	high		102.1		107.8		95.1
	dry		24.5		23.8		
	wet		26.2		25.8		
	bare		26.7		28.8		

variability is due to the greater and more variable interception losses by *P. halepensis* on the plots at Sierra Escalona.

Runoff Hydrographs

A sample of the runoff hydrographs is shown in Figure 22.4. For the high-intensity sprays (a) they show a rapid rise and recession, and in some cases variable discharge during the experiment. The hydrographs also clearly indicate

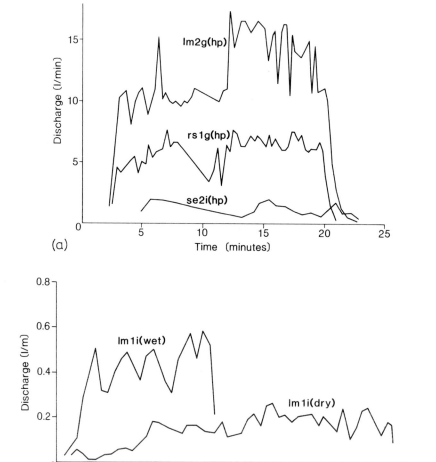

Figure 22.4 Examples of runoff hydrographs showing (a) variations in runoff for the high intensity simulations at La Muela, Rambla Salada, and Sierra Escalona; (b) variations between the dry and wet runs

differences in the magnitude of runoff on the different sites. Typical hydrographs from the low-intensity sprays (b) show a slower time to rise, but still exhibit variable discharges and between plot variations. Three variables of the runoff hydrographs are considered in more detail: the mean flow rate, the steady flow rate after the hydrograph has stabilized, and the time to rise.

The mean runoff is taken as the length of time of the experiment divided by the total accumulated runoff volume, and the values are given in Table 22.4. During the high-intensity runs there is a large difference in the runoff rates between La Muela (mean $= 9.96 \, l \, m^{-1}$), Rambla Salada (mean $= 5.66 \, l \, m^{-1}$), and Sierra Escalona (mean $= 0.87 \, l \, m^{-1}$). Expressed as a percentage of the application rate this gives mean values of 59.2 per cent on La Muela, 31.7 per cent on Rambla Salada, and only 4.9 per cent on Sierra Escalona. This verifies the much greater runoff on La Muela and Rambla Salada compared with Sierra Escalona.

The runoff generated on the metal apron of the Gerlach trough during the low-intensity experiments can be estimated by dividing the average application rate by the area of the impermeable surface. Assuming a catchment area of $0.2 \, m^2$, the amount of discharge due to the spray is about 0.06 to $0.09 \, l \, m^{-1}$. It is evident that the low flows registered at Rambla Salada for the dry runs

Table 22.4 Mean flow rates $(l m^{-1})$ and percentage of application rates at La Muela, Rambla Salada and Sierra Escalona

		La Muela $l \, m^{-1}$	%		Rambla Salada $l \, m^{-1}$	%		Sierra Escalona $l \, m^{-1}$	%
High intensity									
	1g	11.77	71.1	1g	4.95	25.4	1g(d)	0.10	0.9
	1i	8.29	48.9	1i	5.15	30.4	1gw	0.17	1.1
	2g	9.59	65.5	3g	6.96	40.9	1i	0.56	4.8
	2i	10.17	51.2	3i	5.57	30.2	2i	0.97	5.9
							3g	1.97	10.0
							3i	1.42	6.9
Low intensity									
dry	1g	0.59	15.5	1g	0.13	3.3			
	1i	0.14	3.8	1i	0.07	1.5			
	2g	0.19	4.1	3g	0.05	1.2			
	2i	0.29	6.7	3i	0.11	3.7			
wet	1g	1.07	24.9	1g	0.12	2.8			
	1i	0.38	10.1	1i	0.17	3.8			
					(0.19)	(4.4)			
	2g	0.74	14.1	3g	0.39	8.0			
	2i	0.41	9.7	3i	0.15	4.1			
bare	1i	0.41	9.3	1g	0.06	1.4			
	2g	0.69	13.5	3g	0.27	5.1			
	2i	0.43	11.4	3i	0.28	5.5			

are almost entirely due to the spray rather than overland flow across the plots. This is corroborated by field observations that in the first three cases overland flow was not observed.

Some generalizations can be made about the runoff for the low-intensity mean rates. Firstly, the runoff rates are higher at La Muela than Rambla Salada for the various conditions (dry, wet, and after cropping) by an order of 2.6 to 3.4 times; and secondly, runoff values increased by twofold between dry and wet experiments, but cropping under damp conditions did not further increase runoff rates. In the latter case the existence of very rough ground conditions due to the microtopography and stones still provided enough storage and friction to retard flow despite the lack of protection afforded by the vegetation.

The percent runoff for the low intensities emphasizes the increase in the proportion of runoff by twofold between the dry and wet runs and the slightly lower increase between the dry and devegetated runs.

The steady flow rates (Table 22.5) are higher than the mean flow rates, but basically show the same patterns: higher runoff on La Muela compared with Rambla Salada, similar increases in runoff between dry and wet vegetated conditions and dry and wet devegetated conditions for the low-intensity sprays and the differences between the sites for the high-intensity sprays.

The time to rise, defined as the time taken to reach steady-state flow rates, is most rapid for the high-intensity sprays (Table 22.6). The mean response time

Table 22.5 Steady flow rates at La Muela, Rambla Salada, and Sierra Escalona ($l\,m^{-1}$)

		La Muela		Rambla Salada		Sierra Escalona	
High intensity							
	1.	1g	15.459	1g	6.511	1g(d)	0.188
	2.	1i	12.760	1i	7.019	1g(w)	—
	3.	2g	14.760	3g	9.261	1i	—
	4.	2i	11.541	3i	6.674	2i	1.300
						3g	2.373
						3i	2.052
Low intensity							
dry	1.	1g	0.799	1g	0.022		
	2.	1i	0.178	1i	0.070		
	3.	2g	0.303	3g	0.048		
	4.	2i	0.337	3i	0.131		
wet	1.	1g	1.137	1g	0.167		
	2.	1i	0.455	1i	0.180		
					(0.202)		
	3.	2g	0.749	3g	0.365		
	4.	2i	0.442	3i	0.102		
bare	1.	1i	0.598	1g	—		
	2.	2g	1.124	3g	0.356		
	3.	2i	0.577	3i	0.469		

Table 22.6 Estimates of the time to rise at La Muela, Rambla Salada, and Sierra Escalona

		La Muela		Rambla Salada		Sierra Escalona	
High intensity							
	1.	1g	4′18″	1g	2′56″	1g(d)	6′50″
	2.	1i	3′33″	1i	5′34″	1g(w)	8′44″
	3.	2g	3′36″	3g	4′01″	1i	8′44″
	4.	2i	5′51″	3i	3′26″	2i	5′44″
						3g	3′45″
						3i	4′39″
Low intensity							
dry	1.	1g	10′35″	1g	—		
	2.	1i	30′30″	1i	—		
	3.	2g	28′30″	3g	—		
	4.	2i	22′30″	3i	—		
wet	1.	1g	12′15″	1g	19′30″		
	2.	1i	12′30″	1i	5′30″		
					(7′40″)		
	3.	2g	9′52″	3g	10′00″		
	4.	2i	7′45″	3i	7′30″		
bare	1.	1i	17′29″	1g	38′00″		
	2.	2g	18′45″	3g	7′30″		
	3.	2i	13′30″	3i	9′55″		

is marginally greater at Rambla Salada, with 3′59″, than at La Muela with 4′20″, with Sierra Escalona lagging at 6′24″. For the low-intensity sprays the time to rise at Rambla Salada for dry conditions was ignored due to the difficulty of establishing whether or not flow occurred at all. On La Muela the time to rise occurred at a mean value of 23′01″. For the wet run this time was halved to 10′36″, which is very similar to the mean response time at Rambla Salada of 10′02″. When the vegetation was cropped, the time to rise was slightly greater, being on average 16′35″ at La Muela and 18′28″ at Rambla Salada. This underlines the observation that runoff is greater and more rapid for wet antecedent conditions than cropped conditions.

Soil Losses

The transported material consists of both organic and inorganic fractions. These two components have not yet been separated, and the following analysis is based on the total amount. The mean values for sediment concentrations are given in Table 22.7, and shown for high intensities in Figure 22.5. For the high-intensity experiments mean concentrations are similar with $0.51 \, \text{g} \, \text{l}^{-1}$ on Rambla Salada and $0.57 \, \text{g} \, \text{l}^{-1}$ on Sierra Escalona, whereas on La Muela the mean concentration is about fourfold at $2.25 \, \text{g} \, \text{l}^{-1}$. On the assumption of a one-hour storm at the high-intensity rate, total soil losses, assuming a constant sediment concentration,

Table 22.7 Mean sediment concentrations (g l^{-1}) per experimental simulation at La Muela, Rambla Salada, and Sierra Escalona

	La Muela		Rambla Salada		Sierra Escalona	
High intensity						
	1g	3.39	1g	0.519	1g(d)	0.254
	1i	2.33	1i	0.245	1g(w)	0.488
	2g	1.44	3g	0.472	1i	1.054
	2i	1.84	3i	0.803	2i	0.407
					3g	0.586
					3i	0.631
Low intensity						
dry	1g	0.658	1g	—		
	1i	0.341	1i	—		
	2g	0.249	3g	—		
	2i	0.392	3i	0.093		
wet	1g	0.495	1g			
	1i	0.322	1i	0.106		
	2g	0.401	3g	0.085		
	2i	0.384	3i	0.110		
bare	1i	0.667	1g	0.350		
	2g	0.510	3g	0.165		
	2i	0.415	3i	0.214		

would be 1326 g for La Muela, 195 g for Rambla Salada and 27 g for Sierra Escalona.

During the low-intensity runs not all experiments were sampled due to the apparent lack of visible overland flow. However, the data indicate that there is little difference in concentrations between the wet and dry runs, but an increase (by 1.3 times on La Muela and 2.4 times on Rambla Salada) between the vegetated and non-vegetated conditions—this is despite there being no increase in overland flow. Between-site comparison indicated higher soil losses on La Muela than Rambla Salada.

Linear regression analysis between runoff rate and sediment concentration for *all* data give r equal to 0.817 for 32 data pairs, which is significant at the 0.001 level. Regression of percentage vegetation cover against soil loss also has a strong highly significant but negative correlation, with r equal to -0.731.

Further regression analysis of sediment concentrations and runoff indicates slight variations in performance between sites. Whilst the correlation is not significant for Sierra Escalona, at La Muela the rate of increase of concentration with runoff is greater than on Rambla Salada. The regression coefficients (r) are 0.928 on La Muela (significant at 0.001), and 0.763 on Rambla Salada (significant by 0.01).

The sediment concentrations during runoff tend to show a rapid fall from the maximum value of the first reading to lower and relatively stable values.

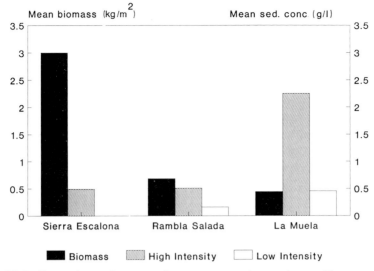

Figure 22.5 Comparisons of mean sediment concentrations and mean biomasses at the different sites

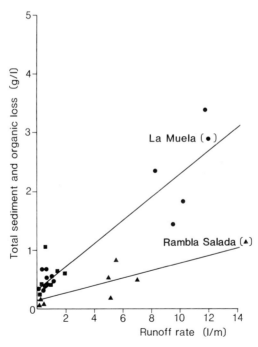

Figure 22.6 Relationships between sediment concentration and water runoff for the three sites at La Muela (●), Rambla Salada (▲), and Sierra Escalona (■). The regression for Escalona was not significant and is not shown

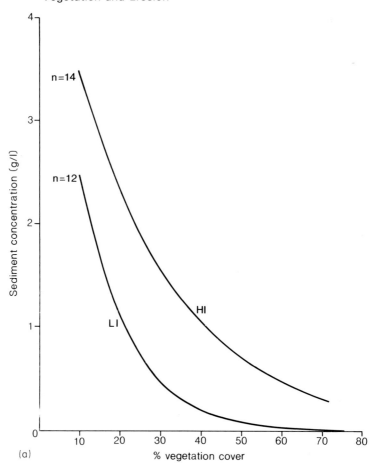

Figure 22.7(a) The relationship between vegetation cover and sediment concentration separating out the high and low intensity runs

on occasions the sediment concentration increases following a peak in runoff or when the run is reinitiated after the half-way stage. The sediment concentrations appear to be more stable for the low-intensity experiments on Rambla Salada than La Muela. That is, there is a greater response on La Muela to changes in runoff rates.

The interaction between runoff, vegetation and soil loss is of course intensity dependent. We have followed the prevailing view and fitted exponential curves, though correlation coefficients differ little between the linear and exponential models. Figure 22.7(a) illustrates the relationship between vegetation cover and sediment concentration separating out the low- and high-intensity simulations, and Figure 22.7(b) the relationship between cover and runoff when fitted by

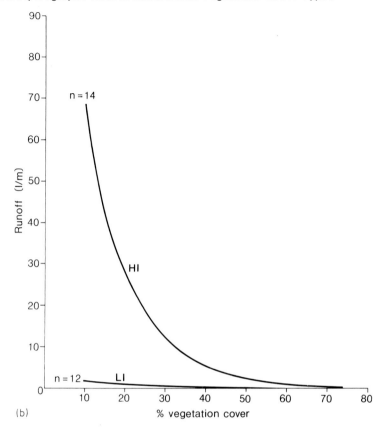

Figure 22.7(b) The relationship between vegetation cover and runoff distinguishing between high and low intensities

exponential models. There is a strong effect of cover on runoff for the intense storm simulations but a relatively small effect for low intensities. The form of the soil loss–vegetation curves, on the other hand, suggests that the effect of vegetation is greater at lower intensities.

CONCLUSIONS

As shown in Figure 22.5, site biomass increases in the order degraded matorral (0.443 kg m^{-2}), shrub matorral (0.687 kg m^{-2}) and high matorral with pines (c. 3.2 kg m^{-2}). Although the degraded and shrub matorral differ insignificantly in biomass, the difference in cover is quite marked (La Muela 35 per cent, Rambla Salada 49 per cent). The simulations undertaken were typical of the extremes of rainfall events in south-east Spain though the drop size distributions were on average too small for the low-intensity events and somewhat broader and flatter for the high-intensity events than occur in nature.

The runoff from these simulations confirm in general terms the results obtained by earlier workers. For high-intensity simulations runoff decreased in the order: high matorral ($9.95 \, l \, m^{-1}$), low matorral ($5.65 \, l \, m^{-1}$) and degraded matorral ($0.87 \, l \, m^{-1}$). Low-intensity simulations were not carried out on the high matorral. For dry soil simulations both degraded and bush matorral produced very little runoff but for simulations on wetted soil the runoff was much higher. In these cases the degraded matorral produced much higher rates of runoff than the shrub matorral. After harvesting the biomass the same strong contrasts were observed. Times to rise were halved when the soil was already wet.

Sediment yields were higher on the shrub matorral than the high matorral, but much higher on the degraded matorral at La Muela, the ratios being of the order of $1:7:48$ for a one-hour storm. Exponential curves fitted to the cover data and extrapolated below 30 per cent cover indicate strong intensity effects with the vegetation control being stronger at the lower intensities.

In general, although the bush matorral has higher erosion rates than that under trees, the difference is principally between the low matorral and the bush matorral. This indicates that bush matorral (in this case about 1 m high plants of *Anthyllis cytisoides*) can provide an extremely effective cover and protection against erosion and should be seriously considered as the major form of cheap erosion protection.

Finally we note that although small plots tend to give rates of erosion which are higher than normal, the size of these plots is sufficient to provide reasonable approximations to the natural variety of vegetation cover within the main types. Although this is offset to some extent by the inadequacy of the drop size distributions, the results obtained for the high-intensity storms are quite similar to those obtained by Romero Díaz *et al.* (1988) for low matorral at a similar location nearby. For such large events they suggest a range of about $70-244 \, t \, km^{-2}$, compared with $132 \, t \, km^{-2}$ found here.

ACKNOWLEDGEMENTS

This research project was funded by the Natural Environment Research Council of Great Britain. The authors are much indebted to the University of Murcia, Spain, for providing logistic support for both laboratory and field experiments. We particularly wish to thank Professor F. López Bermúdez (Departamento de Geografía) for his continual support during two years of experimental work. We are also grateful to Dr F. Alcaraz Ariza (Departamento de Botánica) for identifying plant species, and to Rosemary Thornes, Glenn Watts, Dra M. A. Romero Díaz, and Jose Martínez Fernandez for help with the field work.

REFERENCES

Alcaraz Ariza, F. and M. Peinado Lorca (1987). El sudeste Iberico semiarido. In M. Peinado Lorca and S. Rivas Martínez (eds), *La Vegetacíon de España*, Universidad de Alcala de Henares, pp. 259–81.

Brandt, C. J. (1986). *Transformations of the kinetic energy of rainfall with variable tree canopies*. Unpublished PhD Thesis, University of London.

Capel Molina, J. J. (1984). El clima de las zonas áridas. *Seminario sobre zonas áridas*, Inst. de Estudios Almerienses, 14–45.

Carter, C. E., J. D. Greer, M. J. Braud and J. M. Floyd (1974). Raindrop characteristics in south central United States. *Trans. Am. Soc. Agric. Engrs.*, **17**, 1033–7.

Dargie, T. C. D. (1987). An ordination analysis of vegetation patterns on topoclimate gradients in south east Spain. *Journal of Biogeography*, **4**, 197–211.

Dodge, J. M. (1975). *Vegetational changes associated with land use and fire history in San Diego County*. PhD Dissertation, Riverside: University of California.

Dunne T., W. E. Dietrich and M. J. Brunengo (1978). Recent and past erosion rates in semi-arid Kenya. *Zeitschrift für Geom. Supplmentbd.*, **29**, 130–40.

Dunne, T., W. E. Dietrich and M. J. Brunengo (1980). Simple portable equipment for erosion under artificial rainfall. *Journal Agr. Eng. Res.*, **25**, 161–8.

Elias Castillo, J. and L. Ruiz Beltran (1979). *Precipitaciones máximas en España*. Ministry of Agriculture, Spain, ICONA Monograph 21.

Elwell, H. A. and M. A. Stocking (1976). Vegetative cover to estimate soil erosion hazard in Rhodesia. *Geoderma*, **15**, 61–70.

Esteve Chueca, F. (1972). *Vegetación y flora de las regiones central y meridional de la Provincia de Murcia*. CEBAS, Murcia.

Fisher, G. C., M. A. Romero Díaz, F. López Bermúdez, J. B. Thornes and C. F. Francis (1987). La producción de biomasa y sus efectos en los procesos erosivos en un ecosistema mediterraneo semiárido del SE de España. *Anales de Biología*, **12**, 91–102.

Francis, C. F., J. B. Thornes, M. A. Romero Díaz, F. López Bermúdez and G. C. Fisher (1986). Topographic control of soil moisture, vegetation cover and land degradation in a moisture-stressed Mediterranean environment. *Catena*, **13**, 211–25.

Godron, M., J. L. Guillerm, P. Poissonet, M. Poissonet, M. Thiault and L. Trabaud (1981). Dynamics and management of vegetation. In F. Di Castri, D. W. Goodall and R. L. Specht (eds), *Mediterranean-Type Shrublands*, Elsevier, pp. 317–44.

Gray, J. T. (1983). Nutrient use by evergreen and deciduous shrubs in southern California. I. Community nutrient cycling and nutrient-use efficiency. *Journal of Ecology*, **71**, 21–41.

Kummerow, J., G. Montenegro and D. Krause (1981). Biomass, phenology, and growth. In P. C. Miller (ed.), *Resource Use By Chapparral And Matorral*, Springer-Verlag, Ecological Studies 39, pp. 69–96.

Laws, J. O. and D. A. Parsons (1943). The relation of raindrop size to intensity. *Trans. Am. Geophy. Union*, **24**, 452–9.

Lee, C. R. and J. G. Skogerboe (1985). Quantification of erosion control by vegetation on problem soils. In El Swaify, W. C. Moldenhauer and A. Lo (eds), *Soil Erosion and Conservation*, Soil Conservation Soc. America, pp. 437–44.

López Bermúdez, F. (1973). La vega alta del Segura: Clima, hidrología y geomorfología. Departamento de Geografía, Universidad de Murcia, Murcia.

López Bermúdez, F. (1985). *Sequía, aridez, y desertificación en Murcia*, Academia Alfonso X El Sabio, Murcia.

Marshall, J. S. and W. McK. Palmer (1943). The relation of raindrop size to intensity. *Journal of Meteorology*, **5**, 165–6.

Romero Díaz, M. A., F. López Bermúdez, J. B. Thornes, C. F. Francis and G. C. Fisher (1988). Variability of overland flow erosion rates in a semi-arid Mediterranean environment under matorral cover (Murcia, Spain). *Catena*, Supplement 13, 1–11.

Thornes, J. B. (1988). Erosional equilibria under grazing. In J. Bintliff, D. Davidson and E. Grant (eds), *Conceptual Issues in Environmental Archaeology*, Edinburgh University Press, pp. 32–64.

Whittaker, R. H. and P. L. Marks (1975). Methods of assessing terrestrial productivity. In H. Lieth and R. H. Whittaker (eds), *Primary Productivity of the Biosphere*, Springer-Verlag, pp. 55–118.

Whittaker, R. H. and W. A. Niering (1975). Vegetation of the Santa Catalina mountains, Arizona. V. Biomass, production and diversity along the elevation gradient. *Ecology*, **56**, 771–90.

23 The Influence of Lichens on Slope Processes in Some Spanish Badlands

R. W. ALEXANDER
Department of Geography, Chester College.
and
A. CALVO
Department of Geography, University of Valencia

SUMMARY

Semi-arid badlands represent an extreme environment for plant life and those in south-east Spain typically support only a sparse vegetation that is dominated by crustose and squamulose lichens. The most extensive and diverse lichen communities occur on shaded north-east facing slopes. In order to investigate the influence of lichens on surface processes, rainfall simulation tests were carried out in twelve plots at three sites in Almeria province. A number of parameters were recorded during the simulations and samples of runoff were collected for measurement and analysis of sediment content. Sediment concentration shows a significant positive correlation with slope gradient and is strongly related to surface characteristics. Lichen-covered plots exhibit more rapid ponding and runoff generation but produce lower sediment concentrations than bare plots. These results are interpreted in the light of experimental error, the presence and influence of mineral crusts and surface damage by goats. Their significance in elucidating the relationships between lichen cover and slope development is assessed.

OBJECTIVES

Badland surfaces in semi-arid lands are traditionally considered as areas in which biological activity is absent or nearly absent, but in many places this exists and

Vegetation and Erosion
Edited by J. B. Thornes

conditions the activity of geomorphological processes and landform evolution to a considerable degree. In badlands within the Almería area, lichens are the predominant life form and these are concentrated on north-facing slopes. The aim of this chapter is to examine the influence of lichens on surface wash.

STUDY AREA

The results presented in this chapter were obtained from three badland areas in Almería province, south-east Spain. These three sites (Figure 23.1) are located near to the settlements of Vera, La Herrería and Tabernas, and they represent three different situations of lichen development and other biological activities on badland slopes.

The Vera badlands, cut in Messinian gypsiferous marls (IGME, 1975a), have a restricted vertical extension with a maximum local relief of up to c. 50 m. These are young and have only developed from the dissection of one of the younger surfaces of the Vera basin (Harvey, 1987, p. 211). On these slopes exists a full range of erosional forms from shallow discontinuous gullies to deep linear gullies. Bare surfaces, on south-facing slopes (V1), are rilled and there is local evidence of mudsliding (Harvey, 1982). North-facing slopes (V2) have a denser vegetation cover containing a thin, resistant lichen crust.

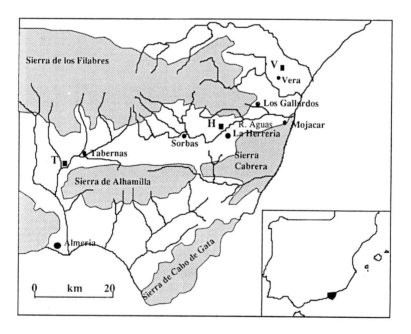

Figure 23.1 Location of study areas within Almería province

The badlands at La Herrería are developed on the Messinian marls in the east of the Sorbas basin (IGME, 1975b), along the central Aguas valley. Similarly to the situation at Vera, the north-facing slopes (H1, H2, H3, H4) have significant vegetation cover and are partially cut by deep gullies. South-facing slopes (H5, H6), however, have only a sparse vegetation and the bare surface is rilled. As in the Vera badlands, the main valley floors are flat and are used as agricultural fields.

The Tabernas badlands, cut in Tortonian mudstones (IGME, 1975c), have the greatest vertical extension, up to c. 200 m, and have been developed over much of the Quaternary during sustained, tectonically induced, dissection (Harvey, 1987). The study area seems to represent typical badlands developed entirely by surface wash and shallow mudslides (Harvey, 1982). On south-facing slopes (T1, T3, T4), rill networks are clear and well developed on the steepest slopes of the gully heads. The gully divides are smooth and the lower part corresponds to an old pediment surface with a cover of stones, vegetation and lichens. North- and east-facing slopes are generally less steep, have fewer rills and support a more continuous lichen cover in their lower parts (T2).

Present climatic conditions are very similar in the three areas. Using data from Elias and Ruiz (1977) for the locations of Vera, Los Gallardos and Tabernas, the annual precipitation ranges from 237 to 268 mm and annual mean temperatures from 22 to 23 °C. The annual average number of days of rain ranges from 17 in Los Gallardos, to 43 in Tabernas.

VEGETATION

The areas investigated support an open steppe vegetation characterized by false esparto grass (*Lygeum spartum*). Cover values of higher plants are very low and the vegetation is frequently dominated by lichens. A characteristic semi-arid terricolous lichen community (Seaward, 1983; Hawksworth and Hill, 1984) containing *Diploschistes dicapsis*, *Psora decipiens*, *Catapyrenium lachneum*, *Fulgensia fulgens*, *Squamarina lentigera* and *Toninia caeruleonigricans*, together with some small mosses (*Tortula* spp.), occurs at all sites. The composition and cover of this community show much variation with some areas being totally devoid of vegetation. Where plant cover is significant, the lichens appear to be the most effective in terms of surface processes and, for the purposes of this investigation, it is the physiognomy (morphology), rather than the floristics, of the lichen community that is of greatest importance.

In terms of their physiognomy, the lichens within the study area can be divided into four groups (Figure 23.2), which appear to differ in their influence on water and sediment movement. Group A contains *Toninia caeruleonigricans*, a *Collema* species and an indeterminate black crust and these form continuous

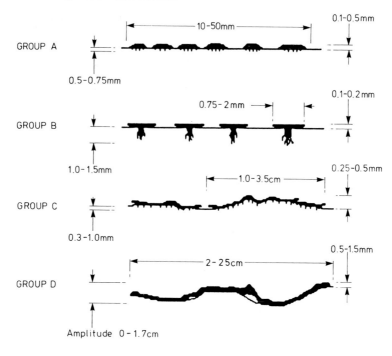

Figure 23.2 Diagrammatic representation of lichen growth form types

patches of thin, porous crust which are likely to provide some degree of surface binding. Group B contains *Psora decipiens* and *Catapyrenium lachneum*, which grow as groups of discrete squamules of 1–2 mm diameter and thus give rise to a discontinuous, non-porous but permeable surface. Some surface binding is provided as the squamules are anchored by rhizinae which penetrate to depths of 1–5 mm. Group C consists of *Squamarina lentigera*, which grows as a series of inbricate squamules, and *Fulgensia fulgens*, a crustose species with raised margins (placodioid). Where extensive (>1 cm) these plants provide a discontinuous, permeable crust with good surface binding. Finally, Group D consists of *Diploschistes dicapsis*, a robust crustose species which provides a non-porous, slightly permeable surface that frequently develops a microtopography with a relief amplitude of up to 1.7 cm.

EXPERIMENTAL DESIGN

In order to examine the response of badland surfaces to rain and the effects of differing lichen types and cover values, twelve plots were selected as representative of the main environments of each study area. The characteristics

Table 23.1   Site characteristics (V1, V2, etc., refers to sites as in text)

	V1	V2	T1	T2	T3	T4	H1	H2	H3	H4	H5	H6
Slope	33	18	25	21	11	26	18	18	18	22	25	26
Aspect	180	325	160	80	180	170	30	350	25	5	180	160
% Sand	17	46	22	25	31	22	20	13	13	15	21	21
% CaCO$_3$	35	27	34	22	25	39	48	48	48	45	45	45
% Lichens A	—	32	—	40	20	—	54	12	15	—	—	—
% Lichens B	—	2	—	5	5	—	—	—	—	—	—	—
% Lichens C	—	8	—	11	6	—	—	—	—	—	—	—
% Lichens D	—	—	—	30	—	—	—	—	—	—	—	—
% Lichens Tot	—	42	—	86	31	—	54	12	15	—	—	—
% Moss	—	15	—	4	—	—	10	4	—	—	—	—
% Lichen + Moss	—	57	—	90	31	—	64	16	15	—	—	—
% Herbs	—	9	—	1	—	—	5	10	10	4	3	1
% All plants	—	66	—	91	31	—	69	26	25	4	3	1
% Moisture (by weight)	1.1	2.5	0.6	0.1	0.3	0.9	2.1	1.3	1.3	1.5	2.4	2.4

of these plots are summarized in Table 23.1 and the photographs in Figure 23.3 show the surface of each plot prior to rainfall simulation.

Mineralogical analysis of clays from Tabernas and Vera (J. Petch, personal communication) shows that both materials are very similar, but the presence of smectite and gypsum in the Vera marls marks the main difference, possibly increasing the infiltration capacity of the Vera badland surfaces.

Simulated storms were carried out using the portable rainfall simulator described by Calvo et al. (1988). The rain was applied at a constant rate during 15–20 minutes and the selected intensity was 60 mm h^{-1}. This is the intensity used by Gilman and Thornes (1985) in the application of the extreme event model of Scoging–Thornes and the recurrence interval of this for the south-east of Spain is 75–150 years. At this value the simulator yields the most even distribution of rain over the plot but, because of strong winds encountered during the field work, the amount of rainfall reaching the ground was reduced to 40–50 mm h^{-1}. The plot used was a circular ring (55 cm in diameter) fitted in the soil, with a small tube at the lower part where water and sediment were collected as samples of one or two minutes at Vera and Tabernas, and as total amount at La Herrería. Before each simulation was carried out, samples of the upper 2 cm of soil were collected from a point immediately adjacent to the plot for the measurement of soil moisture (as percentage by weight). Percentage lichen cover was assessed by eye and is estimated to be accurate to within 5 per cent. Accuracy is greater for species having very small cover values where the diameters of individual thalli were measured using calipers.

Visual observations were also made during the experiments related to the time from the start of rain to the closing of cracks, to surface ponding, and to the beginning of runoff on the surface and at the spout.

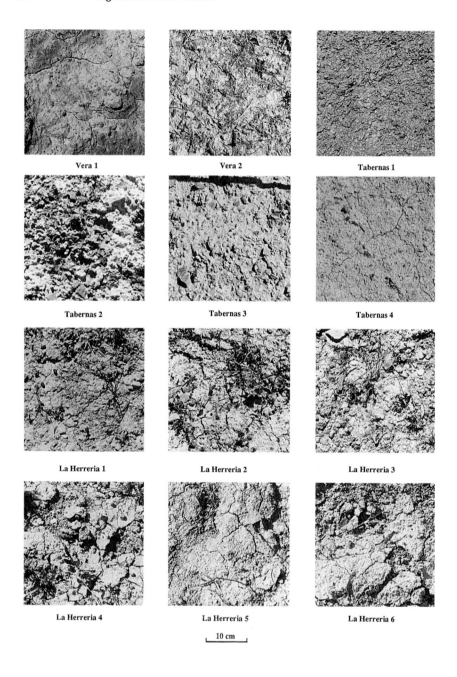

Figure 23.3 Photographs showing surface features of each plot prior to rainfall simulation

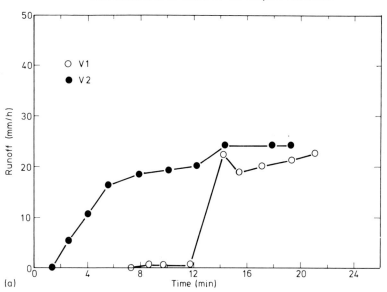

(a)

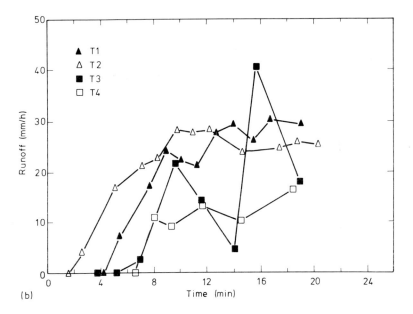

(b)

Figure 23.4 Hydrographs for simulated storms at (a) Vera and (b) Tabernas

RESULTS

The hydrographs for the six simulated storms at Vera and Tabernas badlands are represented in Figures 23.4(a) and (b), and these data have been fitted to:

$$r = A + B/t \qquad (1)$$

where r represents the runoff rate, t is the time elapsed since ponding, and A and B are the parameters (see Scoging and Thornes, 1979). At the La Herrería sites runoff and sediment concentration were measured as total values only and the time distribution of runoff is thus unknown.

This equation form is similar to the Green and Ampt (1911) and Philip (1957) equations and has been fitted to several infiltration tests on soils of the study area by Thornes (1976), Scoging and Thornes (1979), Scoging (1982), Thornes and Gilman (1984) and Gilman and Thornes (1985) with good results. In our case (Table 23.2), the model explains more than 80 per cent of the variance in four curves with data from T3 and T4 having lower values due to data-collection problems caused by wind varying the intensity of rainfall. The fitting of the data to this curve implies a low storage capacity of the soils with a very quick approximation to the final runoff rate, always between 10 and 15 minutes from the beginning of the storm. In general the results of all the experiments show the typical hydrological properties of badland surfaces, that is, very little time from the start of the rain to ponding (between 34 seconds and 5 minutes with an average of $1'52''$).

The influence of lichen cover can be examined in relation to a number of result variables. Figure 23.5(a) shows the relationship between lichen cover and sediment concentration which indicates an exponential correlation, but the large number of bare plots (50 per cent) and the small variance between plots with differing amounts of lichen cover suggest a more guarded interpretation at this stage. The relationship largely reflects the influence of slope which is exponentially correlated with sediment concentration (Figure 23.5b) and the plots with no lichens are also those with the steepest slopes ($>22°$). This interaction was examined using multiple regression which produced an R^2 value of 0.68 (significant at $p < 0.01$) with slope accounting for the major (and significant)

Table 23.2 Parameters A and B derived from fitting runoff data to equation (1) (V1, V2, etc., refers to sites as in text)

	V1	V2	T1	T2	T3	T4
A	37.2382	24.5081	31.9838	29.9487	29.8923	14.7377
B	−258.8569	−33.1597	−52.4988	−53.0613	−130.6042	−16.1523
r^2	0.8025	0.9216	0.9021	0.8965	0.3856	0.3461

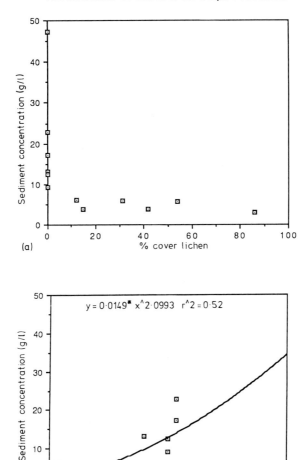

Figure 23.5 Relationships between variables in the twelve sample plots. (a) percentage lichen cover with sediment concentration. (b) Slope with sediment concentration

part of the variance in sediment concentration and lichen cover contributing only a small and in significant amount to the explanation of variance.

Despite the weak relationship between total lichen cover and sediment concentration it is important to note the effects of the different physiognomic types of lichen. The average sediment concentration in plots with lichens of group A only (H1, H2, H3) is $5.2 \, g \, l^{-1}$, the average in those with groups A, B, and C only (T3, V2) is $4.9 \, g \, l^{-1}$, and plot T3, with lichens of all groups including group D, has a sediment concentration of $3 \, g \, l^{-1}$. By contrast, the average in plots with no lichens is $20.3 \, \text{g} \, \text{l}^{-1}$. More data are required in order to investigate these relationships further.

In terms of the other variables related to lichen cover (Figures 23.6a–c) the data referring to plots with lichens exhibit a logical progression, even though the curve fitting is not good and some plots give contradictory results due to the influence of data quantity and the 'noise' produced by factors such as type of lichens present (not expressed in the total percentage), plot morphology and soil properties. Ponding and runoff start times are shorter with greater lichen cover and the total runoff during the first 15 minutes increases linearly in the presence of lichens. With regard to the influence of lichen physiognomy, the differences in porosity described above are especially effective on the ponding times. Whereas it took 2.21 minutes to produce ponds in the sites with no lichens, ponds appeared after 1.66 minutes in the plot with lichens of group A only, after 1.03 minutes in the plots with lichens of groups B and C, and, when the group D lichen was present, the ponding time reduced to 34 seconds (T2).

As an additional approach to investigating the influence of lichens on surface processes, the plots were classified on the basis of the three result variables: ponding time, surface runoff start time and sediment concentration. The analysis was performed on standardized data using Ward's error sum method (Ward, 1963), a polythetic, agglomerative clustering technique based on a distance coefficient, within the CLUSTAN package (CLUSTAN 1C, Release 2.1, January, 1982). The resulting dendrogram (Figure 23.7) can best be examined at the five-cluster level which shows two clear groups of plots with lichens of different characteristics, a group of plots with no lichens, and two single plots also without lichens.

Cluster (i) consists of the plots from Vera and Tabernas that possess lichens (V2, T2, T3). These show a very low sediment concentration and very early ponding and runoff. All three sites have moderate to high cover of lichens in groups A, B and C and T2 contains, in addition, substantial cover of the group D lichen, *Diploschistes dicapsis*.

Cluster (ii) contains three plots from La Herrería (H1, H2, H3), all of which contain group A lichens only. These plots yield low sediment concentrations but have a higher infiltration rate due to the porosity of the group A lichens. Plot T1 is also included in this group as, although it has no lichens, it behaves in a similar way due to the coarse-grained nature of its mineral crust surface.

Cluster (iii) consists of the three remaining La Herrería plots (H4, H5, H6) which have no lichens, a very low cover of herbaceous plants and a mineral crust damaged by goats. The effect of the breakdown of the crust is manifested in later ponding and runoff generation and an increase in the sediment yield, as well as a greater presence of concentrated flow and rill generation.

Finally, two plots with no lichens, V1 and T4, appear as single-member clusters which are quite dissimilar to the others. This is due to their different behaviour resulting from their physical properties. Both are steep, south-facing sites with

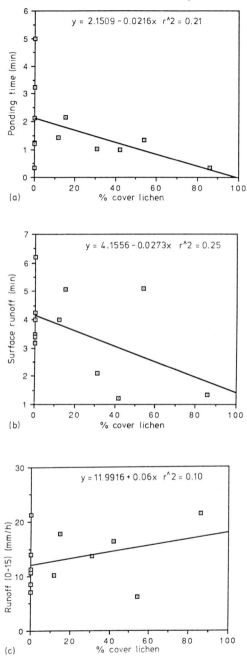

Figure 23.6 Relationships between percentage lichen cover and (a) ponding time; (b) surface runoff start time; (c) runoff during first 15 minutes, in the twelve sample plots

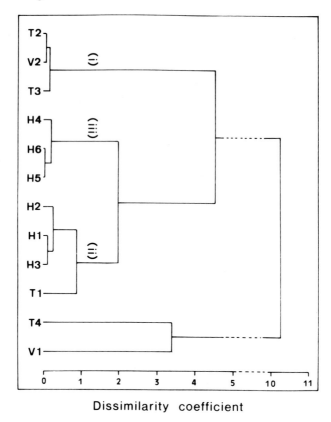

Figure 23.7   Dendrogram resulting from classification of the twelve sample plots in terms of ponding time, surface runoff start time and sediment concentration

low antecedent soil moisture which are late in ponding and in producing runoff, but which ultimately yield the highest sediment concentrations.

CONCLUSIONS

The twelve sites in which the rainfall experiments were carried out represent four of the main situations in interrill areas of badland surfaces in Almería: lichen cover, lichen and herb cover, mineral crust and mineral crust damaged by goats or by surface processes. All of these are overlapped by the control of aspect, in that lichen cover is only well developed on north or near-north facing slopes. Therefore, on south-facing slopes with no vegetation (V1, T1, T4) but covered by a mineral crust with cracks, the experiments show a 'high' initial infiltration followed by surface swelling to close the cracks. Runoff is slow to start but eventually produces high sediment concentration.

On south-facing slopes with some higher plants but no lichens, and a mineral crust partially damaged by goats (H5, H6) (or even on parts of north-facing slopes with similar characteristics (H4)), the times to ponding and to runoff generation are shorter, but initially the runoff goes into deeper cracks and goat-damaged areas that promote temporary storage. The resulting sediment concentration, as well as the total runoff, is lower.

On east- to north-facing sites with moderate slopes and group A lichens only (H1, H2, H3), the presence of a porous but binding lichen crust increases the infiltration rate, thereby causing slow ponding and surface runoff and also a low sediment concentration.

On north- to east-facing sites with moderate to gentle slopes and lichens of groups B, C and D as well as A (V2, T2, T3), the infiltration rates are the lowest, with a very quick ponding and runoff generation but a very low sediment concentration due to the effective cover of lichens.

In summary, north-facing slopes are less steep, less dissected by rills, gullies or piping, not affected by rapid mass movements, and have better developed/conserved hanged micropediments due to the protective effect of the lichen crust (in Tabernas and Vera). South-facing slopes are steeper (especially in the upper part), strongly dissected by rills, gullies, pipes (in Vera) and experience a greater activity of swelling, mass movements and cracking.

In general, badland erosion rates are between 2 and 20 mm yr^{-1} of ground lowering (Campbell, 1989). The erosion rates obtained with the rainfall simulations are clearly lower than others measured for longer periods with other methods which include the interaction of several processes and not only the effect of an individual storm. This is especially important in places such as Vera where swelling work is very effective and produces mass movements (Calvo and Harvey, 1989).

ACKNOWLEDGEMENTS

We are grateful to participants in the 1988 BGRG Almería field meeting for assistance in the field, to Dr A. M. Harvey for his numerous suggestions in the field and in the discussion of the manuscript, and to the referees for their comments and suggestions.

REFERENCES

Calvo, A., J. Gisbert, E. Palau and M. Romero (1988). Un simulador de lluvia portátil de facil construcción. In M. Sala and F. Gallart (eds.), *Métodos y técnicas para la medición en el campo de procesos geomorfológicos*, Soc. Española de Geomorfología, Monografía 1, pp. 6–15.

Calvo, A. and A. M. Harvey (1989). Morphology and development of selected badlands in Southeast Spain. In *Landscape–Ecological Impact of Climatic Change*, Amsterdam (in press).

Campbell, I. A. (1989). Badlands and badland gullies. In D. S. G. Thomas (ed.), *Arid Zone Geomorphology*, Bellhaven Press, pp. 160–83.

Elias, F. and L. Ruiz, (1977). *Agroclimatología de España*, Ministerio de Agricultura, Madrid.

Gilman, A. and J. B. Thornes (1985). *Land-use and Prehistory in Southeast Spain*. The London Research Series in Geography, **8**.

Green, W. H. and G. A. Ampt (1911). Studies on soil physics, I. The flow of air and water through soils. *Journal of Agricultural Science*, **4**, 1–24.

Harvey, A. M. (1982). The role of piping in the development of badlands and gully systems in south-east Spain. In R. B. Bryan and A. Yair (eds), *Badland Geomorphology and Piping*, Geobooks, Norwich, pp. 317–35.

Harvey, A. M. (1987) Patterns of Quaternary aggradational and dissectional landform development in the Almeria region, southeast Spain: A dry-region, tectonically active landscape. *Die Erde*, **118**, 193–215.

Hawksworth, D. L. and D. J. Hill (1984). *The Lichen-Forming Fungi*, Blackie, Glasgow.

IGME (1975a). *Mapa geológico de España E. 1:50,000, hoja de Vera (1014)*. Segunda serie, Instituto Geológico y Minero de España.

IGME (1975b). *Mapa geológico de España E. 1:50,000, hoja de Sorbas (1031)*. Segunda serie, Instituto Geológico y Minero de España.

IGME (1975c). *Mapa geológico de España E. 1:50,000 hoja de Tabernas (1030)*. Segunda serie, Instituto Geológico y Minero de España.

Philip, J. R. (1957). The theory of infiltration, 4. Sorptivity of algebraic infiltration equations. *Soil Science*, **84** (3), 257–64.

Scoging, H. (1982). Spatial variations in infiltration, runoff and erosion on hillslopes in semi-arid Spain. In R. B. Bryan and A. Yair (eds), *Badland Geomorphology and Piping*, Geobooks Norwich, pp. 89–112.

Scoging, H. M. and J. B. Thornes (1979). Infiltration characteristics in a semi-arid environment. IASH publ. 128, 159–68.

Seaward, M. R. D. (1983). Lichens of Malaga Province, S. Spain. *Nova Hedwigia*, **XXXVII**, 325–45.

Thornes, J. B. (1976). *Semi-arid erosional systems: Case studies from Spain*. LSE Geogr. Papers, 7, 106 pp.

Thornes, J. B. and A. Gilman (1984). Potential and actual erosion around archaeological sites in south east Spain. *Catena*, Suppl. 4, 91–113.

Ward, J. H. (1963). Hierarchical grouping to optimize an objective function. *J. Amer. Stat. Ass.*, **58**, 236–44.

24   Spatial Distribution Patterns of Morphogenetic Processes in a Semi-arid Region

J. CARLOS G. HIDALGO, FRANCISCO PELLICER
Department of Geography, Zaragoza University

LEONEL SIERRALTA and M. VICTORIA LOPEZ
Mediterranean Agronomic Institute, Zaragoza

SUMMARY

The spatial distribution of plant cover, and its influence on the morphogenesis are analysed in a semi-arid Mediterranean test area. The analysis shows how the topography, and slope facet in particular, seems to be one of the most important factors for accelerated morphogenesis associated with water processes. This factor determines the type of vegetation by its influences on evapotranspiration rates. In areas of low vegetation cover the low rainfall of this environment nevertheless has high morphogenetic effectiveness. Human activities, including short-term plant cover, increase the fragility of this environment, resulting in an acceleration of processes.

INTRODUCTION

Two groups of morphogenetic processes can be recognized. First, those processes related to potential morphogenesis, often called geological erosion, in which topography, underlying geology and climate, control landform development. Second, morphogenesis resulting from the morphogenetic processes where, besides the factors above, the effects of plant cover and human activities influence the processes.

Vegetation and Erosion
Edited by J. B. Thornes
©1990 John Wiley & Sons Ltd

The distinction made is not only due to methodological criteria, but also to practical ones (Bergsma, 1980; Marques, in press). The power that a society has over the distribution, type and the state of the plant cover suggests that control over morphogenetic processes (especially during the erosion and transport stages), should be focused, among other things, on control of the plant cover as one of the most dynamic factors (Thornes, 1981). In order to do that, it is necessary to identify the extent of human modification of the environment, and, above all, to understand as much as possible about the dynamics of the systems.

In this work plant cover distribution patterns and their influence on morphogenesis are analysed in a semi-arid Mediterranean test area, where the influence of plant cover is higher because of the climatic characteristics of this environment (Tricart and Cailleux, 1960).

The analysis attempts to identify areas with similar conditions in the morphogenetic processes, and has been focused on plant cover spatial distribution as an indicator of the sites (Sanz, 1979).

The patterns studied were slope value, slope facet and human activities, all of them recognized by their effects, direct and indirect, on the plant cover and consequently on processes. In this sense, Douguedroit (1974) has indicated how the slope facet has influence on soil moisture, and then on plant cover. Small (1972) has analysed the texture variation in slopes with different aspect, as a consequence of different environmental and morphogenetic conditions. The spatial variation of plant cover distribution between different facets has been studied by Cotonnec (1971) and Llorente (1985).

The slope value effect has been recognized in many works (e.g. Djorovic, 1980; Pinczes *et al.*, 1981). Here we note the work of Lecarpentier (1974) who has studied the variation of the slope angle with the aspect, reflecting different processes.

Last but not least, human activities have been analysed in many different ways. Among other subjects, Neboit (1979) and Thornes (1980) studied the influence of crop calendar and the temporal variation of plant cover protection; Thornes (1987) identified the influence of grazing on the plant cover; Chisci and Zanchi (1981) have studied tilling systems; and finally Fuentes and Hajek (1979) have offered a global evaluation of human modification over morphogenetic processes in Mediterranean Chile.

STUDY AREA

The field site is in a region strongly changed by human activities by the introduction of irrigation 40 years ago, located on the margins of the Violada basin, situated to the north of the Ebro river, Spain (Figure 24.1).

Topographically, it is an area of gentle slope that varies in relief from 480 m to 380 m a.s.l. (Figure 24.2). From north to south, the area has a series of infilled

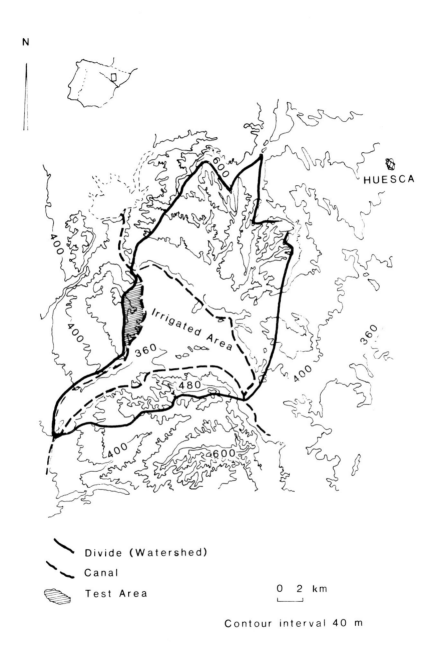

N

HUESCA

Irrigated Area

Divide (Watershed)
Canal
Test Area

0   2   km

Contour interval 40 m

Figure 24.1 Violada Basin

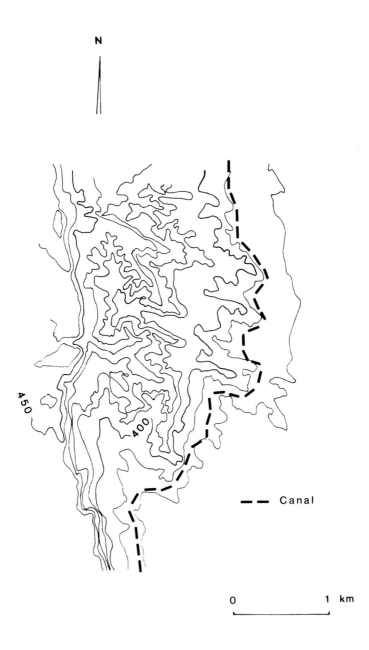

Figure 24.2   Topographical map

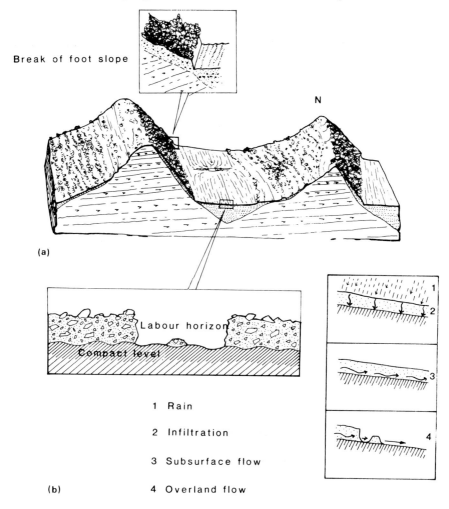

Break of foot slope

N

(a)

Labour horizon

Compact level

1 Rain

2 Infiltration

3 Subsurface flow

(b)   4 Overland flow

Figure 24.3 Valley asymmetry, cross section and rill development

valleys oriented west–east. They are excavated in the loamy and gypsum materials that make up the central part of the Miocene basin. The cross-sections of these valleys show some asymmetry between the slopes facing north and south (Figure 24.3a). This asymmetry is due not only to morphoclimatic conditions, as has been indicated in other areas of Aragon (Mensua and Ibañez, 1977a, b; Zuidam, 1980) and Castilla (Mensua and Plans, 1981), but also to structural control due to the recent geomorphological history of the depression, whose base level has been excavated progressively towards the southwest (Pellicer and Hidalgo, 1987; Hidalgo, 1988).

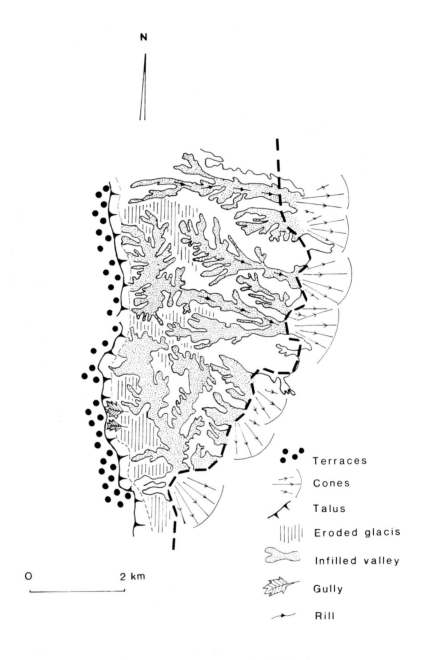

N

Terraces

Cones

Talus

Eroded glacis

Infilled valley

Gully

Rill

O   2 km

Figure 24.4   Cross-section and longitudinal rill development

Along the escarpment there is a slope, of about 15 m width, from which inclined glacis extend to form secondary divides between the valleys. The infilled valleys connect with the bottom of the depression (irrigated area), by alluvial cones overlapping laterally and extending to the middle of the depression; they are, at present, strongly modified by irrigation works (Figure 24.4).

Morphoclimatic conditions, according to Wilson's method (1968, 1969), are semi-arid (Rodriguez Vidal, 1982), but conditions become arid from time to time (Hidalgo, 1988). The average annual rainfall is c. 500 mm (Liso and Ascaso, 1969) and the average temperature 12 °C. There is an annual moisture deficit of c. 200 mm (Table 24.1).

Table 24.1 Climatic data

Month	Rainfall (mm)	Temp. (°C)	Month	Rainfall (mm)	Temp. (°C)
Jan.	36.6	4.45	Jul.	23.7	22.37
Feb.	33.5	6.13	Aug.	32.6	21.69
Mar.	44.6	8.64	Sep.	46.2	18.33
Apr.	49.3	11.56	Oct.	40.6	13.72
May	58.5	14.70	Nov.	46.7	7.98
Jun.	40.1	19.24	Dec.	44.0	6.39

The mean annual total rainfall is uniformly distributed throughout the year, and the seasonal percentages are higher than 20 per cent of the mean annual total rainfall, although the interannual variations are very important. The intensities, expressed as concentrations higher than 10 mm/day, show that spring and autumn are the seasons with highest erosion potential (Hidalgo, 1988).

The vegetation abundance–dominance, sociability class and dynamics in the area indicate an environment with a low natural productivity, due to the semi-arid conditions, and to a poor soil. The soils, where present, are shallow, and often saline. Such conditions are very favourable for Mediterranean shrub with sclerophyllous species, with or without gypsum-tolerant species, depending on the degree of salinity.

This vegetation has undergone progressive substitution by dry land crops, following very clear spatial patterns since the beginning of irrigation in the areas at the bottom of the depression. The cultivated crops were grown first along the bottom of the infilled valleys, and later occupied the lateral areas (Hidalgo, Sierralta and Pellicer, 1987). Human impacts are very strong, and the building of road and livestock activities must be also considered.

METHODS

The degree of the plant cover was analysed through aerial photographs, and tested by field work. The types of cover mapped are: dense natural cover

(>50 per cent), open natural cover (<50 per cent) and cultivated land areas (wheat and barley).

The state of the natural plant cover is analysed through field sampling in 25 m^2 plots, according to Bertrand's methodology (1966). This method consists of recognizing the percentage of the surface covered by plants (abundance–dominance) and how the different plants are spatially distributed (sociability) as used by SIGMA (Station Internationale de Géobotanique Mediterranéenne et Alpine, Montpellier). The sampling is made by strata and finally the results are drawn in a diagram (called a 'vegetation pyramid') in which slope degree, underlying geology, slope facet and soil are marked (see legend of Figure 24.8). Human impact, erosional processes, and other circumstances are noted indicating the state of the community (in regression, progression, or balanced).

In a few words, the method is based on plant cover as indicator of the ecological conditions. The total field samples are 40 (20 in the north facet and 20 in the south facet).

The slope facet measurements were made in the field with a clinometer, grouping the values by slope facet and categories. The soil moisture was analysed at the same sites.

RESULTS

Figure 24.5 shows how the natural plant cover is located in narrow areas along the watershed lines. At the same time, the infilled valley bottoms are planted with crops. These crops very often spread on to the slopes themselves, when the slope gradient is not very high. This is the case in the area near the irrigation channels (east), and in the southern area.

The distribution of the natural plant cover and its present state are very clear. They reflect aspect as well as human influences. Differences between cold facets (north, northeast) with dense cover, and warm facets (south, southeast, southwest) with open cover can be observed by comparison of Figures 24.1, Figure 24.5 and Figure 24.6.

Topographic asymmetry between cold and warm exposure shows higher average slope values in cold (26 per cent) than in warm facets (23 per cent) (Figure 24.7 and Table 24.2), although the most variable values, and the highest, are in the second group, the differences being significant at a level of 0.05 after log transformation of the data.

Moisture values show higher percentages in north facets (mean 18.6 per cent) than in south aspects (mean 13.6 per cent) (Table 24.2), with a difference significant at a level of 0.01.

The state of the plant cover is shown in two pyramids that represent the general situation in cold and warm slopes (Table 24.3). In Figure 24.8, the north facet

DENSE COVER

OPEN COVER

CROPS

0   1 km

Figure 24.5 Plant cover map

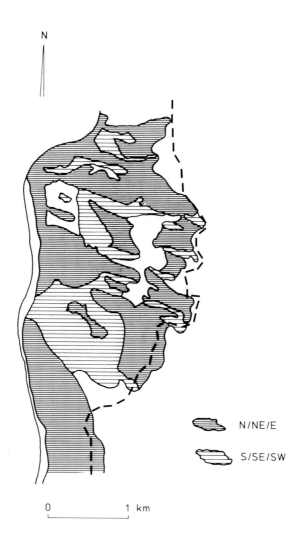

Figure 24.6   Slope facet map

North Facet   **South Facet**

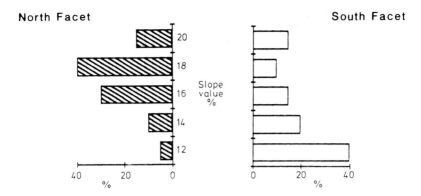

Figure 24.7 Slope facet value

(Table 24.4a) has three strata: bushy, sub-bushy and stratum A. They are quite well consolidated, especially from *Rosmarinus officinalis*. The existence of an incipient soil and the absence of morphogenetic features, considered to be due to the protection of plant cover, can also be observed.

In the second case (Table 24.4b), warm slopes have a very sparse vegetation where there is usually a lack of the bushy stratum. There is no firm evidence of the replacement of the sub-bushy stratum by stratum A. There are no soils and the morphogenetic features are very clear.

The signs of morphogenetic processes are nearly always of water erosion and deposition. The warm slopes and the valley bottoms have benches and small cones of fine material, in the first case deposited on small steps excavated by differential erosion, and it is possible to observe headwater recession and small incisions. In the second case, incisions deeper than 30 cm can be observed in the deposits at the bottom of the valley.

Rills are very frequent between terraces where the slope gradient along the valley is offset by the effect of the old containing walls, and show a cross-section with compact level at about 20–30 cm deep. Cones of material extracted from upstream are formed in contact with the containing wall and can be thicker than 20 cm (Figure 24.9). Every year, new ploughing activities destroy these forms, but they reappear after rainfall. Along the contact between north facet and valley bottom, the ploughing activities have originated a break foot slope where the overland flow is concentrated as shown in Figure 24.3(a). Evident signs of geomorphological processes cannot be observed on the north and north-east facing slopes.

Finally, along the talus slope, thin deposits of fine materials with boulders from the upper surface appear. Rill processes are incipient and towards the south there are excavations of small gully head-waters. These are linear incisions, with abrupt steps, opening on to deposits that cover the concave profile of slopes.

Table 24.2 Moisture, slope, plant cover and erosion rates

		North-facing slope		
Sample	Moisture (%)	Slope (%)	Plant cover (%)	Erosion rate (t/ha/yr)
1	18.4	27	100	2.7
2	18.8	24.8	100	2.4
3	20.2	24.8	100	2.4
4	19.3	25.6	100	2.5
5	17.5	23.1	100	2.3
6	19.2	24.8	75	3.3
7	19.3	27.3	100	2.7
8	18.3	19	100	1.9
9	19.7	22.3	100	2.2
10	18.9	28.2	75	3.7
11	19.8	27.3	100	2.7
12	17.9	27.3	100	2.7
13	18.3	27.3	100	2.7
14	20.2	23.1	100	2.3
15	19.6	21.5	100	2.1
16	19.3	31.7	100	3.1
17	17.2	26.5	75	3.5
18	14.7	29	75	3.8
19	18.1	31.6	75	4.2
20	18.3	31.6	75	4.2
		South-facing slope		
21	12.7	18.2	25	7.2
22	12.1	22.3	25	8.9
23	12.0	23.1	10	23.1
24	14.6	15.8	25	6.3
25	15.2	15.8	25	6.3
26	12.6	24.8	50	4.9
27	13.5	29	50	5.8
28	14.5	19	75	2.5
29	14.8	23.1	50	4.6
30	12.4	27.3	75	3.6
31	12.8	22.3	25	8.9
32	16.1	22.3	25	8.9
33	17.9	18.2	25	7.2
34	12.4	18.2	25	7.2
35	11.2	18.2	25	7.2
36	11.7	36.8	10	36.8
37	13.4	40.5	10	40.5
38	15.0	31.6	10	31.6
39	13.9	18.2	25	7.2
40	13.5	22.3	10	22.3

Table 24.3 Plant cover structure (global results)

	North-facing slopes							South-facing slopes					
	St.A		St.B		St.C			St.A		St.B		St.C	
Sample	A	S	A	S	A	S	Sample	A	S	A	S	A	S
1	5	(5)	3	(3)	—	—	21	2	(3)	+	(1)	—	—
2	5	(5)	2	(2)	—	—	22	2	(1)	—	—	—	—
3	5	(5)	+	(1)	—	—	23	1	(1)	—	—	—	—
4	5	(5)	3	(1)	—	—	24	2	(2)	+	(1)	—	—
5	5	(5)	3	(1)	—	—	25	2	(3)	+	(1)	—	—
6	4	(5)	3	(4)	+	(1)	26	3	(3)	—	—	—	—
7	5	(5)	3	(3)	2	(3)	27	2	(2)	—	—	—	—
8	5	(5)	3	(3)	+	(1)	28	4	(5)	+	(1)	—	—
9	5	(5)	3	(3)	—	—	29	3	(3)	+	(1)	—	—
10	4	(5)	4	(5)	—	—	30	4	(3)	+	(1)	—	—
11	5	(5)	3	(3)	—	—	31	2	(3)	+	(2)	—	—
12	5	(5)	3	(3)	+	(1)	32	2	(3)	1	(2)	—	—
13	5	(5)	3	(3)	—	—	33	2	(3)	+	(1)	—	—
14	5	(5)	4	(3)	—	—	34	2	(3)	+	(1)	—	—
15	5	(5)	2	(3)	2	(3)	35	2	(3)	+	(1)	—	—
16	5	(5)	+	(1)	—	—	36	+	(1)	—	—	—	—
17	4	(4)	2	(1)	—	—	37	+	(1)	—	—	—	—
18	4	(5)	3	(4)	—	—	38	+	(1)	+	(1)	+	(1)
19	4	(4)	3	(3)	—	—	39	2	(4)	—	—	—	—
20	4	(5)	4	(5)	—	—	40	+	(2)	+	(1)	+	(1)

A—Abundance–dominance; S—Sociability;
St.A—Stratum A (<50 cm); St.B—Sub-bushy (50–100 cm);
St.C—Bushy (>50 cm <3 m) (see legend to Figure 24.8)

DISCUSSION

The protection provided by the surface vegetation cover against water processes is one of the factors which condition the accelerated morphogenesis, not only by rainfall interception, but also through its influence on the soil structure, due to organic matter supply, porosity, etc. (e.g. Tricart, 1977, 1979; Evans, 1980; Dudal, 1981; Tricart and Killian, 1981).

In the study area, the existence and distribution of water processes seem to respond to two fundamental facts derived firstly from the degree and nature of the plant cover, and secondly from the presence of cultivated crops with temporal and cyclic characteristics derived from the crop calendar.

Topography seems to be the dominant factor which determines the distribution of plant cover through slope facet. With the same general rainfall conditions, we can see the change from a dense to an open cover, perhaps due to the decrease on the evapotranspiration rates as different values of soil moisture have shown.

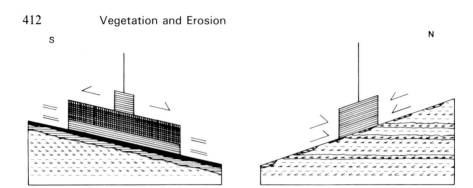

Figure 24.8   Plant cover diagram

Plant cover structure

Strata	Height	Representation
Stratum C (Bushy)	1.0–3.0 m	1.0 cm
Stratum B (Sub-bushy)	0.5–1.0 m	1.0 cm
Stratum A (*)	<0.5 m	0.5 cm

Abundance–Dominance (Percentage of surface cover)

Code	Cover	Representation
5	75–100%	5.0 cm
4	50– 75%	4.0 cm
3	25– 50%	3.0 cm
2	10– 25%	2.0 cm
1	Abundant without appreciable surface covered	1.0 cm
+	Rare	0.5 cm

Sociability (grouping mode)

Code	Mode	Representation
5	Dense patch	
4	Non-extended patch	
3	Grouping	
2	Grouping in 2 or 3	
1	Isolated	

Soil

Parent material

       Loam and gypsum

State

     Progression $\longrightarrow$ Regression $\swarrow$ Balanced $=$

(*) Stratum A includes herbaceous plant and shrubs less than 0.5 m.

This factor can even counteract the effort of slope gradient on the runoff rate, as density of plant cover is highest on the cold slopes, but the evidence of runoff is mainly located at the warm slopes, with lower slope values, but with more sporadic plant cover (Tables 24.2 and 24.3).

Table 24.4   Plant cover structure (particular sample)

(a)

Site no. 1	Stratum A		Stratum B		Stratum C	
	A	S	A	S	A	S
Rosmarinus officinalis	2	4	3	5	+	3
Thymelaea tinctoria	+	1	+	3	—	—
Artemisia herba-alba	2	2	—	—	—	—
Asphodelus albus	1	1	—	—	—	—
Thymus vulg.	1	2	—	—	—	—
Gramineous non dif.	2	3	—	—	—	—
Total by strata	3	3	3	5	+	3

Slope: *c.*18°
Aspect: North
Microclimate: Protection of NW–SE winds; moisture *c.* 20%
Parent material: Gypsum and loam
Humus: Non-existent
Soil: Carbonates at the surface. Thin. Not well differentiated horizontally
Geomorphological processes: Little steps by headwater recession along the contact between parent materials
Human activities: Grazing (?). Fire evidence
Global dynamic: Stability, although the progression could be broken by soil thickness

(b)

Site no. 2	Stratum A		Stratum B		Stratum C	
	A	S	A	S	A	S
Rosmarinus officinalis	+	3	1	3	—	—
Genista scorpius	+	1	1	2	—	—
Thymus vulgaris	1	1	—	—	—	—
Artemisia herba-alba	1	3	—	—	—	—
Gramineous non dif.	1	3	—	—	—	—
Total by strata	1	3	1	3	—	—

Slope: *c.* 12°
Aspect: South
Microclimate: High insolation
Parent material: As above
Humus: Non-existent
Soil: Non-existent (parent material at surface)
Geomorphological processes: Evidence of overland flow, rill, etc.
Human activities: Grazing
Global dynamic: Regressive

Secondly, human activity is focused on the bottoms of the infilled valleys, and, from there, it impinges on areas where ploughing is not prevented by topography. These areas coincide with the south-facing slopes.

The factors which create favourable conditions for the initiation of erosion processes derived from runoff are: the indiscriminate character of these activities occupying areas where, *sensu stricto*, there is no soil; ploughing parallel to the

414

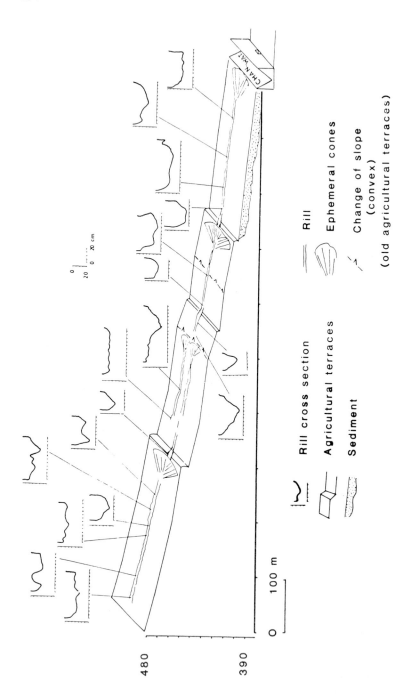

Figure 24.9 Infilled valley, longitudinal profile with rills, cones development and agricultural terraces

slope gradient; utilizing heavy machinery, thereby compacting the soils; and finally the crop calendar means that the highest morphogenetic activity coincides with the unprotected surface (Thornes, 1981).

The runoff type is combination of wash and splash ('battant', Tricart, 1977; Tricart and Killian, 1982) which forms a pavement and seals the surface resulting in loss of water supply. This can be explained by two factors: first, the clay and silts of the deposits at the bottom of the valleys have a low porosity rate at a compact level; and second, the decrease in the number of terrace-containing walls from 1945 to 1985, increased the length of terrace between them in order to enable mechanization. As a result the runoff can move easily and rapidly, helped by the tilling direction (Figures 24.3b and Figure 24.9).

Finally, trying to compare the different effects of slope value and natural plant cover on theoretical rates of erosion, Musgrave's equation $E = f(q*s)$ (Thornes, 1987), has been modified in the following form:

$$E = \frac{1}{PC} * S$$

where PC is plant cover; E is erosion rate; $1/PC$ is overland flow; S is slope angle, given that overland flow in Horton's model can be considered as the inverse of plant cover.

The results, Table 24.3, show relative differences, not real and absolute rates, between the north and south facets, with average values higher in the south facet (12.5 t/ha/yr) than in the north facet (2.8 t/ha/yr), the coefficient of correlation between plant cover and erosion rate being best in the north facet (-0.88), compared with that in the south facet (-0.65), because of a more uniform profile of north-facing slopes than in south-facing slopes, where there is a more complex topography along the profiles.

CONCLUSIONS

The analysis of the distribution of the natural plant cover in a semi-arid region is related to topography and controls the distribution of accelerated morphogenesis by water processes.

The topography, and the slope facet in particular, determine the type of the existing vegetation by the influence on evapotranspiration rates. Under such conditions, the low rainfall rates characteristic of this environment can still acquire geomorphic effectiveness when the surface protection is poor.

Human activities, such as the introduction of a temporary plant cover, can break the stability of these fragile environments, creating new situations which show poor adaptation to the conditions and result in an acceleration of erosive processes.

REFERENCES

Bergsma, E. (1980). Methods of a reconnaissance survey of erosion hazard near Merida, Spain. In De Boodt and D. Gabriels (eds), *Assessment of Erosion*, John Wiley & Sons, Chichester. pp. 55–66.

Bertrand, G. (1966). Pour un étude géographique de la végétation. *Rev. Géogr. des Pyrénées et du Sud-Ouest*, 128–43.

Chisci, G. and C. Zanchi (1981). The influence of different tillage systems and different crops on soil losses on hilly silty-clayey soil. In R. P. C. Morgan (ed.), *Soil Conservation, Problems and Prospects*, John Wiley & Sons, Chichester, pp. 211–17.

Cotonnec, M. M. (1971). Comparaison des deux vesants du Vallon de la Côte Blanche (Forêt de Beynes, Yvelines). *Bull. de l'association de Géograph. Française*, **387–88**, 169–73.

Djorovic, M. (1980). Slope effect on run-off and erosion. In De Boodt and D. Gabriels (eds), *Assessment of Erosion*, John Wiley & Sons, Chichester, pp. 215–25.

Douguedroit, A. (1974) Le rôle de l'humidité des sols dans l'opposition de la végétation entre adrets et ubacs. *Bull. de l'association Géograph. Française*, **415–16**, 130–40.

Dudal, R. (1981). An evaluation of conservation needs. In R. P. C. Morgan (ed.), *Soil Conservation, Problems and Prospects*, John Wiley & Sons, pp. 3–12.

Evans, R. (1980). Mechanics of water erosion and their spatial and temporal controls: An empirical viewpoint. In M. J. Kirkby and R. P. C. Morgan (eds), *Soil Erosion*, John Wiley & Sons, Chichester, pp. 109–28.

Fuentes, E. and E. R. Hajek (1979). Patterns of landscape modification in relation to agricultural practice in Central Chile. *Environmental Conservation,* 4, 265–71.

Hidalgo, J. C. (1988). *Geomorfologia y procesos morfogenéticos en las márgenes de la depresión de la Violada. Una aproximación metodológica.* MSc Thesis, C.I.H.E.A.M., Zaragoza.

Hidalgo, J. C., L. Sierralta and F. Pellicer (1987). Changes in the spatial distribution between shrubs and dryland crops in the neighbourhood of an irrigation area: A not expected effect. Communication in *Mediterranean Ecosystems Conference*, Montpellier.

Lecarpentier, M. (1974). Analyse numérique de la topographie et mesure des pentes. *Cahiers de Géograph. de Québec*, **18**(45), 483–94.

Liso, M. and A. Ascaso (1969). *Introducción al estudio de la evapotranspiración y clasificación climática de la Cuenca del Ebro.* Anales de la Est. Exp. Aula Dei, Zaragoza, vol. 10, 1–2, 505.

Llorente, J. M. (1985). *Los paisajes adehesados salmantinos.* Centro de Estudios Salmantinos, Salamanca.

Marques, M. A. (in press). Procesos de erosión hidrica y desertización. In Porta, ed.), *IV Curso de Ordenación del Territorio.*

Mensua, S. and M. J. Ibañez (1977a). Los valles asimétricos de la orilla derecha del Ebro. *Actas II Reunión del Grupo de Estudios del Cuaternario*, 113–22.

Mensua, S. and M. J. Ibañez (1977b). *Mapa de terrazas fluviales y glacis en el sector central de la Depresión del Ebro.* III Reunión Nacional del Grupo de Trabajo del Cuaternario, Zaragoza.

Mensua, S. and P. Plans (1981). La disimetría de los valles del Páramo Leonés. *Estudios de Geografía, Homenaje A. Floristán*, 17–29.

Neboit, R. (1979). Les facteurs naturels et les facteurs humains de la morphogenèse. Essai de mise au point. *Ann. de Géograph.*, **490**, 649–70.

Pellicer, F. and J. C. Hidalgo (1987). Depósitos del Somontano Oscense en la Depresión de la Violada. *Actas X Conq. Nac. de Geografía.* Dep. Geografía y Ordenación del Territorio Univ. de Zaragoza, Zaragoza, 47–56.

Pinczes, Z., A. Kerényi, K. Erdós and P. Csorba (1980). Judgement of the danger of erosion through the evaluation of regional conditions. In R. P. C. Morgan (ed.), *Soil Conservation*, John Wiley & Sons, Chichester, pp. 87–104.

Quirantes, J. (1978). *Estudio sedimentológico y estratigráfico del Terciario Continental de los Monegros*. Inst. Fernando el Católico, C.S.I.C., Zaragoza.

Rodriguez, J. (1982). Distribución morfoclimática en la Depresión media del Ebro: Procesos dominantes y modelado actual. *Estudios Geológicos*, **38**, 43–50.

Sanz, C. (1979). La vegetación como medio de información geoecológica. *Estudios Geográficos*, **156–7**, 465–9.

Small, T. W. (1972). Morphological properties of driftless area soils relative to slope aspect and position. *Prof. Geographers*, **24**(4), 321–6.

Thornes, J. B. (1980). Erosional processes of running water and their spatial and temporal controls: A theoretical viewpoint. In M. J. Kirkby and R. P. C. Morgan (eds), *Soil Erosion*, John Wiley & Sons, Chichester, pp. 129–82.

Thornes, J. B. (1981). Conservation practices in erosion models. In R. P. C. Morgan (ed.), *Soil Conservation, Problems and Prospects*, John Wiley & Sons, Chichester, pp. 265–71.

Thornes, J. B. (1987). Erosional equilibria under grazing. In J. Bintliff, D. Davidson and E. Grant (eds), *Conceptual Issues in Environmental Archeology*, Elsevier, pp. 193–210.

Tricart, J. (1977). *Géomorphologie Dynamique*, Ed. Sedes, Paris.

Tricart, J. (1979). L'analyse de système et l'étude intégrée du milieu naturel. *Ann. de Géographie*, **88**, 700–14.

Tricart, J. and A. Cailleux (1960). *Le modelé des régions sèches. Le milieu morphoclimatique. Les mécanismes morphogénetiques des régions sèches*, C.D. Universitaire, París.

Tricart, J. and J. Killian (1982). *La Ecogeografía y la Ordenación del Medio Natural*, Ed. Anagrama, Barcelona.

Wilson, L. (1968). Morphogenetic classification. In Fairbridge (ed.), *Encyclopedia of Geomorphology*, Reinhold, pp. 717–31.

Wilson, L. (1969). Les relations entre les processus geomorphologiques et le climat moderne comme méthode de paleoclimatologie. *Rev. Geogr. Phys. Geol. Dyn.* **IX**(3), 303–14.

Zuidam, R. van (1980). Un levantamiento geomorfológico de la región de Zaragoza. *Geographicalia*, **6**, 103–34.

25 Soil Moisture, Runoff and Sediment Yield from Differentially Cleared Tropical Rainforest Plots

S. NORTCLIFF
Department of Soil Science, University of Reading

S. M. ROSS and J. B. THORNES
Department of Geography, University of Bristol

SUMMARY

Experimental plots were subjected to complete and partial clearance under dry tropical forest on slopes varying from 2 to 13 degrees and in different slope positions. Soil moisture is controlled mainly by slope position whereas runoff and sediment yields are affected mainly by treatment. The effects of soil, slope and topography lead to significant variations in response.

INTRODUCTION

In recent years quantitative proof has been provided of what has been taken for granted for many decades: that removal of vegetation cover significantly enhances erosion. Elwell and Stocking (1976), Dunne, Dietrich and Brunengo (1978), Lee and Skogerboe (1985) and in a related paper, Francis and Thornes (Chapter 22 of this volume) have all demonstrated that both runoff and sediment yield fall exponentially as the percentage of vegetation cover increases in a wide range of environments. Moreover, it is evident that the major changes occur between 0 and 30 per cent cover, with changes of cover above this having a relatively small impact. What is now needed is a more careful evaluation of how these general (and, under transport-limited removal,

Vegetation and Erosion
Edited by J. B. Thornes
©1990 John Wiley & Sons Ltd

perhaps steady-state) conditions are reached and how they are likely to vary through time, with different topographic, edaphic and vegetation cover types.

We have attempted to address some of these questions in the context of the problem of the clearance of dry tropical forest by selective felling and successive monitoring of the evolving conditions. In particular, we are interested to know if the state of clearance (degree and nature of vegetation cover) is more important in determining runoff and erosion rates than the magnitude and variability of slope and the nature and characteristics of the underlying soils. We also wish to examine how these sources of variation interact to influence soil moisture conditions and soil erosion. In addition we shall be concerned to explain the dynamics of clearance effects through changing soil, hydrological and surface litter characteristics over time.

In this chapter we describe the experimental design and discuss the differential response of runoff and soil loss to different clearance treatments and test the basic hypothesis that the components of variation due to clearance are less important than those due to slope and soil conditions. Later papers will be concerned with the nature and effects of progressive litter breakdown on the partially cleared sites, the character and quantity of transported organic matter and the detailed investigation of changes in the surface and subsurface hydrological budget.

The second major interest in this chapter is regionally specific. It is frequently claimed that in the Amazonian forests extensive removal of the plant cover leads not only to a reduction of soil fertility accompanied by specific chemical imbalances, producing, for example, aluminium toxicity, but also to massive soil erosion. It is widely assumed that clearance inevitably and irreversibly leads to deep soil loss and associated off-site effects such as increased sediment yields and ecological degradation of fresh-water environments and their quality. The fact that vigorous regeneration of cover occurs rapidly following clearance suggests that the case may have been grossly exaggerated. Mature and experienced observers (Fearnside, 1985, 1987) recognize that the problem is appreciably more complex than this oversimplified conservationist view implies. In this experiment we sought to compare the effects of complete clearance with those of only partial clearance, a condition which might be analogous in many if not all respects to partial recovery after clearance.

FIELD EXPERIMENT

The field experiment was conducted over a 300-day period in dry tropical forest on Maraca Island, Roraima Territory in northern Brazil. The site is a sloping hill side of about 13° draining fairly smoothly down to a small stream. The slope is generally convexo-concave, draining broad flat interfluves to a narrow floodplain zone (Figure 25.1), and was covered over its length with tall mixed terra firma forest characteristic of the eastern side of Ilhada Maraca. The change towards the floodplain is marked by an increase in palms and 'thicket' type riparian vegetation.

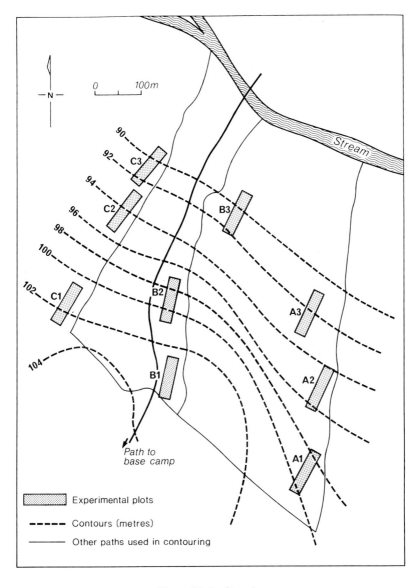

Figure 25.1   Site plan

In general terms the soil sequence comprises a coarse sandy loam overlying a clay-loam and dense clay, which in turn merge into a strongly indurated lateritic layer. Along the profile the sandy loam thinned downslope to a mid-slope position, and then thickened sharply in the lower part of the slope close to the stream. Typical soil profiles for the mid-slope position are given in Figure 25.2(b).

In order to capture the effects of slope position, transects were set up normal to the contours and these are indicated in Figure 25.1. On each transect three plots of 6 by 15 m were surveyed, in top, mid-slope and bottom slope, positions. These positions essentially relate to sites dominated by a well-drained sandy-loam over clay sequence (top), a thin sandy loam clay sequence overlying the lateritic layer at 90–120 cm depth (mid-slope) and thick sands near to the water table (bottom sites). There are variations from transect to transect of this pattern, mainly in the thickness of the sands and loams, but differences between positions within transects are greater than the differences between the transects for the same position. Slope profiles for each transect, together with soil thicknesses are shown in Figure 25.2(a)

Each plot comprised a 15 m by 6 m area and examples of three of the surveyed plots are shown in Figure 25.3, with contours accurately surveyed but relative to a datum within the plot. Internal subplots, 5 m by 2 m, were used to measure runoff and were trenched around to lead away upslope runoff. Plot runoff drained into metal troughs 2 m long and thence into large collection barrels set into the ground. These barrels were cleared of sediment and water periodically and after large rainfall events. On each site a nest of tensiometers was installed, with instruments at 15, 30, 60, 90 and, where possible, 120 cm depth. In the mid-slope plots a second set of tensiometers was installed, just downslope of the first to assist in the computation of subsurface fluxes. These will subsequently be referred to as A2.1 and A2.2, B2.1 and B2.2, and C2.1 and C2.2 respectively. Each plot also has a neutron-probe access tube and a simple raingauge. Within the sites subplots were set aside for periodic sampling of litter and subsurface organic matter. We also had access to the base-camp rainfall data for the entire period, enabling us roughly to estimate the canopy throughfalls for different treatments against central site values.

All the sites in transect A were left as undisturbed as possible, the only interference being the installation of the runoff troughs and the trampling along access paths to the probes and tensiometers. The B sites had all vegetation above breast height removed and the debris of this removal carefully extracted from the plots. This was done with the minimum possible disturbance to the ground layer, and on plot B1 a very large tree trunk had to be left across a part of the site. Overhanging branches were removed and in some cases large off-plot trees had to be cut down to ensure the plots received all the possible rainfall and sunlight energy. The plots along transect C were all completely cleared.

Trees and undergrowth, though not roots, were felled or cut out and litter was gently raked or hand-picked from the surface of the plots. These plots were carefully cleared of vegetation regrowth and fallen litter during the period of data collection.

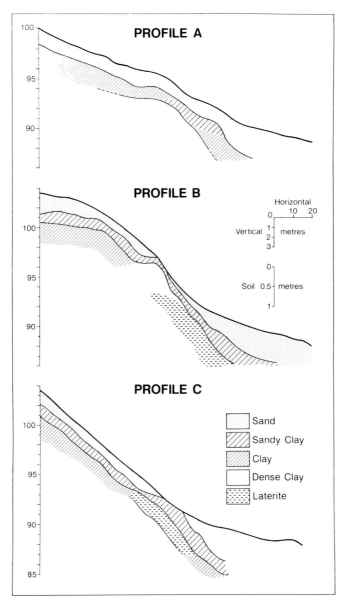

Figure 25.2(a)   Slope profiles (note vertical exaggeration)   *(continued on next page)*

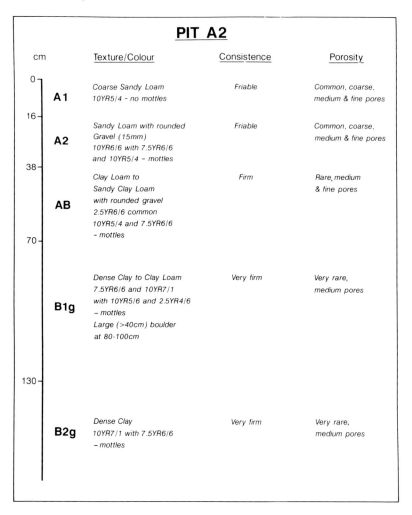

Figure 25.2(b) Schematic profile at mid-slope site   *(continued opposite)*

RESULTS

Rainfall

Over the period of 300 days the mean daily temperatures were of the order of 35°C and relative humidities typically about 90 per cent. The major variations were in rainfall (Figure 25.4). In April and early May there was heavy rainfall such that between 12 April (day 71) and 17 May 1988 (day 106) some 540 mm of

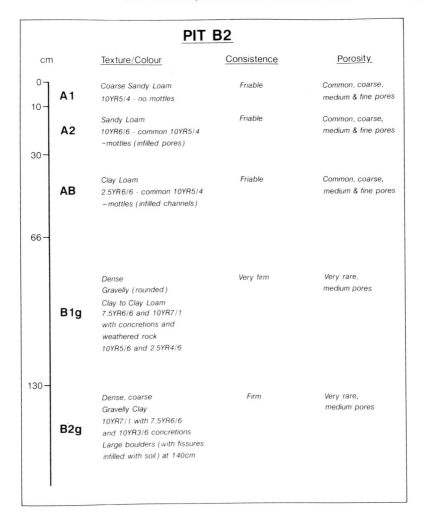

PIT B2

cm		Texture/Colour	Consistence	Porosity
0	**A1**	Coarse Sandy Loam 10YR5/4 - no mottles	Friable	Common, coarse, medium & fine pores
10	**A2**	Sandy Loam 10YR6/6 - common 10YR5/4 -mottles (infilled pores)	Friable	Common, coarse, medium & fine pores
30	**AB**	Clay Loam 2.5YR6/6 - common 10YR5/4 - mottles (infilled channels)	Friable	Common, coarse, medium & fine pores
66	**B1g**	Dense Gravelly (rounded) Clay to Clay Loam 7.5YR6/6 and 10YR7/1 with concretions and weathered rock 10YR5/6 and 2.5YR4/6	Very firm	Very rare, medium pores
130	**B2g**	Dense, coarse Gravelly Clay 10YR7/1 with 7.5YR6/6 and 10YR3/6 concretions Large boulders (with fissures infilled with soil) at 140cm	Firm	Very rare, medium pores

Figure 25.2(b)

rain fell, with average daily rainfall per rain day in the range 20–40 mm. There then followed a long and relatively dry period punctuated only by the large storms (115.1 mm) on 15/16 October (day 258) and a smaller storm (38.9 mm) on 4 November (day 277). After this the rains further diminished and the system continued to dry out until the last observations on this subproject were made in early December (Figure 25.4). The period from 20 October to 1 December was especially dry. Rainfall intensities also diminished in this early winter period when rainfall per rain day was usually in the range 3–6 mm.

426

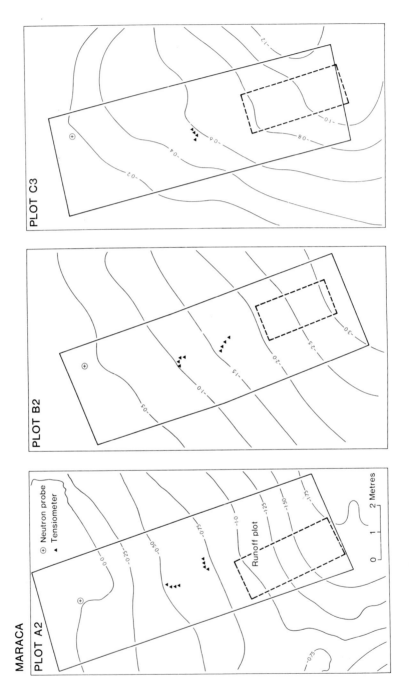

Figure 25.3 Plans of plot layouts for three plots

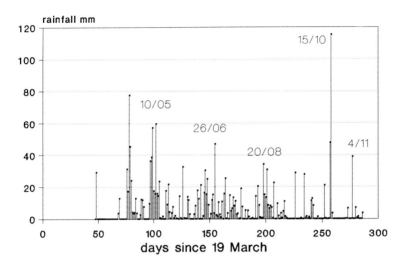

Figure 25.4   Rainfall for survey period

Soil Tensions

Because of difficulties with the neutron probe data we have chosen to discuss soil moisture conditions in terms of soil tension which, though spatially more variable, is more readily and more accurately measured. It is important to bear in mind that the soils on each site are somewhat different and therefore a simple substitution between soil moisture and tension cannot be made, especially since we are working on the upper limb of the moisture–tension curve, where rates of change are quite steep. Differences in the characteristic curve could be responsible for some of the observed differences, even though the moisture values were the same. However, the large contrasts observed here suggest that differences in soil texture are overwhelmed by differences in treatment and in soil structure. Measurement of soil tensions commenced on 11 May following plot clearance by about one week and the very wet period of early May. Figure 25.5 shows characteristic tension traces for three tensiometers chosen to indicate the range and variety of conditions at the field sites. In Figure 25.5(a), data are given for site B2.2 at 15 cm. They show a typically large range (0–80 kPa) and a quick response to even quite small rainfalls, reflecting the near-surface position and rapid draining and drying. Figure 25.5(b) is for the top-slope forested site (A1) at 60 cm and

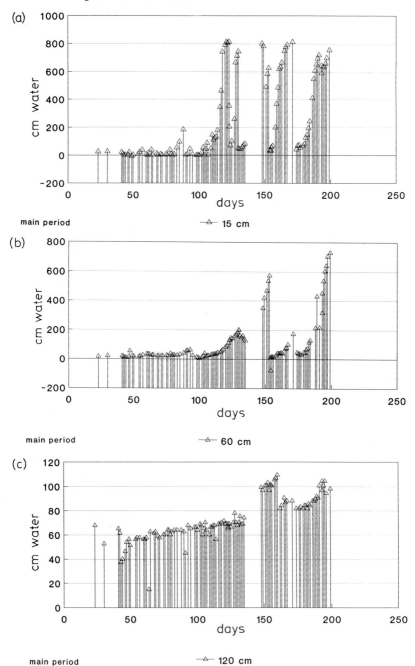

Figure 25.5   Soil tension to illustrate variations with site and depth: (a) site B2.2, 15 cm, (b) site A1, 60 cm, (c) site C3, 120 cm

shows a similar range but with slower and smoother responses. Figure 25.5(c) is for the completely cleared lower slope site (C3) at a depth of 120 cm. Here the range of values is small (2–11 kPa) and the response really quite smooth.

For the purposes of summarizing the tension conditions and comparing them between transects and plots within transects, we have identified the characteristic periods shown in Table 25.1. This illustrates three sets of soil conditions with very wet (periods 1 and 2), medium-wet (periods 3 and 4) and very dry soils (periods 5 and 6) when soil moisture tension values were relatively stable. They are separated by periods of rapid change and for the purposes of analysis we have regarded each set as statistically independent. However, there is an overall shift to drier conditions during the observation period and autocorrelation between daily values is very significant in the first wet period up to day 120. After that autocorrelations for lags of more than three days are not significant.

For each of these periods and for three tensiometer levels (30, 60 and 90 cm) we have carried out an analysis of variance with treatment (forest, partially cleared and completely cleared) and slope position (top, middle and bottom) as the factors. Each cell contains the readings for the tensiometer over the period at a specific site on a particular treatment and for a given depth. Table 25.2 shows typical mean values for the three periods, by treatment, for a depth of 60 cm. This table shows that during the wet period, immediately after clearance, there were no real differences between the sites in soil tensions; as time since clearance increased *and* as the climatic conditions became drier, all the soils became drier; and by the end of the observation period the virgin forest sites had much higher tensions (drier soils) than the partially cleared sites, and these again had higher tensions than the completely cleared sites. We interpret this as reflecting the much higher transpiration rates for the uncleared forest than for the partially cleared and, in turn, the completely cleared plots. This is true both in the driest periods of the study after the effects of clearance, including progressive change of surface conditions such as organic matter content and soil stability, have had time to work through.

Table 25.1   Periods used for analysis

Period	Date	Elapsed days	Prior 7-day rainfall (mm)
Very wet			
1	28–31 May	117–120	63.3
2	26 June–3 July	149–153	80.6
Medium-wet			
3	16–20 August	193–196	44.8
4	28 September–4 October	236–238	42.1
Very dry			
5	9–14 October	251–256	1.4
6	21–29 November	294–297	0.4

Table 25.2 Soil tension mean values for wet, medium and dry periods for the three clearance treatments at a depth of 60 cm (values in kPa)

Soil condition	Clearance treatment		
	Virgin forest (A plots)	Partially cleared (B plots)	Completely cleared (C plots)
Wet	1.8	2.9	4.3
Medium	63.3	20.8	9.6
Dry	73.6	56.4	19.3

Table 25.3 shows the mean values of soil tension (kPa) by treatment, by site and by depth for the final dry period. The data indicate the expected variations with depth, the deeper parts of the profile generally exhibiting lower tensions. The upper part of the table indicates that for the upper part of the profile, treatments A and B are more alike than treatment C. This can be interpreted as indicating that near the surface the uncleared and partially cleared plots are losing water from shallow-rooting shrubs and herbs at a rate which is more important than the bare-soil evaporative losses on the cleared sites. Deeper in the profile the treatments are more similar, though C and B are now more alike than A. It is tempting to attribute this to the lack of large trees on B and C but with such small plots the roots of large trees off the plot may be extending beneath them.

The lower part of the table indicates that downslope, the soils become wetter at the surface, whilst remaining relatively uniform at depth. Hence topographic and slope effects on soil drainage may be more important in the surface horizons. Investigations of the subsurface hydrology reveal very strong subsurface flows through the upper profile, especially in the mid-slope sections as the plateau is drained following heavy rain (Nortcliff and Thornes, in press).

In order to evaluate the relative contribution of the different sources of variance (clearance treatment, slope position and soil type) on soil moisture

Table 25.3 Mean soil tensions by site, treatment and depth for period 6 (very dry soil conditions) (Values in kPa)

Clearance treatment	Soil depth		
	30 cm	60 cm	90 cm
Transect A (virgin forest)	52.7	38.4	35.2
Transect B (partially cleared)	57.0	24.1	16.7
Transect C (completely cleared)	35.0	12.6	9.8
Top slope (sites No 1)	77.6	22.7	21.0
Middle slope (sites No 2)	41.3	30.5	27.1
Bottom slope (sites No 3)	25.0	16.8	13.4

tensions, we have used a two-way crossed classification with replication and tested as systematic effects (Huitson, 1971). Separate analyses were carried out for each of the three levels (treatment, position and depth) for each period. The detailed analysis of variance reveals the extent to which these trends can be regarded as statistically significant. Essentially they indicate that in statistical terms for the *wettest periods* (1 and 2), differences in slope position are a more important control on soil moisture tension than are differences in treatments, with much higher F ratios in the first, wet period than in the second, medium-wet period (3 and 4). For the *dry periods* (5 and 6), contrasts between clearance treatments are more important than contrasts between sites. Two trends may account for these results: first, the progressive changes in moisture regime resulting from the different treatments; second, the more marked impact of treatment than position when the system is much drier. Both are probably responsible for the observed changes.

Runoff

Runoff depends on storm intensity as well as position (reflecting slope and soil conditions) and clearance treatment. In order to separate out rainfall characteristics, regressions between rainfall and runoff have been obtained and the characteristics of the regression curves are used to evaluate treatment and slope effects. Rainfall has been determined by ground-level gauges on the top sites for each treatment but the rainfall was so clearly affected by the treatment that it cannot be used as a control. In other words, by using *site* rainfall we are measuring canopy throughfall in the case of transect A, understorey throughfall in transect B and true rainfall in transect C. Significantly, the correlation with runoff decreased in that order from r values of 0.9 to values of about 0.65.

We decided therefore to correlate runoff and rainfall for storm events using the rainfall recorded at the base camp. In this case the lowest Pearson correlation coefficient is 0.76 and values are typically over 0.8. The models are linear and express runoff in litres against rainfall in millimetres. Regression coefficients average 0.60 for forested sites, 0.632 for partially cleared sites and 1.614 for totally cleared sites indicating that, after compensating for plot area, on average about 6 per cent of the rain occurs as runoff on the forested and partially cleared plots whilst about 16 per cent runs off from the completely cleared plots. These figures are much smaller than those for the forested sites studied by Leopold, Franken and Matsui (1985) for the Barro Branco (19.3 per cent) and the Taruma-Acu basin (25 per cent), and for the forested sites this presumably reflects the sandy topsoils of Maraca when compared with the latosols of the area around Manaus. In part, however, it may reflect the transient character of the soil water status during the transition period, at least for the completely cleared transects. Moreover, the rainfall intensity effects have not been accounted for in our analyses because of the smaller data set.

A more detailed inspection of the regression coefficients indicates that, on the completely cleared plots, the middle site (C2) had the heaviest runoff coefficient (22.6 per cent) and C1 the least heavy (7.8 per cent). This is also true of the partially cleared plots where the mid-slopes site (B2) had a significantly higher runoff coefficient (11.4 per cent) than either B1 (7.5 per cent) or B3 (7.2 per cent). This almost certainly reflects the combined effects of a steeper slope and the lateritic horizon at shallow depth in the mid-slope positions. The mid-slope soils act as a major conduit for the plateau drainage, as revealed by throughflow measurements, and runoff here must largely reflect impeded infiltration due to the higher moisture levels at depth. For top-slope positions the mean runoff coefficient is 7.7 per cent, for the mid-slope positions 13.9 per cent and for the bottom-slope positions 11.8 per cent. The differences resulting from treatment appear therefore to be more important than differences due to position.

The total measured runoff volumes from the different sites for the entire observational period reflects the importance of treatment and slope as shown in Figure 25.6(a), the differences in treatment being more significant at the mid-slope position. At the slope foot sites the role of saturated overland flow appears to play a much less significant role than at Reserva Ducke, Manaus (Nortcliff and Thornes, 1988).

To summarize, there are major differences between the uncleared and partially cleared plots on the one hand and the completely cleared plots on the other. There are also observable but less strong differences due to slope position, probably reflecting differences in soil structure. The first of these contrasts suggests that partially cleared sites behave in runoff terms more or less like virgin forest, despite the evident differences in soil moisture regimes in the wet season between the two treatments referred to in the previous section.

Soil Loss

Total receipts of sediment for all sites varied according to antecedent rainfall conditions, with collections on 28 August and 16 October 1987 being much larger than the rest. The former reflects the storms of 16–19 August, the latter mainly the 162.8 mm that fell on the previous two days.

More important, however, for the purposes of this chapter, are the relative contributions of slope and clearance treatment to soil loss. These are summarized in Table 25.4 and indicated in Figure 25.6(b). Above all, these results reveal the great contrast between virgin forest and partially cleared plots on one hand and the completely cleared plots on the other. This illustrates the wisdom of the traditional assertion that complete clearance of sites will result in very high soil losses. Note, however, the more surprising and perhaps

Maraca runoff

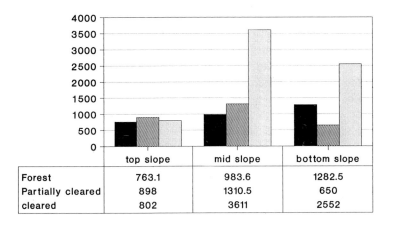

	top slope	mid slope	bottom slope
Forest	763.1	983.6	1282.5
Partially cleared	898	1310.5	650
cleared	802	3611	2552

Maraca sediment yield

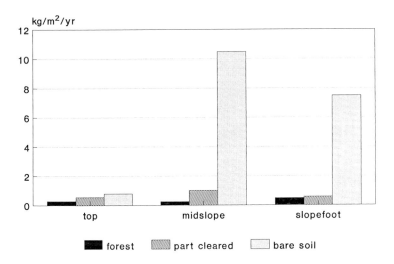

Figure 25.6   Upper, runoff differences between sites and treatments (litres, 27/05–22/11/87). Lower, sediment yield differences between sites and treatments

in some respect more important result, that the removal of the canopy in the B sites has not resulted in erosion at levels higher than under natural forest. This supports the assertion by Wiersum (1985), Brandt (1986) and others that it is the clearance of the understorey and litter that are more important than the clearance of the canopy. It also reinforces the comments made at the beginning of this chapter that it is probably the rate of regeneration of ground cover to about the 30 per cent level that is critical for understanding erosional problems, rather than the destruction of the forest itself. In this respect, the related studies of forest regeneration in artificially cleared forest gaps at Maraca will prove to be very instructive. Clearly the conditions of regeneration on sloping sites will themselves depend very much on the erosion rates, however, (Thornes, 1988; and Chapter 4 of this volume).

It is also evident from Table 25.4 that within the treatments there are other factors causing variation in soil loss. One factor we have already mentioned is antecedent rainfall. Regressions of yield against antecedent rainfall for each site do not reveal a consistent strength of relationship, with correlation coefficients varying very significantly. Of the nine sites, five have correlations higher than 0.7. Even here the number of observations is small. On transects B and C, the mid-slope positions have appreciably higher yields of eroded soil than the other positions. This reflects not only the steeper slopes, but also the higher levels of runoff generation. Further analysis of these controls will be undertaken in a later paper. Finally we note that of the completely cleared sites, C1 is anomalous in sediment yield as it is in most other respects. On re-inspection of the maps and contour data for the runoff subplot it appears that this may be due to loss of both runoff and sediment from the side of the subplot, which was set slightly obliquely to the contours.

CONCLUSIONS

The results show unequivocally that soil moisture conditions are significantly affected by both the nature of clearance and slope position.

Table 25.4 Total observed soil loss (kg) by site and treatment (25/07–26/10/87)

Clearance treatment	Slope location			
	Top (1)	Mid (2)	Bottom (3)	Total
Forest (A)	1.44	1.28	2.42	5.14
Partially cleared (B)	2.72	5.17	2.37	10.26
Completely cleared (C)	3.93	52.40	37.50	93.83
Total	8.09	58.85	42.29	

Slope position is most important in the wettest period at the beginning of the experiment immediately following clearance and the treatment is most important under moisture stress at the end of the observation period. These results reflect both the direct effects of different treatments on the moisture budgets and the progressive changes during the transition to new stable equilibria following clearance.

Both runoff and sediment losses are dominated by treatment effects, the latter very strongly. Our findings confirm the general concern over forest clearance in that complete clearance results in very large increases in sediment yields. They do indicate, however, that more emphasis needs to be made on the rate of vegetation recovery in so far as even a simple understorey and ground litter (which, on Maraca, is very thin, even under full forest) provide a level of protection not very different to that of the forest. This point is also made in a related paper by Francis and Thornes (Chapter 22 of this volume) relating to semi-arid vegetation removal.

The major differences between these results and those found for example in the temperate zone are of degree rather than kind. At the simplest level we can interpret them in terms of the exponential reduction in erosion as cover increases, referred to at the beginning of the chapter.

Two caveats remain to be mentioned. First, the plots are still relatively small and subsurface rooting, and boundary effects play their part. Second, the subsurface hydrology has here been compounded with slope and soil structure, in the term 'position'. Dynamically it relates to the wider topographic setting. Despite these caveats it is clear that the impact of clearance is intimately related with soils, slope and topography and in predicting the impact of clearance these must be taken into account.

ACKNOWLEDGEMENTS

The authors wish to express their appreciation of the support given for this work by the Royal Geographical Society during the 1987/8 Maraca Expedition and during the subsequent analysis by the Leverhulme Trust. They also wish to express their deep appreciation for the logistical support in the field of Steve Bowles, Fiona Watson and Sarah Latham.

REFERENCES

Brandt, J. (1986). *Transformations of the kinetic energy of rainfall with variable tree canopies*. Unpublished PhD Thesis, University of London.

Dunne, T., W. E. Dietrich and M. J. Brunengo (1978). Recent and past erosion rates in semi-arid Kenya. *Zeitschrift fur Geomorphologie, Supplementbd.*, **29**, 130–40.

Elwell, H. A. and M. A. Stocking (1976). Vegetative cover to estimate soil erosion hazard in Rhodesia. *Geoderma*, **15**, 61–70.

Fearnside, P. M. (1985). Agriculture in Amazonia. In G. T. Prance and T. E. Lovejoy (eds.), *Amazonia*, Pergamon, Oxford.

Fearnside, P. M. (1987). Causes of deforestation in the Brazilian Amazon. In R. E. Dickenson (ed.), *The Geophysiology of Amazonia*, John Wiley & Sons, Chichester.

Huitson, A. (1971). *The Analysis of Variance*, Griffin Statistical Monograph Series, **18**.

Lee, C. R. and J. G. Skogerboe (1985). Quantification of erosion control by vegetation on problem soils. In S. A. El Swaify, W. C. Moldenhauer and A. Lo (eds), *Soil Erosion and Conservation*, Soil Conservation Society of America, pp. 437–44.

Leopold, P. R., W. Franken and E. Matsui (1988). Hydrological aspects of the tropical rain forest in the Central Amazon. In J. Hemming (ed.), *Man's Impact on Forests and Rivers*, Manchester University Press, pp. 91–108.

Nortcliff, S. and J. B. Thornes (1988). The dynamics of a tropical floodplain environment with reference to forest ecology. *Journal of Biogeography*, **15**, 49–59.

Nortcliff, S. and J. B. Thornes (in press). Forest clearance and subsurface hillslope hydrology. Proceedings of First Maraca Symposium, *Acta Amazonia*, Manaus.

Thornes, J. B. (1988). Erosional equilibria under grazing. In J. Bintliff, D. Davidson and E. Grant (eds), *Conceptual Issues in Environmental Archaeology*, Edinburgh University Press, pp. 32–64.

Wiersum K. F. (1985). Effects of various vegetation layers of an *Acacia auriculiformis* forest plantation on surface erosion in Java, Indonesia. In S. A. El Swaify, W. C. Moldenhauer and A. Lo (eds), *Soil Erosion and Conservation*, Soil Conservation Society of America, pp. 79–89.

26 The Use of Tree Mounds as Benchmarks of Previous Land Surfaces in a Semi-arid Tree Savanna, Botswana

YVAN BIOT

School of Development Studies, University of East Anglia

SUMMARY

The tree root exposure technique has been used successfully to derive rates of erosion in East Africa. In regions where annual soil loss is little and/or the turnover of the woody vegetation very rapid, roots are rarely exposed but the soil surface under trees is slightly raised. The use of these 'tree mounds' as benchmarks of the soil surface at tree germination is evaluated in a case study situated in the Hard Veld of Botswana. It is found that the rate of denudation as calculated using tree mound heights and tree age is 10 to 15 times higher than rates of erosion derived using SLEMSA and an ongoing water balance study in the region, and that tree mound heights can be explained by a difference in bulk density between 'mound' and 'flat' soils alone. It is concluded that, in the study area, tree mounds cannot be used as benchmarks of the position of the land surface at the time of germination of the tree. The tree mounds are not formed by a lowering of the surrounding surface but by a raising of the land surface in the immediate vicinity of the tree stem because of high root density and termite/ant activity. Trapping of sediment at the tree stem could possibly also contribute to these tree mounds. The results from this study confirm the need for careful utilization of the tree mound technique for a rapid appraisal of rates of erosion.

Vegetation and Erosion
Edited by J. B. Thornes
©1990 John Wiley & Sons Ltd

INTRODUCTION

Measuring rates of sheet and rill erosion without interfering with the processes which produce it is almost impossible (Stocking, 1987). Indirect field monitoring involves the determination of stream runoff, sediment loads and reservoir sedimentation rates and is based on the concept of sediment delivery ratios which are notoriously imprecise (Task Committee on Sedimentation, 1970; Rapp *et al.*, 1972). Examples of direct measurement are: the measurement of the position of the land surface around erosion *pins* (Schumm, 1967; Haigh, 1977); the collection of sediment in troughs (Gerlach, 1967) and from standard erosion plots (e.g. Wendelaar and Purkis, 1979); the measurement of the position of benchmarks of former land surfaces such as exposed tree roots (Eardley, 1967; Lamarche, 1968); and the survey of the redistribution of radioactive fall-out (Reece and Campbell, 1984). Of all these methods only the erosion pins, the tree root exposure technique and the measurement of the redistribution of radioactive fall-out can claim to avoid effectively the interference of the measuring procedure with the erosion process.

Based on the experience of Dunne, Dietrich and Brunengo (1978, 1979), El Swaifi, Moldenhauer and Lo (1985) and Stocking (1984) recommend the use of the tree root exposure method as a 'rapid rural appraisal technique'. This recommendation was followed in a survey of sheet and rill erosion rates in a region of the Hard Veld of Botswana. This work was carried out as part of a study of the sustainability of cattle production on communal rangeland in the Hard Veld (Biot, 1988a, b, c).

PHYSICAL GEOGRAPHY OF THE STUDY AREA

Climate, Geology, Geomorphology and Soils

The study area is located in Central District, Botswana, between 26°30' and 27°00' longitude East and 22°45' and 23°15' latitude South (Figure 26.1). Climate is semi-arid (Bsh according to Köppen, 1923) with an average annual rainfall of 450 mm. The main rock types (migmatites, granite, ortho- and paragneisses and grano-dioritic dykes) are part of the Pre-Cambrium Basement Complex (Ermanovics, 1980).

The present-day landscape is believed to be the result of several cycles of pediplanation whose origin can be traced to previous tectonic upheavals and climatic fluctuations (King, 1967; Jennings, 1962). The bulk of the region belongs to a 'Late Tertiary Pediplain', with scattered remnants of an older 'African Surface' and a younger 'Quaternary Surface' along the main rivers. Occasional runoff waters are drained through the mostly dry Mhalatschwe River, which flows in the Limpopo east of Mahalapye. It is within the headquarters of the

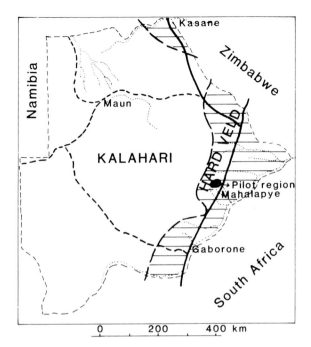

Figure 26.1 Location of the study area

Mhalatschwe River, where slopes are steepest (1.5 to 5 per cent) and signs of water erosion most obvious, that the field study was conducted.

The land surface consists of scattered inselbergs, pediment slopes, and occasional levees (Figure 26.2). The soil cover was formed and has been redistributed during wetter climatic periods in the past (Siderius, 1973; Biot, 1988b). Past erosion processes are probably responsible for a dense gravel horizon which covers the parent rock and for the now sand-filled and stabilized old gully scars easily recognizable as wide and shallow, grass-covered depressions. Present-day abrasion is mainly caused by sheet and rill erosion in the upper and middle pediment slopes. Incision through gullying occurs in the lower pediment slopes, where soils exhibit high sodium contents.

Soils have been mapped and classified by Siderius (1973) and Remmelzwaal (1987). The main mapping units are classified within the Regosol, Arenosol, Acrisol, Luvisol and Planosol groups of the FAO/UNESCO system (FAO, 1974). The physico-chemical soil characteristics are typical for a semi-arid area: high base saturation and cation exchange capacity in the subsoil and low organic matter, nitrogen and phosphorous levels in the topsoil. Sodium saturation and (petro-) calcic horizons can occur in the lower pediment slopes.

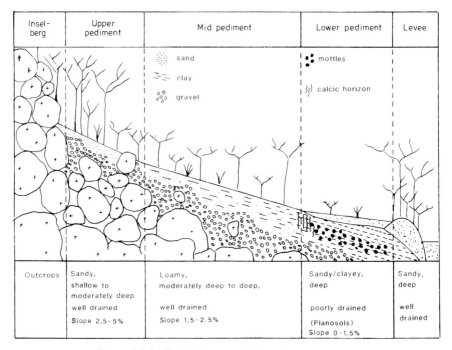

Figure 26.2 Typical geomorphological cross-section

Vegetation and Land Use

The low and erratic rainfall is responsible for frequent crop failure and the farming system has an important animal production component. Traditionally livestock is managed from 'cattle posts' which are scattered within the communal range. The rangeland in this area mainly consists of tree savanna (Weare and Yalala, 1971) with the density and species composition of the tree cover depending on the soil type and geomorphological position. Average canopy cover is 33 per cent and dominant species are *Acacia tortilis*, *Combretum apiculatum*, *Dicrostachys cinerea*, *Euclea undulata*, and *Grewia flava*.

At the time of the survey, grass cover was low (29 per cent on average) because of drought. Dominant grass species are (in order of importance): *Digitaria milanjiana*, *Aristida congesta*, *Brachiaria nigropedata*, *Tragus berteronianus*, *Eragrostis rigidior* and *Urochloa trichopus*.

TREE MOUNDS

Tree root exposures were not found in the study area. This may be an indication of the low rates of past and present erosion and/or a high rate of turnover of the woody layer. Each tree and bush, however, is situated on a low mound, on average 5.5 cm higher than the surrounding flats. At first sight, the height and shape of these tree mounds appear to vary with tree species, age and density. Two such mounds are illustrated in Figure 26.3.

It was assumed, initially, that the tops of the tree mounds were benchmarks of the land surface at the time of germination of the tree. If this were substantiated, annual rates of denudation would be equal to the ratio between tree mound height and tree age, referred to hereafter as 'annual tree mound height increments' or 'tree mound increments'.

MATERIALS AND METHODS

Tree mound heights and tree ages were measured in three areas on a total of 16 sites representative of the main landscape units of the study area: upper, middle and lower pediment slopes over granitic and granodioritic rocks, wide shallow depressions and levees. Each site was sampled in triplicate within a 2500 m^2 area. Mound height was taken from the top of the mound at the tree base to the edge of the tree canopy or the lowest point under the canopies of neighbouring trees. The height of the tree mounds above the flats was measured in four directions: upslope, downslope and in the two directions orthogonal

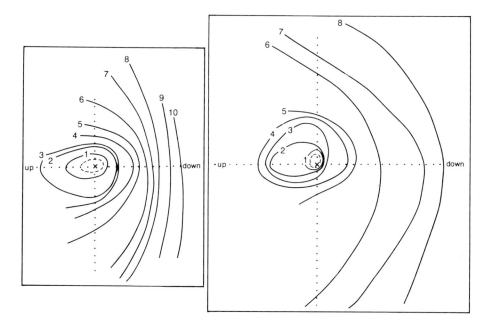

Figure 26.3 Typical tree mounds (heights in cm relative to top of tree mound)

to the direction of steepest slope. Tree age was determined from a careful tree-ring count on slices of the tree trunk sampled between 30 and 50 cm height. The five trees closest to the centre of each replicate were selected.

Not all tree species are suitable for the purpose of erosion monitoring using the tree root exposure and tree mound technique. The following set of requirements must be fulfilled:

1. The tree must be a one-stem tree. Many bushes are composed of a bunch of semi-perennial branches which die off after 3–5 years and are replaced by other branches from the base (e.g. *Grewia* spp., *Tarchonanthus camphoratus*), which is unsuitable for the purpose of erosion monitoring.
2. The tree species must occur in all sample sites.
3. The tree age must be established with some acceptable degree of certainty.

Only the different *Acacia* species fulfilled the first two requirements, and the following were used (in order of importance): *Acacia tortilis*, *A. nilotica*, *A. karroo*, *A. fleckii*, *A. robusta*, *A. gerardii*.

RESULTS

Tree Age

Dunne *et al.* (1978) mention the problem of determining the exact age of trees using tree ring counts in an environment with more than one growing season. In the study area, only one growing season occurs, but as it is of a rather disparate nature there are often unpredictable flushes of growth.

Tree core samples taken during a reconnaisance visit revealed only two specimens of *Acacia tortilis* with an observable ring count on the core. Other attempts failed because of the lack of observable rings in the cores of most other species and difficulties in sampling caused by the dense wood.

The dendrochronological potential of 108 tree species in Southern Africa has been studied and discussed by Lilly (1977). According to this study, the *Acacia* species rate fairly poorly mainly due to the occurrence of false ring margins and discontinuous rings—a common feature among angiosperms of the Southern Hemisphere.

The use of a tree corer as a means of obtaining reliable tree-ring counts in Southern Africa has been discussed by Curtis, Tyson and Dyer (1978). Apart from the obvious problems related to the dense wood in a dry environment, tree coring has also been found to be unsatisfactory owing to the occurrence of localized missing rings associated with ring convergence. A new sampling method was recommended and followed in the present study, which consists of the sampling of a whole cross-section of the tree.

In view of the uncertainties associated with tree age determination through ring counts, tree age was calibrated against ring counts using samples from two areas with known time of return to fallow (from aerial photography and local confirmation with farmers). The following results were obtained:

3-year fallow: average tree ring count = 4.2 ± 0.8 (6 samples)
16-year fallow: average tree ring count = 16.3 ± 1.3 (5 samples)

Although the sample size was insufficient with regards to the complete range in tree age, it is tentatively concluded that the tree-ring count exercise is accurate to within 10 per cent around the mean for the older trees and 20 per cent for the younger trees. The age of the younger trees tends to be overestimated by one year.

The age of the 240 sampled *Acacias* was 8 to 10 years on average, with a maximum of 18 years. This maximum age, however, is underestimated as it was not possible to fell the few very large trees using the equipment available at the time of the survey.

Kruskal–Wallis analysis revealed no significant differences in tree age between areas. Differences between sample sites within each sample area were slightly significant, revealing the following trends: younger trees on the wide and open depressions and shallow soils, and older trees on the levees. The wide and shallow depressions are characterized by an impervious layer of silt at about 90 cm depth and the concentration of runoff waters from the surrounding areas after heavy showers. According to Tinley (1982) such conditions make it difficult for trees to develop deep roots and compete with the herbaceous layer which thrives in the additional supply of soil moisture. The levees are well drained and trees benefit from a reliable supply of moisture from the river and stored runoff waters from the pediment slopes.

The relationship between tree trunk diameter and age is good, and little affected by tree species (within the genus *Acacia*):

$$\text{age} = 0.46 \cdot \text{diameter}^{0.74} \qquad R^2 = 0.87 \qquad \text{(Figure 26.4)}$$

Tree Mound Heights

The typical shape of the tree mounds has been illustrated in Figure 26.3. Average tree mound height is 5.5 cm, ranging from 2 to 18.5 cm. Kruskal–Wallis analysis revealed no significant differences between the tree mound heights of different sample areas. Sites within each area were significantly different, with lower mounds in the wide and open depressions and higher mounds on the levees.

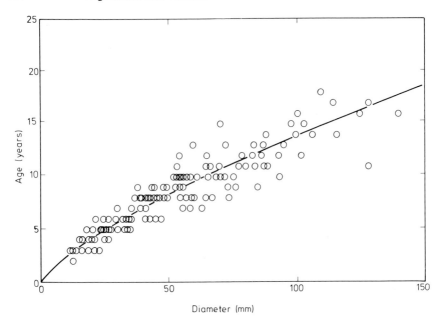

Figure 26.4 Relationship between tree age and tree trunk diameter

Tree Mound Height Increments and Rates of Erosion

Tree mound height increments were calculated for a subset of the total sample consisting of the sample sites on the upper and middle pediment slopes over granite. The samples in this subset represent different stages in the erosion process as indicated by soil depth (from > 120 cm down to a few centimetres). The tree mound height measurements taken perpendicular to the direction of steepest slope were used in the subsequent enquiries. The distribution of the tree mound increments was found to be positively skewed, so a logarithmic transformation was applied for further statistical analysis. Analysis of variance revealed no significant effect of species and sample area on tree mound increments.

Average tree mound height increment is 6.1 ± 1.8 mm/year, which, on the basis of a bulk density of 1.65 g/cm^3 (Biot, 1988b) results in a rate of erosion of 91 t/ha. This value is about 10 to 15 times higher than that forecast using SLEMSA (Elwell and Stocking, 1982) for the observed slope (Abel and Stocking, 1987; Biot, 1988b) and the results of a water balance study (Biot, 1988b, d), and it is concluded that a lowering of the surrounding surface by soil erosion is unlikely to be the (sole) process responsible for the creation of tree mounds.

Dunne *et al.* (1978) have discussed alternative tree-mound forming processes, including the following:

1. Termites are known to bring up subsoil to the top, and Biot (1986) has found significantly more signs of termite activity under tree canopy than on the flats.
2. The tree mound might have been developed on top of an existing tree mound connected to a neighbouring tree.
3. The tree might have germinated on an existing tree mound.
4. Trees are obstacles to overland flow and wind where sediment can accumulate.
5. Certain trees are known to produce tree mounds and even expose their roots on sites protected against denudation.

In order to examine other possible explanations for the occurrence of tree mounds, disturbed and undisturbed soil samples were taken on the tree mounds and on the surrounding flats. Bulk density, organic carbon and infiltration capacity of the tree mounds were found to be significantly different from the flats (Table 26.1).

The different bulk densities have a significant effect on calculated rates of erosion. This can be illustrated using the relationship between bulk density and total porosity of the soil:

$$TPS = \left(1 - \frac{BD}{SG}\right)100$$

where TPS = total pore space, BD = bulk density, SG = specific gravity of the soil particles (= 2.65 g/cm^3). Using the values of bulk density given above we find:

TPS under tree canopy 49 per cent
TPS on flats: 38 per cent

For a given volume of soil (say 5 cm $\times$ 5 cm $\times$ 5 cm = 125 cm^3), this means that under tree canopy 61.25 cm^3 and on the flats 47.50 cm^3 is porosity—a difference of 13.75 cm^3. When spread equally over the whole surface of the given volume

Table 26.1 Average infiltration capacity, ground cover, organic matter content (OC) and bulk density (BD) under tree canopy and in unshaded areas (flats)

	Infiltration (min/cm)	Ground cover (%)	OC (%)	BD (g/cm^3)
Canopy	36	27	1.07	1.34
Unshaded	162	15	0.62	1.64

Bulk density and infiltration under canopy measured at 10 cm from base of tree.
Infiltration: mean time in min/cm for 2 cm infiltration in 5 cm diameter ring (19 samples, 3 replicates).
Bulk density: using 3 $\times$ 5 cm rings (16 samples, 3 replicates).
Ground cover: visual estimate within 1 $\times$ 0.25 m sample frame (16 areas, 5 samples per area).
All differences significant at probability = 0.001.

of soil (i.e. 25 cm²), this gives rise to a 0.55 cm layer of soil on the tree mound which can be attributed to the difference in bulk density alone. Assuming that the tree germinated on a surface with uniform bulk density equal to the bulk density of the soils on the flat, tree mound height measurements must be corrected for this contribution of soil porosity to the soil's bulk density if we want to determine the exact position of the land surface at the time of tree germination. Depending on the depth down to which the difference in bulk density prevails, the correction of tree mound height for a higher porosity under tree canopy differs (Table 26.2, assuming an average tree mound height of 4.88 cm and a tree age of 8 years for the subset considered).

It is obvious from these calculations that, unless the exact differences in bulk density and the depth down to which these apply are known, it is impossible to use the results of the tree mound survey to calculate absolute rates of erosion in the survey area. An alternative would have been to use an indication on the tree stem of the level of the soil surface at the time of germination. Not enough was known about tree stem morphology at the time of the survey to do this, and the few attempts yielded unreliable results.

Following the disappointing results obtained from this tree mound survey, a systematic analysis of the impact of 30 variables characterizing the soil, soil surface, type and intensity of erosion, topography and herbaceous and tree vegetation on tree mound height increments was carried out. The following significant (0.05 level) Spearman correlation coefficients were obtained:

micro-topography	-0.571
type of litter	-0.525
amount of litter	-0.442
vegetation index	-0.418
percentage bare soil	0.390
canopy cover	-0.350

Table 26.2 Correction to be applied to the tree mound height depending on the depth down to which a difference in bulk density of 0.30 g/cm³ prevails and rates of erosion

Depth (cm)	Height due to higher porosity (mm)	Corrected mound height (mm)	Annual increment (mm)	Erosion (t/ha/yr)
5	5.5	43.3	5.4	81
10	11.0	37.8	4.7	71
15	16.5	32.3	4.0	60
20	22.0	26.8	3.4	50
25	27.5	21.3	2.7	41
30	33.0	15.8	2.0	30
40	44.0	4.8	0.6	9
50	55.0	-6.2	-0.8	-12
60	66.0	-17.2	-2.2	-32

where micro-topography: a number derived from the index used to characterize the position of the herbaceous cover sample within the micro-topography. A high number is related to a dominance of tree mound positions, a low number to flats. Thus the number can be considered as a measure of the density of tree mounds per ha of land,

type of litter: large number—woody litter, small number—herbaceous litter,

vegetation index: large number—tree shade, small number—no tree shade.

All other characteristics of the environment were not related significantly to tree mound increment.

Stepwise multiple linear regression using backward elimination on the variables isolated above yielded the following relationship:

$$\text{tree mound increment} = 8.37 - 2.49 \text{ vegetation index}$$
$$r^2 = 14.4, \text{ significant at } \alpha = 0.05$$

The results of this analysis are interpreted as follows: with an increasing number of trees, the number of tree mounds increases but their average height is lower. This could be due to the protective role of trees, but also to other reasons:

1. At high tree density it is likely that the tree mounds merge into each other.
2. At low tree density it is likely that termites concentrate on the few trees available and build higher structures.

Both phenomena were observed during the survey: at high tree density it was often impossible to find isolated trees, and it was obvious that the lowest point of the soil surface was influenced by the proximity to a neighbouring tree. On the sites with sparse canopy cover, intense termite activity was observed at the tree base.

CONCLUSION

The results above clearly demonstrate that tree mound height increments cannot be trusted as a measure of erosion in the case of the tree savanna in the Hard Veld of Botswana. Tree mound height can be explained by a lowering of the bulk density in the direct vicinity of the tree stem, probably as a result of higher organic matter content and higher porosity caused by decaying root material and termite activity. Thus the tree mounds are caused by a raising of the local surface rather than a lowering of the surrounding surface caused by erosion. This raising of the local surface could possibly be enhanced by the deposition of sediment at the tree base. In the case of a dense tree cover, mounds tend to merge and the mound measurement technique used in this survey is inadequate.

This conclusion is contrary to the claims made by Dunne *et al.* (1978, 1979) and Stocking (1984) for environments which are characterized by higher rates of erosion. Dunne *et al.* (1978) discussed briefly the difficulties described in this study, and the present findings confirm the need for careful utilization of this technique to measure rates of erosion, especially in environments which are characterized by low rates of erosion. Only if the position of the soil surface at tree germination can be identified on the tree stem will this technique offer a possible means of establishing rates of erosion. However, given an average tree age of 10 years, a mean rate of erosion equal to 5 t/ha/year and a bulk density of 1.65 g/cm^3, an average lowering of the soil surface of 3 mm will have been experienced on average since the time of tree germination. It is highly unlikely that a sufficiently accurate positioning of the previous soil surface can be established on the tree stem to allow the measurement of such small height differences.

ACKNOWLEDGEMENTS

Field and laboratory work for this study were supported by a grant from the International Livestock Centre for Africa (ILCA), Addis Ababa.

REFERENCES

Abel, N. and M. Stocking (1987). A rapid method for assessing rates of soil erosion and sediment yields from rangelands. *Journal of Range Management*, **40**(5), 460–6.

Biot, Y. (1986). *Modelling of the Degradation of Rangeland in Botswana: Report on the Findings of the Reconnaisance Visit*, School of Development Studies, University of East Anglia.

Biot, Y. (1988a). *Modelling productivity losses caused by erosion*. Paper presented at the Vth Int. Soil Cons. Conference, Bangkok.

Biot, Y. (1988b). *Forecasting productivity losses caused by sheet and rill erosion in semi arid rangeland: A case study from the communal areas of Botswana*. PhD thesis, University of East Anglia.

Biot, Y. (1988c). Calculating the residual suitability of agricultural land from routine land resorces surveys. In J. Bouma and A. Bregt (eds), *Land Qualities in Space and Time*, Proceedings of a Symposium organised by the ISSS, Wageningen, pp. 261–4.

Biot, Y. (1988d). *The Morale Water Balance Study: Objectives, Materials and Methods*, International Livestock Centre for Africa (ILCA), Addis Ababa, and Animal Production Research Unit (APRU), Gaborone.

Curtis, B. A., P. D. Tyson and T. G. F. Dyer (1978). Dendrochronological age Determination of *Podocarpus falcatus*. *South African Journal of Science*, **74**, 92–5.

Dunne, T., W. E. Dietrich and M J. Brunengo (1978). Recent and past erosion rates in semi-arid Kenya. *Z. Geom., Suppl. Bd.*, **29**, 130–40.

Dunne, T., W. E. Dietrich and M. J. Brunengo (1979). Rapid evaluation of soil erosion and soil lifespan in the grazing lands of Kenya. In International Association of Hydrological Sciences, *The Hydrology of Areas of Low Precipitation*, IAHS Publ. 128.

Eardley, A. J. (1967). Rate of denudation as measured by bristlecone pines, Cedar Breaks, Utah. *Utah Geol. and Miner. Surv. Spec. Study*, 21.

El Swaifi, S. A., W. C. Moldenhauer and A. Lo (eds.) (1985). *Soil Erosion and Conservation*, Soil Conservation Society of America.

Elwell, H. A. and M. A. Stocking (1982). Developing a simple yet practical method of soil loss estimation. *Trop. Agriculture*, **54**(1), 43–8.

Ermanovics, I. F. (1980). *The Geology of the Mokgware Hills Area*. Geol. Surv. Dep. Bull. no. 13, Rep. of Botswana.

FAO (1974). *Soil Map of the World, Volume I: Legend*, FAO/UNESCO, Paris.

Gerlach, T. (1967). Hillslope troughs for measuring sediment movement. *Revue Géomorphologique Dynamique*, **17**, 173.

Haigh, M. J. (1977). *The Use of Erosion Pins in the Study of Slope Evolution*. Brit. Geom. Res. Group. Techn. Bulletin no. 18, pp. 31–49.

Jennings, C. M. H. (1962). *Note on Erosion Cycles in the Bechuanaland Protectorate*. Unpublished paper.

King, L. C. (1967). *South African Scenery: A textbook of geomorphology*, 3rd edition, Oliver and Boyd, Edinburgh.

Köppen, W. (1923). *Die Klimate der Erde*, Walter der Gruyter, Berlin.

Lamarche, V. C. (1968). Rates of slope degradation from botanical evidence, White Mountains, California. *US Geol. Surv. Profess. Paper*, 352.

Lilly, M. A. (1977). *An Assessment of the Dendrochronological Potential of Indigeneous Tree Species in South Africa*. Department of Geography and Environmental Studies, University of the Witwatersrand, Occasional Paper no. 18.

Rapp. A., D. H. Murray-Rust, C. Christianssen and L. Berry (1972). Soil erosion and sedimentation in four catchments near Dodoma, Tanzania, *Geogr. Ann.*, **54**, A, 255–318.

Reece, P. H. and B. L. Campbell (1984). *The Use of ^{137}Cs for Determining Soil Erosion Differences in a Disturbed and Non-disturbed Semi-Arid Ecosystem*. Proc. Second Int. Rangelands Congress, Adelaide.

Remmelzwaal, A. (1987). *Soils of Central District, Botswana. First Draft: Field Document*, FAO/ Gov. of Botswana, FAO/BOT/80–003.

Schumm, S. A. (1967). Erosion measured by stakes. *Rev. Géom. Dyn.*, **17**, 161–2.

Siderius, W. (1973). *Soil Transitions in Central East Botswana*. Doctoral thesis, State University of Utrecht.

Stocking, M. (1984). Rates of erosion and sediment yield in the African environment. *Challenges in African Hydrology and Water Resources*. IAHS Publ. no. 144.

Stocking, M. (1987). Measuring land degradation. In P. Blaikie and H. Brookfield, *Land Degradation and Society*, Methuen, pp. 27–48.

Task Committee on Sedimentation (1970). Sediment sources and sediment yields. *J. Hydraul. Div. American Society of Civil Engineers*, **96**(HY6), 1283–1329.

Thornes, J. (1988). Erosion equilibria under grazing. In J. Bintcliff, D. Davidson and E. Grant (eds.), *Environmental Archaeology*, Edinburgh University Press.

Tinley, K. (1982). The influence of soil moisture balance on ecosystem patterns in Southern Africa. In B. Huntley and B. Walker (eds.), *Ecology of Tropical Savannas*, Springer-Verlag, Berlin, pp. 175–92.

Weare, P. R. and A. Yalala (1971). Provisional vegetation map of Botswana. *Botswana Notes and Records*, **3**, 131–48.

Wendelaar, F. E. and A. N. Purkis (1979). *Recording Soil Loss and Runoff from 300 m² Erosion Research Field Plots*. Research Bulletin no. 24, Dep. of Conserv. and Extension, Salisbury (Harare).

27 Vegetation and Fluvial Geomorphic Processes in South-east Asian Tropical Rainforests

TOM SPENCER
Department of Geography, University of Cambridge

IAN DOUGLAS, TONY GREER
School of Geography, University of Manchester

and

WAIDI SINUN
Danum Valley Field Studies Centre, Yayasan Sabah

SUMMARY

Natural disturbance of tropical rainforests leads to spatially and temporally irregular exposure and detachment of soil on slopes, creating a series of potential sediment sources and sinks greatly influencing denudation processes in the humid tropics. Tree fall sometimes provokes long-lasting surface soil exposure, especially where bare ground is exploited by animals. Spatial variations in litter fall affect rates of erosion. Debris dams store sediment during low-magnitude, moderate-frequency events, and release it in high-magnitude, rare events, thus accentuating the geomorphic importance of extreme events. Debris dams are more closely spaced on smaller streams and are rare on those over 20 m in width; nevertheless, storm runoff carries large numbers of floating logs down major rivers. Debris dams have lower residence times than those of humid temperate forest streams. The interaction between natural disturbance processes and the magnitude and frequency of hydrometeorological events has to be built into any model of tropical rainforest fluvial geomorphic processes.

Vegetation and Erosion
Edited by J. B. Thornes
©1990 John Wiley & Sons Ltd

INTRODUCTION

New theories to explain the species diversity of tropical forests stress the dynamism of forest structure and function and the close relationships between geomorphological, hydrological and ecological processes. While phases of forest regeneration and decay activate pulses of water, sediment and nutrient flux, the physical presence of trees, litter and ground-cover plants on forest slopes and in forest streams provides obstacles to the movement of water and materials into and through the drainage system. The decay and decomposition of these obstacles introduce mechanical weakness which may result in failure and sudden release of material. Observations in humid temperate forests suggest that vegetation and litter may prevent the development of organized, concentrated runoff on slopes and reduce the energy of flowing water in streams. Large obstacles may persist for several decades while seasonal accumulations of debris on slopes may provide effective protection against erosion from one year to the next. This chapter asks whether the continuous year-long decay and decomposition in equatorial rainforests make such organic obstacles as effective and long-lived as at extra-tropical latitudes.

A THEORY OF NATURAL DISTURBANCE IN TROPICAL RAINFORESTS AND ITS GEOMORPHOLOGICAL CONTEXT

Old ideas of the rainforest as a stable ecosystem, ordered by fixed, multiple canopy levels (e.g. Richards, 1952, 1969) have recently been replaced by the notion of a dynamic forest mosaic, first recognized by Aubréville (1938), in which interacting processes of tree growth and decay create vegetation pattern.

A theory of irregular episodes of forest growth and disruption, including the formation of gaps and their subsequent closure on colonization and succession, has been used to explain rainforest diversity (e.g. Whitmore, 1978; Denslow, 1980; Hartshorn, 1980; Lang and Knight, 1983; Brokaw, 1985). This theory suggests that the forest is a mosaic of patches of different ages, with recently opened gaps colonized by pioneer species adapted to the enhanced light levels now reaching the forest floor; patches of growing forest species in the building stage, with a mass of young, fast-growing, slender, tall-stemmed trees; and patches of mature forest with widely spaced, old trees with spreading crowns interspersed with some saplings. To this temporal view must be added a spatial dimension, significant in terms of microclimate and thus species composition, from small disturbances of the forest floor at the 10^{-2} to 10^{-1} m^2 scale, through gaps around trees which remain upright on death (e.g. Lieberman and Lieberman, 1987) to the large-scale clearances associated with landslides (for review see Spencer and Douglas, 1985), windthrow (e.g. Whitmore, 1974), fire and human activity (e.g. Lovejoy et al., 1983). In addition, cutting across these vegetation time and space scales are the irregular disturbances made, or enhanced, by insects and animals, especially the translocation and comminution of material by termites, the tracks and foraging areas created by deer, and the wallowing and trampling zones of pigs and larger quadrupeds.

Clearly such a theory of forest structure and dynamics has important implications for the nature of sediment sources and sinks, nutrient storages (e.g. Anderson and Swift, 1983) and hydrological pathways beneath a tropical forest cover, and suggests a tropical geomorphology different from that associated with the traditional static and passive model of forest stability and maintenance.

Forest-growth cycle theory suggests that a disturbance regime should locally accelerate the work of slope and channel fluvial geomorphic processes. However, as yet, there is little information on the impacts of forest canopy disruption on such processes in the humid tropics. However, by analogy, the wide range of observations on humid temperate forests reported in the literature enables a pattern of vegetational influences on slope and channel processes to be established. Observations made in the context of study of tropical rainforest disturbance in the Ulu Segama district of eastern Sabah, Malaysian Borneo, can then be used to see whether tropical forest processes operate in the same way at similar rates.

VEGETATION AND HILLSLOPE PROCESSES IN TEMPERATE FORESTS

Although many writers on forest influences suggest that the forest canopy breaks the fall of raindrops and reduces surface soil erosion, Mosley (1982) noted that in temperate rainforests in southern New Zealand, the kinetic energy of raindrops under the canopy was greater than those in the open. Bare soil areas in the forest suffered more erosion than those in the open. Raindrops tend to flow towards the leaf tip and coalesce, forming larger drops than in the open (Chapman, 1948; Ovington, 1954), and in forests of large trees the foliage is high enough above the ground for drops to reach their velocity.

Organic material on the floors of humid temperate forests varies considerably, but where it is reasonably thick, it operates in two important ways:

1. it absorbs the impact of raindrops;
2. it has an extremely high permeability which allows water to move into the soil at a high rate (Walsh and Voigt, 1977).

Tree roots regularly act as traps for fine material moved downslope. The soil on the upslope side of a lateral root may be 5 to 20 cm higher than that on the downslope side. Many tree trunks have bigger differences in ground level on their up- and downslope sides, with soil accumulating on the upslope side and being washed away, probably by stemflow running down the trunk on the downslope side.

ORGANIC DEBRIS IN HUMID TEMPERATE FOREST STREAMS

The forests of the Pacific Northwest of the USA have long been recognized as having rainfalls and sediment and solute yields comparable with those of some humid tropical areas (Douglas, 1969). Detailed investigations in the Pacific Northwest show that on minor streams a debris dam occurs every 1.5 to 8 m of channel (Heede, 1981), while in the central Oregon coast range third-order streams have 3.97 log steps per kilometre (Marston, 1982). Even though actual treefall may be most frequent along first- and second-order streams in these steeplands, log steps are not formed because the V-shaped incision of the low-order channels prevents the logs from lying on the streambed.

Storage of sediment behind log steps may exceed the mean annual sediment yield. In the Oregon coast range, the volume of sediment stored behind log steps in third-, fourth- and fifth-order streams amounted to 123 per cent of the total mean annual suspended and bed load discharge (Marston, 1982); in western Oregon debris loading in old-growth Douglas fir forest exceeds $40 \, kg \, m^{-2}$ (Keller and Swanson, 1979). In the redwood forests of northern California (Keller and Tally, 1979), the morphology on first- and second-order streams is largely controlled by logs and branches exceeding 10 cm diameter, while in third- and fourth-order streams, large organic debris tends to become concentrated in debris dams or jams that locally influence channel morphology and the pattern of erosion and deposition. The largest redwood logs may remain where they fall for long periods of time, but the major rivers are able to evacuate even the biggest log material supplied to them to the sea.

The Californian and Oregonian studies have led to the following general conclusions (Klein, 1984; Keller and Tally, 1979; Swanson, Lienkaemper and Sedell, 1976; Keller and Swanson, 1979; Lienkaemper and Swanson, 1987):

1. By exaggerating the variability in local channel gradient, a high proportion of a stream's potential energy is expended at localized points along the profile.
2. Energy dissipation by vertical waterfalls over these organic steps result in less available energy for channel erosion and sediment transport. This condition allows streams to maintain steeper overall gradients than would otherwise be possible.
3. The loading of forest streams with organic debris generally decreases with increasing stream order.
4. Large organic debris moves through the fluvial system primarily by flotation at high flows or perhaps, in very steep reaches, by debris torrents. Debris length is important: movement is greatest where pieces are shorter than bankfull width.
5. The minimum residence times for large tree trunks in debris dams range from 20 to 220 years.
6. Debris dam spacing ranges from more than 3 per 100 m of channel to less than 1 per 100 m in catchments of up to $5 \, km^2$.

These general conclusions are supported by other observations in the Pacific coast ranges and islands of North America where annual rainfalls are comparable with those of the humid tropics (Table 27.1). This chapter examines whether or not the same conditions operate in north-east Borneo where rainfall and lithology are similar, but overall relief is somewhat less.

A TROPICAL RAINFOREST STUDY IN SABAH, MALAYSIAN BORNEO

The Tertiary sediments of the Ulu Segama district of Sabah (Figure 27.1) are composed of shales, mudstones and sandstones (Leong, 1974) similar to those of the Pacific coast ranges, but the steep slopes are shorter. In this environment, the bulk of the runoff and sediment transport occurs in high-magnitude, short-duration storm events, in which fourth-order streams may rise from 20 cm depth to more than 3 m depth in less than one hour. The energy of peak flows is thus much greater than that prevailing between storms and may be capable of lifting fallen organic debris and removing it by flotation. In this environment, at the Danum Valley Field Studies Centre (Figure 27.1), hydro-meteorological events are being monitored in three catchments: two, of 10 and 1 km^2, in natural forest, and one, of 0.5 km^2, on a stream where commercial logging began in early 1989. Each monitoring station provides continuous water level, pH, temperature and conductivity records, with an automatic water sampler being triggered by a float switch during storm events. The small natural forest streams

Table 27.1 North American Pacific coast organic debris dam data

Catchment	Area (km^2)	Debris dams per 100 m	Age range of fallen logs (years)	Source
Little Lost Man Creek, Lower Reach	4.9	1	20–220	Keller and Tally, 1979
Little Lost Man Creek, Upper Reach	1.1	3.5	70–200	Keller and Tally, 1979
Oregon Coast Range 3rd-order streams		0.397		Marston, 1982
Government Creek, Queen Charlotte Islands, British Columbia	3.9	1.3		Hogan, 1987
Government Creek, Lower Reach	6.9	1.6		Hogan, 1987
H. J. Andrews Experimental Forest, Cascade Range, Oregon Watershed 2	0.8		100 +	Swanson et al., 1976

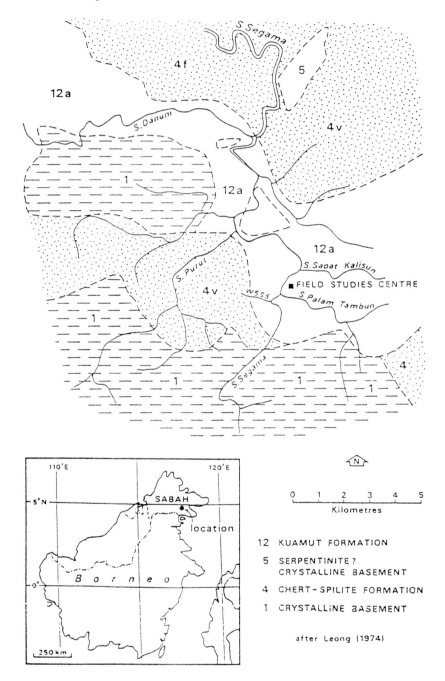

12 KUAMUT FORMATION

5 SERPENTINITE?
  CRYSTALLINE BASEMENT

4 CHERT-SPILITE FORMATION

1 CRYSTALLINE BASEMENT

after Leong (1974)

Figure 27.1 Location of field area and local geology Ulu Segama district, eastern Sabah

rise rapidly during storms, reaching sediment concentrations of 800 to 1500 mg l^{-1} in the largest storms. Much of the sediment is derived from bank erosion of the weak mudstones and shales, or by the re-mobilization of previous flood deposits along the channels.

Erosion plots and throughfall gauges are also operating to provide understanding of slope processes. As in the Pacific north-west (e.g. Swanston and Swanson, 1976), the effects of road construction associated with timber removal are environmentally severe. The dramatic impact of this activity is clear from the contrast in runoff and sediment concentration between the small undisturbed catchment (W8S5) and the catchment affected by road construction (Baru) in the storm of 26 March 1989 (Figure 27.2). Increase in runoff above baseflow was much greater, while peak sediment concentration was four times greater in the Baru catchment. However, the role of vegetation in hillslope and channel processes in the natural catchments has been less clear, although the role of channel debris in storing, releasing and routing sediments has been shown to be of great importance in humid temperate forest sediment budgets (Dietrich and Dunne, 1978; Swanson et al., 1982). Observations of slope erosion and surveys of organic debris dams in the W8S5 catchment are now being made at intervals to provide comparable tropical forest data.

Trees and Surface Erosion

On the friable, easily dispersed mudstones, thin, weak sandstones and occasional resistant quartz outcrops of the Miocene Kuamut formation (Figure 27.1), soils and weathering mantles are often thin, slopes are steep and channel heads are often characterized by deep, steep-sided amphitheatre-like hollows. Tree root systems are overwhelmingly shallow, often with lateral buttresses or near-surface roots. Tall, mature trees fall to the ground, either by a break in their stems, 2 to 4 m above ground level, or by uprooting their surface roots and tearing out a large amount of soil. Every such tree fall may pull down a considerable mass of vines and, through the inter-tree connectivity of climbers, may also bring down several smaller trees (cf. Lieberman et al., 1985). Geomorphology and the nature of the tree do not appear to influence whether stems are broken or the tree is uprooted; buttressed trees are just as likely to be uprooted as to have their stems broken.

Tree falls create temporary breaks in the canopy on the 10^1–10^2 m scale, with areas of bare soil of varying extent. Initially, the foliage of the fallen vegetation withers and raindrop splash and erosion are enhanced. However, within a few weeks, vines and creepers opportunistically invade the well-lit area, and a dense vegetation cover close to the ground is established. The initial acceleration of erosion is thus replaced by reduction in the energy of raindrops reaching the ground and most probably (although this is yet to be confirmed) by slower rates of erosion than existed under the original mature forest.

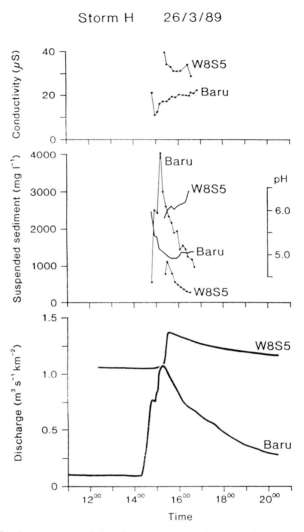

Figure 27.2 Discharge, suspended sediment concentration, specific conductance and pH of the W8S5 and Baru streams during the storm of 26 March 1989

Where trees have been uprooted, small vertical bare soil faces of 1 to 3 m height may be created in the slope. Such faces sometimes have associated bare horizontal, or gently sloping surfaces downslope, which may be kept bare by animal activity, particularly by pigs (*Sus barbatus*), but perhaps even by the rarely seen rhinoceros (*Dicerorhinus sumatrensis*). Rainsplash is the main cause of soil loss on such bare areas, particularly if canopy height and structure are favourable (e.g. Brandt, 1988), but the remaining *in situ* roots often act as barriers to soil movement downslope, turning the exposed area into a series of steps, or an alternate pattern of near-vertical faces of soil loss and near-horizontal zones of sediment accumulation. Some of the higher vertical faces tend to be undercut with emergent interflow, leading to soil loss by sapping mechanisms. Burrowing animals and insects probably also contribute to the maintenance of these exposed vertical faces.

The increase in erosion rate in the 10^1–10^2 m^2 gaps may be restricted to a few weeks if vegetation can invade easily. Erosion on bare soil persists for longer, but leaf litter from surrounding trees may inhibit splash and detachment. Within a few months or a year, a formerly actively eroding site may become almost completely covered with leaf litter. Such debris accumulation is probably episodic and the later removal of leaf litter and subsequent renewed acceleration of erosion may occur. Such litter dynamics are comparable to those described by Ruxton (1967) from Papua.

Where bare patches persist on steep slopes, further destabilization of vegetation may occur as soil is removed from around the root systems of small trees. In extreme cases the surface root systems of such trees are undercut, making them vulnerable to windthrow. The relative magnitudes of the rate of soil removal and the rate of vegetation invasion or litter accumulation determine whether bare areas persist or are eliminated.

As in cyclone-prone montane rainforest (Herwitz, 1988), buttressed trees play an important role in slope processes at the 1–10 m^2 scale in equatorial lowland rainforest. Quite small buttresses can influence topography. Thus, in one case, an 18 cm diameter young tree on a ridge crest has a 7 cm diameter buttress which has created a topographic step 50 cm high. Larger buttresses produce greater differences in level between upslope and downslope sides of trees, such effects being most marked on the steep slopes above first- and second-order stream channels. Long lateral roots on steep slopes, especially where quartzite outcrops are within 1 m of the ground surface, act as barriers to soil movement, producing a succession of topographic steps 20 to 50 cm in height.

At the 10^{-2}–10^{-1} m^2 scale, the effects of funnelling of water down the stems of trees is readily apparent. Many trees have bare soil on their downslope side. In March 1987, along a 0.5 km transect along a ridge crest, 80 per cent of a sample of 70 trees had bare patches on their downslope sides, while 20 per cent had no clear signs of any surface wash adjacent to the trees; 7 per cent of the trees had roots exposed by surface erosion. A year later, such bare patches were less apparent, over 50 per cent of the trees being surrounded by leaf litter.

In some cases, trees appear to stand on a soil pedestal, the soil unprotected by roots having been washed away by stormflow, splash or slopewash. Again, presence or absence of litter is important. Sites exposed at one time may be litter covered a few months later. Further investigations of litter supply and litter removal rates are required to establish how short-lived such bare areas are under the canopy.

At a scale of 10^{-3}–10^{-2} m, individual leaves and seedpods cap 5–10 cm high soil pillars in areas where soil is loose, silt-sized and easily detached by raindrops falling from the canopy 15 to 25 m overhead. In a few localities, 80 to 120 such pillars have been found over 1 m², but such phenomena are infrequent. More often the leaf litter provides a continuous cover, although this may be only one or two leaves thick. Rarely is there more than 120 cm of humic matter on the forest floor. The spatial variation of litter thickness and its relationship to other biotic activity and slope processes remain to be explored.

Organic Debris Dams and Extreme Events

The interrelationships between ecology and geomorphology are particularly strong on stream margins and river banks (Figure 27.3). On streams up to a size which can be blocked by a single fallen log (c. 20 m channel width), tree fall is a major influence on channel dynamics. Debris dams are prominent on first-, second- and third-order streams, but decrease in frequency as channel width increases. Many dams are created by the fall of trees undercut by the

Figure 27.3 Riparian vegetation, Sungei Sapat Kalisun, Ulu Segama, September 1988

stream itself; others develop where a tree has fallen downslope and its crown has crashed into the stream (Figure 27.4).

Most of the debris dams are in the form of prominent organic steps with falls of 2 to 3 m height. Such dams clearly act as energy-dissipating structures. Sand and gravel are deposited upstream, eventually creating a continuous low-angle bar, flush with the top of the log. (Figure 27.5). Below the obstacle there may be scour and channel widening.

Three surveys of the W8S5 channel indicate that there may be considerable change in the number of organic steps in 13 months (Table 27.2). a major storm at the end of March 1988 washed out many of the organic debris dams, especially those at 155, 241, 264 and 302 m upstream of the gauging station. Daily rainfall in this storm was 180 mm, a figure comparable to the annual daily maxima recorded at other raingauges in this part of Sabah (Table 27.3). Even though official records are patchy and short, it is probable that this storm, which washed out most of the organic steps, was of the order of magnitude of the mean annual flood. The exceptions to the general elimination of barriers are locations where the channel bed was flush with the top of the logs (at 570 m upstream;

Figure 27.4   Remains of debris dam/tree fall, Sungei Palum Tambun, September 1988

Figure 27.5   Debris dam on Bole River Tributary, Ulu Segama, September 1988. Stream flows over gravel impounded behind rear log, then cascades under trunk in foreground

Table 27.2). Two processes influence changes in channel morphology at such locations; the rate of rotting of the log and lateral movement of the channel. At 570 m upstream, bank erosion has allowed the channel to move to the right of the obstacle and scour an alternative channel which may eventually pass around the log. Such 'autodiversion' is well known from humid temperate forest stream channels (Keller and Melhorn, 1973; Keller, Melhorn and Gardner, 1976; Hickin, 1984). In the year following the March 1988 storm, new organic debris dams have begun to form at 144 and 386 m upstream, the latter being a blockage of major dimensions.

Parallel observations on other Segama tributaries confirm that the March 1988 storm removed the majority of debris dams in second- and third-order streams. Those in small headwater channels were less disturbed. Log jams thus collect and store sediment in minor and moderate runoff events and release it in major and extreme events. They thus tend to increase the importance of high-magnitude, low-frequency events in the removal of fluvial sediment.

The debris dams composed either of fallen foliage and tree canopies or of several small trees tend to trap coarser floating plant material and gradually to build up a complex mat of decaying organic matter across the stream. Sometimes tree fall brings a mass of vines and creeper into the stream; this filters out floating debris until broken either by another falling tree or an extreme runoff event. It is too early to suggest whether or not the rate of change in organic debris dams observed in Ulu Segama in 1988–89 is characteristic of streams in equatorial environments, but the evidence is strong enough to show that the number of organic steps found in a rainforest stream at any one time depends on antecedent hydrometeorological events. Two factors make it likely that organic steps in humid tropical rivers do not last as long as those in temperate streams:

1. The logs and branches decompose more rapidly (although the exact dating of logs by dendrochronology applied in humid temperate forests is not possible).
2. Rains are more intense and runoff per unit area in low-order streams is consequently higher.

COMPARISON OF TEMPERATE AND TROPICAL EVIDENCE

The tropical forest floor offers less impedance to raindrop impact than the thick litter layers of temperate forests. Litter decomposition is rapid, but locally individual leaves may cap small earth pillars, while elsewhere the leaf litter mat continuously covers the ground, reducing the opportunity for splash erosion. Under these conditions some localized surface wash may occur, but roots, buttresses and fallen logs restrict the downslope movement of material. The decomposition of such obstacles, with active movement of material by ants and termites, may be more rapid than in humid temperate forests, so that, although vegetative obstacles are always present on tropical rainforest slopes, their locations change.

Figure 27.6 Sungei Segama and primary equatorial lowland rainforest, Danum Valley Field Studies Centre. Note log debris in coarse marginal channel bars

Table 27.2 Changes in log jams and debris dams W8S5 stream 1988–89

Distance upstream from gauging station (km)	Situation in March 1988	Situation in August 1988	Situation in April 1989
144			Some in channel vegetation, not blocking flow but holding back some debris
155	Major tree forms bridge over main stream but log jam on side stream	Log jam on side stream completely washed out. No remaining vegetation obstacles	Channel still clear
241	Log trapped by boulders in channel with gravel bar extending upstream	Log washed out. Gravel still held behind boulders. No debris dam	Logs remain boulders. No sediment
264	Fallen log creates gap in forest around stream. Much gravel trapped behind debris dam	No debris dam	No debris dam. Log stump half across channel
273	Fallen log	No debris dam	No debris dam
302	1 m diameter log forms major debris dam with much trapped gravel	No debris dam	No debris dam
386			Large tree and accompanying strangler fig across main and tributary channel. As yet no sediment held back

450	Large boulders block stream with major gravel accumulation and log and vegetation litter debris	Large gravel spread behind rocks, but no remains of log and vegetation litter	No debris dam
453	Fallen log causes major debris dam	No debris dam	No debris dam
570	Large gravel bar behind rotted fallen log which acts as a local base level	Part of rotted log remains, gravel bar is flush with top of log so that water passes over, rather than pushes against log	Organic step remains but stream is enlarging channel on right bank to eventually by-pass obstacle
603	Large tree fall in gap by stream. Debris accumulating	No debris dam	
670	Major log jam caused by tree which has fallen as a result of under-cutting by stream. Fallen log 8 m further upstream	Log jam still intact. Log 8 m further upstream breaks force of water	Broken log. No jam but evidence of gravel bars behind log
766	Major log jam blocking stream	Major log jam blocking stream	Major log jam blocking stream

Table 27.3 Maximum one-day rainfalls (mm) around the Ulu Segama area, Sabah

| Danum Valley Field Studies Centre gauge | | 173 | 29 March 1988 |
| 67 km gauge | | 184 | 29 March 1988 |

Other locations Year	Ulu Kuamut	Lahad Datu	Kuamut	Bukit Garam
1981	83.1	177.8	n.d.	n.d.
1982	100.7	67.3	147.5	n.d.
1983	n.d.	84.8	149.0	121.0
1984	258.0	n.d.	87.0	211.5
1985	128.0	n.d.	106.0	127.0
Highest on record	258.0	266.7	149.0	211.5
(date)	(8 Dec. 84)	(14 Feb. 74)	(28 Nov. 83)	(29 Dec. 84)

n.d. signifies incomplete or missing record.
Sources of other station data:
Drainage and Irrigation Division (1983). *Hydrological data: rainfall records 1975–80*, Publications Unit, Ministry of Agriculture, Kuala Lumpur.
Drainage and Irrigation Division (1987). *Hydrological data: rainfall and evaporation records for Malaysia 1981–85*. Publication Unit, Ministry of Agriculture, Kuala Lumpur.

The first four of the six general conclusions from Californian redwood and Oregonian Douglas-fir forest studies listed earlier apply to the streams of the Ulu Segama. The fifth conclusion on a minimum residence times for large tree trunks of at least 20 years seems unlikely to be valid in Sabah. Minimum residence times would appear to be shorter, possibly related to the mean annual flood recurrence interval. Referring to the sixth conclusion, organic debris dam spacing in the rapidly changing tropical streams varies with hydrometeorological events. Spacing at the three dates in Table 27.2 was 1.3, 0.4 and 0.5 organic debris dams per 100 m. Smaller channels have more frequent organic debris dams, produced by smaller logs than those in W8S5. The balance between the natural disturbance of tree fall and gap formation and the erosive energy of the stream is ever present. However, lateral erosion to avoid an obstacle may create new obstacles by causing tree fall by undercutting. Indeed, stream bank erosion may locally increase the rate of gap formation. This would be an impact of the environment on the forest itself, one way in which Jordan's conclusion (1988) that the landscape and its climate have little impact on ecosystem function can be shown to need reassessment.

CONCLUSIONS

The ease with which material is removed from a tropical rainforest ecosystem is, as in temperate hardwood forests, a function of both geological substrate and biomass. Whether or not the kinetic energy of moving water within the ecosystem is strongly coupled to detachment and transport of sediment, or is dissipated in other ways, is often determined by the biotic conditions of the ecosystem (Bormann and Likens, 1979).

In the Ulu Segama, ecological processes create opportunities for both erosion and sediment storage. Small hydrometeorological events may merely add sediment to some stores and remove it from others. An irregular passage of sediment through a series of slope and stream sediment stores, held back by roots or log jams, may be envisaged. Rare, high-magnitude events may, however, open up all these stores and flush great volumes of sediment through the system. Natural disturbance and magnitude and frequency thus have to be built into any conceptual model of forest fluvial geomorphic processes. However, to that must be added the irregularity of litter fall, rates of decay of fallen vegetation and the roles of insects and animals. In all this spatial and temporal scales are interwoven. Preliminary, tentative findings suggest that the rate of change in tropical rainforest streams is at least as rapid, and probably more rapid than that in wet humid temperate areas. The processes of ecosystem—denudation system interaction—are similar in all forests, but their timing may differ. Our present observations may only indicate the types of processes operating; we will need to study the forest for many more years to be sure of the time scale of these needs.

REFERENCES

Anderson, J. M. and M. J. Swift (1983). Decomposition in tropical forests. In S. L. Sutton, T. C. Whitmore and A. C. Chadwick (eds), *Tropical Rain Forest: Ecology and Management*, Blackwell Scientific Publications, Oxford, pp. 287–309.

Aubréville, A. (1938). La forêt coloniale: les forêts d'Afrique equatoriale française. *Annals Academie Science Coloniale*, 9, 1–245.

Bormann, F. H. and G. E. Likens (1979). *Pattern and Process in a Forested Ecosystem*, Springer, New York.

Brandt, J. (1988). The tranformation of rainfall energy by a tropical rain forest canopy in relation to soil erosion. *Journal of Biogeography*, 15, 41–8.

Brokaw, N. V. L. (1985). Treefalls, regrowth and community structure in tropical forest. In S. T. A. Pickett and P. S. White (eds), *The Ecology of Natural Disturbance and Patch Dynamics*, Academic Press, New York, pp. 53–69.

Chapman, G. (1948). Size of raindrops and their striking force at the soil surface in a red pine plantation. *Transactions, American Geophysical Union*, 29, 664–70.

Denslow, J. S. (1980). Gap partitioning among tropical rainforest trees. *Biotropica*, 12, 47–55.

Dietrich, W. E. and T. Dunne (1978). Sediment budget for a small catchment in mountainous terrain. *Zeitschrift für Geomorphologie, Supplbd.*, 29, 191–206.

Douglas, I. (1969). The efficiency of humid tropical denudation systems. *Transactions Institute of British Geographers*, 46, 1–16.

Hartshorn, G. S. (1980). Neotropical forest dynamics. *Biotropica*, 12 (Suppl.), 23–30.

Heede, B. M. (1981). Dynamics of selected mountain streams in the western United States of America. *Zeitschrift für Geomorphologie*, 25, 17–32.

Herwitz, R. R. (1988). Buttresses of tropical rainforest trees influence hillslope processes. *Earth Surface Processes and Landforms*, 13, 563–7.

Hickin, E. J. (1984). Vegetation and river channel dynamics. *Canadian Geographer*, 28, 111–26.

Hogan, D. L. (1987). The influence of large organic debris on channel recovery in the Queen Charlotte Islands, British Columbia, Canada. *International Association of Hydrological Sciences Publication*, **165**, 343–53.

Jordan, C. F. (1988). The tropical rainforest landscape. In H. A. Viles, (ed.), *Biogeomorphology*, Blackwell, Oxford, pp. 145–65.

Keller, E. A. and W. N. Melhorn (1973). Bedforms and fluvial processes in alluvial stream channels: Selected observations. In M. Morisawa (ed.), *Fluvial Geomorphology* (Proceedings of the 4th Annual Geomorphology Symposium Series), State University of New York, pp. 253–283.

Keller, E. A., W. N. Melhorn and M. C. Gardner (1976). Effects of autodiversion (logjams) on stream channel morphology. *Geological Society of America, Abstracts with Programs*, **8**(6), 950.

Keller, E. A. and F. J. Swanson (1979). Effects of large organic material on channel form and fluvial processes. *Earth Surface Processes*, **4**, 361–80.

Keller, E. A. and T. Tally (1979). Effects of large organic debris on channel form and fluvial processes in the coastal redwood environment. In D. D. Rhodes and G. P. Williams (eds), *Adjustments to the Fluvial System*, Kendall Hunt, Dubuque, Iowa, pp. 169–97.

Klein, R. D. (1984). Channel adjustments following logging road removal in small steepland drainages. In C. L. O'Loughlin and A. J. Pearce (eds), *Symposium of Effect of Forest Land Use on Erosion and Slope Stability*, East–West Center, University of Hawaii, Honolulu, pp. 187–95.

Lang, G. E. and D. H. Knight (1983). Tree growth, mortality, recruitment and canopy gap formation during a 10-year period in a tropical moist forest. *Ecology*, **64**, 1075–80.

Leong, K. M. (1974). The geology and mineral resources of the Upper Segama Valley and Darvel Bay area, Sabah, Malaysia. *Geological Survey of Malaysia Memoir* 4 (revised).

Lieberman, D. and M. Lieberman (1987). Forest tree growth and dynamics at La Selva, Costa Rica. *Journal of Tropical Ecology*, **3**, 347–58.

Lieberman, D., M. Lieberman, R. Peralta and G. Hartshorn (1985). Mortality patterns and stand turnover rates in lowland wet tropical forests in Costa Rica. *Journal of Ecology*, **73**, 915–24.

Leinkaemper, G. W. and F. J. Swanson (1987). Dynamics of large woody debris in streams in old-growth Douglas-fir forests. *Canadian Journal of Forest Research*, **17**, 150–6.

Lovejoy, T. A., R. O. Bierregaard, J. M. Rankin, and H. O. R. Shubart (1983). Ecological dynamics of tropical forest fragments. In S. L. Sutton, T. C. Whitmore and A. C. Chadwick (eds), *Tropical Rain Forest: Ecology and Management*. Blackwell Scientific Publications, Oxford, pp. 377–84.

Marston, R. A. (1982). The geomorphic significance of log steps in forest streams. *Annals Association of American Geographers*, **72**, 99–108.

Mosley, M. P. (1982). The effect of a New Zealand beech forest canopy on the kinetic energy of water drops and on surface erosion. *Earth Surface Processes and Landforms*, **7**, 103–7.

Ovington, J. D. (1954). A comparison of rainfall in different woodlands. *Forestry*, **27**, 41–53.

Richards, P. W. (1952). *The Tropical Rain Forest*, Cambridge University Press, Cambridge.

Richards, P. W. (1969). Speciation in the tropical rainforest and the concept of the niche. *Biological Journal of the Linnean Society*, **1**, 149–54.

Ruxton, B. P. (1967). Slopewash under mature primary rainforest in Northern Papua. In J. N. Jennings and J. A. Mabbutt (eds), *Landform Studies from Australia and New Guinea*, A.N.U. Press, Canberra, pp. 85–94.

Spencer, T. and I. Douglas (1985). The significance of environmental change: diversity, disturbance and tropical ecosystems. In I. Douglas, and T. Spencer (eds), *Environmental Change and Tropical Geomorphology*, Allen and Unwin, London, pp. 39–73.

Swanson, F. J., R. J. Janda, T. Dunne and D. N. Swanston (eds) (1982). Sediment budgets and routing in forested drainage basins. *USDA Forest Service, General Technical Report*, PNW-141, 1–65.

Swanson, F. J., G. W. Lienkaemper and J. R. Sedell (1967). History, physical effects and management implications of large organic debris in western Oregon streams. *USDA Forest Service, General Technical Report*, PNW-56, 1–15.

Swanston, D. N. and F. J. Swanson (1976). Timber harvesting, mass erosion, and steepland forest geomorphology in the Pacific Northwest. In D. R. Coates (ed.), *Geomorphology and Engineering*, Dowden, Hutchinson and Ross Inc., Stroudsberg, Pa., pp. 199–221.

Walsh, R. P. D. and P. J. Voigt (1977). Vegetation litter: An underestimated variable in hydrology and geomorphology. *Journal of Biogeography*, **4**, 253–74.

Whitmore, T. C. (1974). *Change with time and the role of cyclones in tropical rain forest on Kolombangara, Solomon Islands*. Commonwealth Forest Institute Paper, 46.

Whitmore, T. C. (1978). Gaps in the forest canopy. In P. B. Tomlinson and M. H. Zimmerman (eds), *Tropical Trees as Living Systems*, Cambridge University Press, Cambridge, pp. 639–55.

28 The Geomorphological Role of Vegetation in Desert Dune Systems

DAVID S. G. THOMAS
Department of Geography, University of Sheffield

and

HAIM TSOAR
Department of Geography, Ben-Gurion University, Israel

SUMMARY

Despite its inclusion as a major controlling variable in an early desert dune classification scheme, vegetation has largely been ignored in geomorphological studies of desert dunes. Yet in drylands, dune sands can provide a valuable moisture source for plant growth, especially when less mobile dune types are considered. In the light of recent studies, this chapter examines the interactions between vegetation and geomorphology in desert dunes. Vegetation–dune geomorphology interactions fall into three broad groups: *vegetation as a surface stabilizer*, *vegetation as an accretion focus*, and *vegetation as a determinant of dune morphology*. Following a consideration of these three interactions, it is concluded that the stabilizing role of vegetation on dune surfaces has often been overestimated, whereas that of vegetation as a positive contributor to dune processes has frequently been understated.

DESERT DUNE FORMATION AND FORM—SOME BASIC CONSIDERATIONS

The formation of desert sand dunes is dependent on the availability of suitable sediment and winds strong enough to entrain and transport it. For entrainment

Vegetation and Erosion
Edited by J. B. Thornes
©John Wiley & Sons Ltd

to occur, and therefore ultimately for dunes to develop, wind velocities must not only be sufficient to overcome the resistance exerted by the particles themselves, which varies according to particle size, shape, sorting and mineralogy (Sarre, 1987), but also the effects of other ground surface properties, including vegetation (Buckley, 1987), which may influence aerodynamic roughness. For dry, bare, dune sand with particles of 0.25 mm diameter, the fluid threshold shear velocity (U_{*t}) is approximately $20 \, cm \, s^{-1}$, which corresponds to a threshold wind velocity U_t of $6 \, m \, s^{-1}$ at 1 m above the ground surface.

Even when conditions favouring sand entrainment and subsequent transport by saltation and creep are met, dune formation requires initiation through accretion. In general, dune genesis remains one of the least understood components of aeolian geomorphology, but once it has occurred the interactions between a dune and the atmospheric boundary layer will tend to result in the perpetuation of the dune form unless there are detrimental changes in sediment availability or climate. Overall, the necessary combination of factors for dune development are far from prevalent throughout the earth's deserts, and all told dunes occupy only about 12 per cent of arid environments (Thomas, 1989a).

Dune Form

Where the conditions which allow desert dune formation to occur are satisfied, the form of dunes is determined by several factors. Wind regime, both in terms of directional variability and the frequency of velocities above the threshold for entrainment, is widely regarded to assume major importance in the determination of dune morphology (e.g. Fryberger, 1979) and forms the basis of the most widely applied dune classification scheme (McKee, 1979). Sediment supply, which is in part related to wind regime, is regarded as a further major influence (Wasson and Hyde, 1983). In general terms, it is usually accepted that transverse forms (including barchans) develop under the influence of simple, often unidirectional, wind regimes in areas of limited sediment availability; linear and seif (also called longitudinal) dunes occur in locations with bidirectional formative winds; and star dunes where wind regimes are most complex and sediment supply is greatest.

This picture is complicated by the development of compound and complex forms (McKee, 1979), and the growth of other, less frequently occurring, desert dune types. Additionally, dune formation and dune forms cannot be explained in terms of wind and sediment characteristics alone. This was recognized by John T. Hack (1941), whose study of dunes in Navajo Country, Arizona, has been considered to contribute 'an important part of our understanding of sand dune formation' (Greeley and Iversen, 1985). Hack considered vegetation to be a third major variable which could influence dune development (Figure 28.1), yet is is only recently that geomorphological investigations have again included vegetation as a positive component of desert dune dynamics.

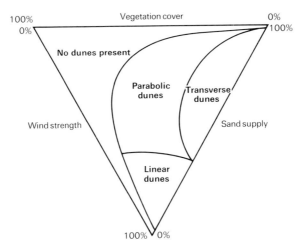

Figure 28.1   Hack's (1941) schematic diagram of the relationships between vegetation, wind strength, sand supply and dune type. Although no consideration is given to the role of wind directional regime, now known to be an important factor in the determination of dune type, and wind strength is not considered in a meaningful way, Hack did recognize the importance of variations in ground cover by plants, which was largely ignored in subsequent studies of desert dune development until the early 1980s

Vegetation and Desert Dunes

The geomorphological importance of vegetation lies in its situation at the interface between the atmospheric boundary layer and the surface of potentially deflatable sediments. Yet it has commonly been ignored in geomorphological studies of desert dune development. This is probably due to the assumption that deserts are too dry to support significant plant growth, but also because many of the more significant studies of desert dunes and aeolian processes have been conducted in extremely arid regions such as the Libyan and Namib Deserts; have concentrated on types of dune which are essentially devoid of vegetation when active; and increasingly, involve the use of laboratory simulations.

Vegetation is, however, often found in conjunction with desert dunes, with plant densities and species varying according to moisture availability, climatic regime and wind strength, dune type and shape, and increasingly the degree of grazing pressure (Hack, 1941; Tsoar and Møller, 1986). In general, plant species with shallow roots are less successful on mobile dune types, where species with anchoring tap roots, such as the sage bush (*Artemisia filifolia*) and rabbit brush (*Chrysothamus* spp.) of the southwestern United States are able to tolerate both deflation and accumulation of sand. Grasses are generally better suited to the less active parts of dunes, such as the plinths and lower flanks of linear dunes, as for example in the southwestern Kalahari (Thomas, 1988a), or the crests of dunes which display only limited mobility. Dunes may in fact be favourable sites for plant growth because of the moisture-retaining properties of sand (e.g. Bowers, 1986; Eriksson *et al.*, 1989) even in very arid areas, as observed in parts of the Negev and Sinai deserts receiving less than 50 mm precipitation per year (Tsoar and Møller, 1986).

The presence of vegetation on desert dunes will clearly influence their geomorphological dynamics, by modifying the aerodynamic roughness of the ground surface. These influences can be grouped into three categories, under which they will now be examined: vegetation as a surface stabilizer; vegetation as an accretion focus; and vegetation as a determinant of dune morphology.

VEGETATION AS A DUNE SURFACE STABILIZER

Hack's (1941) dune classification scheme incorporated the not unreasonable concept that as vegetation cover increased on the ground surface, so dune development became less likely (Figure 28.1). The widespread identification of desert dunes, both within today's arid lands and in extra-arid locations, possessing a partial or near-complete vegetation cover, has commonly been accompanied by their interpretation as inactive, or fossilized, features (e.g. Flint and Bond, 1968; Sarnthein, 1978; Lancaster, 1981). The vegetation cover is therefore considered to act as a buffer between potentially effective sand-moving winds and potentially entrainable sediment.

Climatic Factors and Dune Stabilization by Vegetation

Considerations of the stabilizing effects of dune vegetation have been intimately linked with delimitations of the mean annual precipitation values associated with both 'fossilized' and active desert dunes (Table 28.1). In regions where both categories of dunes are found, it has also been assumed that the difference in mean annual precipitation amounts to the increase which has occurred since the fossilized forms were active (e.g. Goudie, 1977; Lancaster, 1981; Thomas, 1984; Muhs, 1985).

Although some studies have inferred dune stability and relict status from a range of factors, including the presence of surficial stone and Iron Age artefacts, soil development and duneform degradation, the inferences of palaeoclimate, through the stabilizing role of vegetation, have recently come under scrutiny. Studies in the Australian (Ash and Wasson, 1983) and Thar (Wasson *et al.*, 1983) dunefields have demonstrated that in arid and semi-arid areas clear-cut boundaries between dune activity and inactivity are absent. Although the transition between these two states of desert dune dynamics is likely to be gradational, it is equally likely that Ash and Wasson's (1983) observation for the Australian desert dunefields of 'no simple correlation between rainfall and dune mobility, modulated by vegetation cover' has wider applicability.

Table 28.1   Examples of published values of mean annual precipitation limits for 'active' and 'fixed' desert dunes (data in mm)

Location	Active dunes	Fixed dunes	Source
Linear dunes			
Mauritania	25–50	—	Sarnthein and Diester-Hass, 1977
Southern Africa	150	972	Grove, 1969; Lancaster, 1981; Thomas, 1984
Southern Sahara	150	750–1000	Grove, 1958
Australia	100	—	Mabbutt, 1971
W. Australia	200	1000	Glasford and Killigrew, 1976
Parabolic dunes			
Colorado	—	282–464	Muhs, 1985
Parabolic and linear dunes			
Arizona	238–254	305–380	Hack, 1941
N.W. India	200–275	800	Goudie *et al.*, 1973

The range of values given in Table 28.1 demonstrates that factors other than precipitation must contribute to dune stability. In fact dune vegetation, the medium through which precipitation has been viewed to impart desert dune stability, is affected by climatic factors other than simply annual precipitation values. Temperature regimes not only affect evapotranspiration, which contributes to the determination of the amount of effective rainfall, but also the physiology of plants. Wind regimes, through sand movement, will affect plant colonization and growth, whilst the high intensity of rainfall events in some deserts will influence infiltration rates and the subsequent levels of available moisture within dune sands.

Factors Affecting Dune Stabilization by Vegetation

The impact of vegetation upon aeolian sediment transport is affected by factors relating both to vegetation communities and the dune environment (Table 28.2). Dunefield vegetation communities vary at two scales: *regionally* according to broad environmental gradients (e.g. Ash and Wasson, 1983; Thomas, 1988a), and *across dune profiles*. The latter is primarily a consequence of habitat variations induced by differences in wind conditions (Tsoar, 1985) and sediment properties (Thomas, 1984) between dune crests and inter-dune areas, which in turn may have a profound effect on moisture availability (e.g. Tsoar and Møller, 1986). Such dune-scale vegetation differences will complicate any dunefield scale relationships between vegetation and dune stability, as well as effecting differing

Table 28.2 Factors affecting the ability of vegetation to influence aeolian sand transport

---

Vegetation attributes
1. Plant densities
2. Vegetation community species composition
3. Seasonal growth cycles
4. Plant mechanical attributes (height, breadth, foliage density, etc.)
5. Within-dunefield community variations
6. Community variations across dune profiles
7. Stage of plant succession

Factors affecting effectiveness of a given plant community
1. Wind regimes and strengths
2. Sediment supply
3. Dune type
4. Grazing and fire pressures

---

degrees of surface stability at different points on dune profiles, as reported by Grove (1969), Tsoar and Møller (1986) and many others.

Vegetation Characteristics

The most important attributes of a plant cover which influence aeolian sediment movement on dunes are community composition, plant densities and plant shapes. Whilst greater plant densities are likely to provide greater surface stabilization, including through the binding effect of roots (e.g. Hesp, 1981), effective sand movement can occur with up to 35 per cent of the ground surface sheltered by plants (Ash and Wasson, 1983). This is more likely to occur at or near the crests of dunes, because of the higher crestal wind velocities associated with dune-induced airstream modifications (e.g. Tsoar, 1985; Watson, 1987).

Grasses mitigate more effectively against aeolian entrainment than trees and shrubs, as most (90 per cent, according to Heathcote, 1983) wind-transported sand is carried in the lowest 0.5 m of the atmosphere. A grass cover therefore raises the height of the zone of zero velocity above the sediment surface (Bagnold, 1941; see Figure 28.2). However, the clumped nature of desert grasses may limit the effectiveness of ground cover protection.

Plant shape also contributes to the nature of airflow modifi-cation by vegetation (Marshall, 1970; Ash and Wasson, 1983; Buckley, 1987; Thomas, 1988b), further complicating a simple relationship between vegetation cover and dune stability. Indeed, wind velocities may be enhanced as airflow is streamlined around clumped grasses and shrubs, so that aeolian scour and sand transport may be increased around the sides of plant obstacles (Figure 28.3).

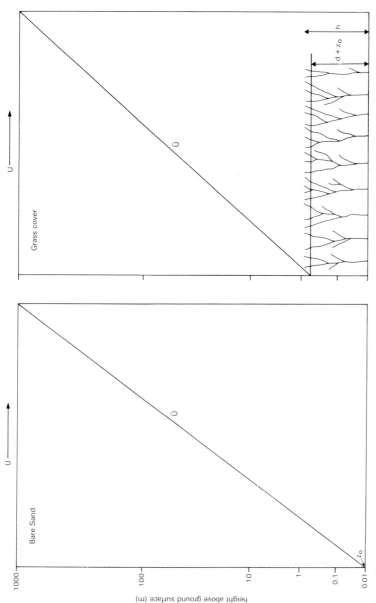

Figure 28.2 Ground surface roughness creates a zone of zero wind velocity. The height of this zone above the ground surface (z_0) is related to the size of the surface characteristics which constitute the roughness elements (Bagnold, 1941); the left-hand diagram constitutes the case for a surface of sand devoid of any vegetation: z_0 is equal to 0.033 of the mean grain diameter. An even vegetation cover (e.g. of grass, diagram on right) has the effect of substantially increasing the height of z_0 by d, the displacement distance. For a grass cover 0.25 to 1.0 m high, z_0 is 0.04–0.1 m and d is <0.66 m (two-thirds the height of the vegetation: Tsoar and Pye, 1987).

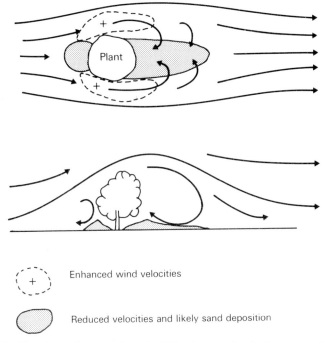

Enhanced wind velocities

Reduced velocities and likely sand deposition

Figure 28.3   The clumped nature of much arid land vegetation, be it grass or shrubs, creates a complex effect whereby zones of enhanced and reduced windspeeds are created around isolated plants. If the plant has a single main stem or is relatively impervious to wind flow, windward and leeside sand mounds or dunes may develop. If sand is trapped within the plant structure, a coppice dune will form, as in Figure 28.4.

Other Environmental Factors

Vegetation cover cannot be considered in isolation from other environmental factors in a consideration of aeolian sand transport. For a given vegetation cover, higher wind velocities will reduce the protection afforded against sediment movement. Conversely, despite vegetation densities of less than 10 per cent over much of the Australian dunefields, significant sand movement occurs only locally today because of low wind velocities.

Factors which reduce the vegetation cover on dune surfaces will obviously influence dune stability. The *spinifex* grasses of the Australian arid zone are particularly fire susceptible (Wasson and Nanninga, 1986), with notable increases in dune mobility recorded in severely burnt areas. Overgrazing has been widely noted to lead to enhanced dune movement and changes in dune morphology in the Negev (Tsoar and Møller, 1986) and Kalahari (Thomas, 1988a) deserts.

Dune Stabilization by Biocrusts

Higher-order plants are not the only organisms which can contribute to dune surface stabilization. The tendency for biocrusts composed of unicellular algae to develop away from vegetation (Shields, Mitchell and Drouet, 1957) may be particularly significant. Algal and fungal crusts have been noted to stabilize dune surfaces in coastal (Van Den Ancker *et al.*, 1985) and desert (Shields *et al.*, 1957; West and Skujins, 1978) situations. In the White Sands dunefield, New Mexico, blue-green algae cement gypsum particles to form a protective crust on dune surfaces (Shields *et al.*, 1957) where algal crusts, which may undergo lichenization, cover up to 80 per cent of the surface, wind erosion may be substantially reduced, as recorded in the Great Basin, Utah and Nevada (West and Skujins, 1978).

Dune Type and Vegetation Stabilization

The relationships between dune type, wind regime and sand movement (e.g. Fryberger, 1979; Wasson and Hyde, 1983) mean that the presence of dune vegetation will differ in geomorphological significance according to the type of dune under consideration. This was first observed by Hack (1941), but has largely been ignored since. Dunes which are migratory or experience significant morphological changes will be less conducive to plant growth than those which are more stationary and therefore afford greater opportunity for plant colonization.

Transverse dunes (including barchans) develop in environments dominated by unidirectional sand-moving winds. They are, especially in the case of barchans, mobile, migratory forms, affording little opportunity for the development of even partial plant covers, though to some extent this will be affected by the rate of migration, which tends to decrease as dune size increases (Norris, 1966; Hastenrath, 1987). Seif dunes form in environments with bidirectional wind regimes, extending downwind parallel to the resultant sand drift direction (Tsoar, 1978). Whilst their plinths are largely stable, the crests of seifs are sinuous and highly mobile, due to the alternating seasonal or diurnal influences of winds from opposite sides of the dune (Tsoar, 1982, 1983; Livingstone, 1988). The movement of seif crests does not favour plant colonization either.

In contrast, linear ridges are more stable forms, reported to extend parallel to the dominant wind direction (Tsoar and Møller, 1986; Thomas, 1989b) in bimodal wind regimes (see below). Although sand movement occurs along the subdued crests of linear ridges (Ash and Wasson, 1983; Thomas, 1988a), the morphological stability of this dune type favours plant growth, especially as the largely immobile but porous dune body may retain appreciable quantities of moisture from desert rainfall events (Tsoar and Møller, 1986). Thus the distinction between dunes which are morphologically mobile and those which are stationary is probably very important for interpreting the significance of a dune vegetation cover: even a partial cover on a usually migratory form or mobile crest is likely to indicate dune stabilization; whereas it can be a normal component of active sand-passing forms.

VEGETATION AS AN ACCRETION FOCUS

Vegetation can contribute to the genesis and development of dunes by acting as a focus for sand accumulation. This may occur where the vegetation cover of a surface is incomplete, as a facet of the plant-induced airflow modifications discussed in the previous sections, or where aeolian transport moves sediment from a vegetation-free surface to one where plants are present. The former is probably responsible only for the development of minor dune forms, whereas the latter, analogous to the accretion of coastal dune systems, may contribute to the formation of more substantial features, such as lunette dunes on the downwind sides of unvegetated desert pan depressions (e.g. Goudie and Thomas, 1986).

Plant Obstacle and Coppice Dunes

The airflow streamlining affect of clumped plants may induce sand deposition in the leeside cavity of reduced wind velocity (Hesp, 1981), or if the plant is sufficiently large or of low wind 'porosity', on the windward side (Figure 28.3). These plant obstacle dunes, and *coppice* dunes (sometimes called 'rebdon' or 'nebkha': Cooke and Warren, 1973), which develop through the trapping of sand within the body of the plant (Figure 28.4), are rarely major features, often being no more than one or two metres high (Gunatilaka and Mwango, 1987).

The morphology of plant-induced dunes is dependent on the characteristics of the plants involved, wind regime and sediment availability (Bagnold, 1953; Hesp, 1981). Coppice dunes are normally rounded, without slipfaces, and can be found on the crests of partially vegetated linear dunes (Purdie, 1984, p. 20), in inter-ridge areas (Thomas, 1988a), on pan surfaces (Gunatilaka and Mwango, 1987) or near ephemeral or dry river valleys. In all these locations, plants are able to tap subsurface moisture which supports their growth in otherwise dry environments (e.g. Seely, 1987).

It is feasible that, given a sufficient sand supply, coppice dunes may develop into other larger dune forms, with the plant acting as dune initiator but diminishing in geomorphological significance over time as the mound of sand grows. The development of small nebkhas near the margins of saline pans in Kuwait is achieved through sand accumulation around isolated halophyte *Nitraria retusa* plants (Gunatilika and Mwango, 1987). Whilst their growth

Figure 28.4 A coppice dune, created by the sand-trapping effect of the nara (*Acanthosicyos horrida*), which draws upon subsurface moisture using long tap roots. Gobabeb, Namibia

initially keeps apace with the sand accumulation, the halophytes eventually die as root contact with the pan capillary zone ceases. Sand is then transferred from the windward side of the obstacle to the lee shadow, and the dune may grow into an elongated or pyramidal form, possessing low slipfaces.

Sand Sheets

Sand sheets, characterized by low-angle sediment laminae, develop in environments which favour aeolian activity but not dune development (Thomas, 1989b), often on the margins of desert dunefields (Kocurek, 1986). Vegetation, including biocrusts (Fryberger, Schenk and Krystinik, 1988), is now identified as one of the major controls on the development of accretional sand sheets (Kocurek and Nielson, 1986). This was also recognized by Bagnold (1941), who suggested (p. 183) that 'a uniform but thin' covering of grass was the most likely vegetation cover to lead to sand sheet formation, storing incoming sand but inhibiting its further onward movement. A provision necessary for this situation to persist is that the accretion rate does not exceed that which permits grass growth to continue.

While vegetation can favour the growth of sand sheets, the character of sand sheets can also favour plant growth. Sand sheets are often composed of poorly sorted bimodal sands, and in areas of some rainfall this favours moisture retention, and therefore plant growth (Fryberger *et al.*, 1984; Kocurek and Nielson, 1986).

Silt and Clay Additions to Vegetated Dune Surfaces

The stabilizing effects of biogenic crusts and higher-order plant communities on dune surfaces can result in an increase in the silt and clay content of the dune sands which, with the exception of dunes composed of clay pellets (e.g. Bowler, 1973) is commonly less than 3–5 per cent of the sediment by weight. This has been recorded in the Negev (Tsoar and Møller, 1986), where crusted sands contain up to 10 per cent silt and clay.

Atmospheric silt and clay do not normally accumulate in mobile dune sands because fine particles coming to rest on a dune surface are readily remobilized by saltating sand (Tsoar, 1976; Pye, 1982). A surface algal crust can lead to the incorporation of fines into the upper layers of sand because of the significant protection against sediment entrainment that even a weak crust affords (Gillette et al., 1980, 1982). This can then contribute to further stability, moisture retention and soil fertility which can lead to more substantial vegetation growth and plant succession (Danin, 1978; Tsoar and Møller, 1986). Higher-order plant communities on dunes can similarly trap atmospheric dust, by raising the height of the zone of zero wind velocity above the dune surface (Tsoar and Pye, 1987).

VEGETATION AS A DETERMINANT OF DUNE MORPHOLOGY

The complex interaction of environmental variables which contribute to the development and characteristics of desert dunefields means that it is possible to consider the role of vegetation as more than simply inhibiting sand movement or encouraging its deposition (Table 28.3). This is clearly demonstrated in its role in the development of parabolic dunes (Hack, 1941; Wasson et al., 1983) where spatial variations in plant cover allow adjacent communities to act as a surface stabilizer on dune arms, to permit deflation in the central corridor (blow out) by its absence, burial or destruction, and to trap sand at the dune nose. The widespread occurrence of parabolic dunes in some deserts, notably in the Thar in India and in North America, attests to the importance of vegetation as a determinant of dune form.

As some dune types develop in environments where the velocities and directions of potential sand-moving winds vary seasonally or diurnally, the geomorphological impact of dune vegetation can differ according to the component of the wind regime taking effect. This is especially important, for example, in the cases of seif dunes and linear ridges, discussed above. Partially vegetated linear ridges are a major component of the dunefields of Australia (Madigan, 1936; Twidale, 1972), the southwestern Kalahari (Grove, 1969; Thomas, 1988a) and parts of the Negev (Tsoar and Møller, 1986). They are up to about 20 m high and often asymmetrical in cross-profile, sometimes displaying a small slipface on the steeper side of the rounded crest. Vegetation

Table 28.3   Examples of the influences of vegetation upon dune morphology

Dune type	Description of influence	Reference
Parabolic	Anchoring dune arms; permitting scour between arms due to absence	Hack, 1941; Mabbutt, 1977; Wasson *et al.*, 1983
Nebkha, evolving into pyramidal 'shadow dune'	Vegetation binding nebkha dies and dune elongates downwind	Gunatilaka and Mwango, 1987
Barchan, passing downwind into sand mounds	Mobile barchans advancing into vegetated area, which traps sand and disrupts dune form	Bagnold, 1941
Linear ridge	Vegetation restricts sand movement to strongest ridge-parallel winds only	Tsoar and Møller, 1986
Linear ridge, evolving into braided ridges	Ridge vegetation destroyed, oblique superimposed ridges develop under influence of secondary winds	Tsoar and Møller, 1986
Linear ridge, evolving into sinuous seif dune	Ridge vegetation destroyed, dune able to evolve under influence of both components of bimodal wind regime	Tsoar and Møller, 1986; Thomas, 1988a
Nested parabolic, superimposed on linear ridge	Wind scour occurs on ridge crests on bare sand beneath trees	Eriksson *et al.*, 1989

Figure 28.5   The crest of a partially vegetated linear dune ridge, S.W. Kalahari Desert. Aeolian activity is evident in the crestal zone in the form of sand accumulation around clumped grasses and the patches of bare rippled sand.

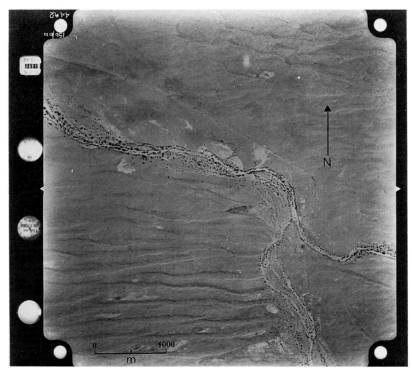

Figure 28.6 Air photographs showing the change in dune morphology associated with the de-vegetating of linear dune ridges. (a) Linear dune ridges, comparable with that in (b) *(opposite)* Sinuous-crested and braided ridges, following destruction of the crestal vegetation. Negev Desert

is least well developed in the crestal zone, which may display evidence of aeolian sand movement in the form of rippled surfaces and coppice sand mounds (Ash and Wasson, 1983; Thomas, 1988a; see Figure 28.5).

Although seif dunes and linear ridges are found in areas with bimodal potential sand moving winds, their alignment is often parallel to the strongest dominant wind direction (Tsoar and Møller, 1986), as the partial vegetation cover inhibits sand movement by weaker winds. A reduction in the vegetation cover of these dunes, by natural means or by overgrazing, has been noted to result in significant changes in the morphology of the dune crestal zones and in dune alignment (Tsoar and Møller, 1986; Thomas, 1988a). Less surface vegetation allows the secondary sand-moving modal wind direction to act dynamically on the dune (Figure 28.6). This results in a more active, sharper, sinuous crest zone and the transformation of the ridge to a seif dune, indicative of the effects of a bimodal wind regime (Tsoar, 1978, 1983); or a braided form, with secondary superimposed dunelets aligned normal to the component

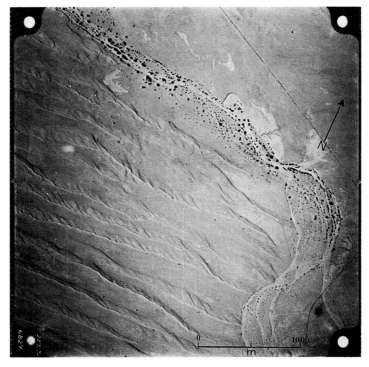

Figure 28.6(b)

of the wind regime which is now able to effectively transport sand (Tsoar and Møller, 1986).

Eriksson *et al.* (1989) have observed a further change in dune morphology through the impact of vegetation on linear ridges in the southwest Kalahari. The higher dune ridge intersections (Y junctions: Thomas, 1986) tend to be favourable locations for the growth of trees (especially the Shepherd's Tree, *Boscia albitrunca*), probably as the greater volume of sand in these locations retains a higher concentration of moisture. The trees support an important micro-ecosystem (Eloff, 1984), leading to overgrazing and trampling in their vicinity. Wind scour has ensued, resulting in the development of nested, dominant wind-aligned, parabolic dunes, superimposed on the linear ridges (Eriksson *et al.*, 1989).

DISCUSSION AND CONCLUSION

Many of the world's arid zones support at least a partial vegetation cover. It is only in hyper-arid locations, or where surface conditions mitigate against it (soils with a high salinity or where bedrock exposures predominate, for example), that vegetation is commonly absent. Although some of the more spectacular desert dunefields, such as the Namib sand sea, the Lybian desert and the great dunefields of Arabia, are largely devoid of vegetation, many do support at least a partial and often more substantial plant cover. Where vegetation is absent or almost absent from desert dunes, it can be accounted for by one of three explanations: hyper-aridity; the types of dune present; or the destruction of plants by fire or overgrazing.

The presence of desert dune vegetation has frequently been assumed to be an indication of dune stabilization and, consequently, vegetated dunes have been interpreted as indicators of greater former aridity. Whilst duneforms which support a dense, in some cases woodland, vegetation, as for example in western Zimbabwe (Thomas, 1984), and those which possess characteristics such as surface soil development, can be considered to be relict features, there are now very good reasons to discard the blanket assumption that the presence of vegetation automatically indicates dune stability. There is no simple relationship between rainfall, vegetation cover and dune stability, whilst significant sand transport can occur with up to 35 per cent of the ground covered by vegetation. Indeed, dune surfaces may provide very suitable environments for plant growth in arid environments, because of their moisture-retaining properties (Tsoar and Møller, 1986). This suitability is especially so in the case of sand-passing dune forms, where sand is transported along the length of the dune, and changes in dune morphology, which may be destructive to plant growth, are restricted to the crestal zone.

As vegetation is becoming recognized as a 'natural' component of many desert dunes, so its role in the development of dune form is being identified. It is evident that further studies in this area are called for, so that dynamic interactions between plant life and aeolian geomorphology can be more fully appreciated.

REFERENCES

Ash, J. E. and R. J. Wasson (1983). Vegetation and sand mobility in the Australian desert dunefield. *Zeitschrift für Geomorphologie, Supplementbd*, **45**, 7–25.
Bagnold, R. A. (1941). *The Physics of Blown Sand and Desert Dunes*, Methuen, London.
Bagnold, R. A. (1953). The surface movement of blown sand in relation to meteorology. *Research Council of Israel special Publication 2*, 89–93.
Bowers, J. E. (1986). *Seasons of the Wind*, Northland Press, Colorado.
Bowler, J. M. (1973). Clay dunes: Their occurrence, formation and environmental significance. *Earth-Science Reviews*, **9**, 315–38.
Buckley, R. (1987). The effect of sparse vegetation on the transport of dune sand by wind. *Nature*, **325**, 426–8.
Cooke, R. U. and A. Warren (1973). *Desert Geomorphology*, Batsford, London.
Danin, A. (1978). Plant species diversity and plant succession in a sandy area of the northern Negev. *Flora*, **167**, 409–22.

Eloff, F. C. (1984). The Kalahari ecosystem. *Koedoe* Supplement, **27**, 11–20.

Eriksson, P. G., N. Nixon, C. P. Snyman and J. duP. Botha (1989). Ellipsoidal parabolic dune patches in the southern Kalahari desert. *Journal of Arid Environments*, **16**, 111–24.

Flint, R. F. and G. Bond (1968). Pleistocene sand ridges and pans in western Rhodesia. *Bulletin, Geological Society of America*, **79**, 299–314.

Fryberger, S. (1979). Dune forms and wind regime. In E. D. McKee (ed.), *A study of global sand seas*. US Geological Survey Professional Paper 1052, 137–69.

Fryberger, S. G., A. M. Al-Sari, T. J. Clisham, S. A. R. Rizvi and K. G. Al-Hinai (1984). Wind sedimentation in the Jafurah sand sea, Saudi Arabia. *Sedimentology*, **31**, 413–31.

Fryberger, S. G., C. J. Schenk and L. F. Krystinik (1988). Stokes surfaces and the effects of near surface groundwater-table on aeolian deposition. *Sedimentology*, **35**, 21–41.

Gillette, D. A., J. Adams, A. Endo, D. Smith and R. Kihl (1980). Threshold velocities for input of soil particles into the air by desert soils. *Journal of Geophysical Research*, **85**, 5621–30.

Gillette, D. A., J. Adams, D. Muhs and R. Kihl (1982). Threshold friction velocities and rupture moduli for crusted desert soils for the input of soil particles into the air. *Journal of Geophysical Research*, **87**, 9003–15.

Glasford and Killigrew (1976). Evidence for Quaternary westward extension of the Australian Desert into southwestern Australia. *Search*, 394–6.

Goudie, A. S. (1977). *Environmental Change*, Oxford University Press, Oxford.

Goudie, A. S. and D. S. G. Thomas (1986). Lunette dunes in southern Africa. *Journal of Arid Environments*, **10**, 1–12.

Greeley, R. and T. D. Iversen (1985). *Wind as a Geological Process on Earth, Mars, Venus and Titan*, Cambridge University Press, Cambridge.

Grove, A. T. (1958). The ancient ergs of Hausaland and similar formations on the south side of the Sahara. *Geographical Journal*, **139**, 243–57.

Grove, A. T. (1969). Landforms and climatic change in the Kalahari and Ngamiland. *Geographical Journal*, **135**, 191–212.

Gunatilaka, A. and S. Mwango (1987). Continental sabkha pans and associated nebkhas in southern Kuwait, Arabian Gulf. In L. E. Frostick and I. Reid (eds), *Desert sediments ancient and modern*, Geological Society Special Publication 35, Blackwell, Oxford, pp. 187–203.

Hack, J. T. (1941). Dunes of the western Navajo Country. *Geographical Review*, **31**, 240–63.

Hastenrath, S. (1987). The barchan dunes of southern Peru revisited. *Zeitschrift für Geomorphologie*, **NF31**, 167–78.

Heathcote, R. L. (1983). *The Arid Lands: Their Use and Abuse*, Longman, Harlow.

Hesp, P. A. (1981). The formation of shadow dunes. *Sedimentary Petrology*, **51**, 101–12.

Kocurek, G. (1981). Significance of interdune deposits and bounding surfaces in aeolian sands. *Sedimentology*, **28**, 753–80.

Kocurek, G. (1986). Origin of low-angle stratification in aeolian deposits. In W. G. Nickling (ed.), *Aeolian Geomorphology*. Allen and Unwin, Boston, pp. 177–93.

Kocurek, G. and J. Nielson (1986). Conditions favourable for the formation of warm-climate eolian sand sheets. *Sedimentology*, **33**, 795–816.

Lancaster, N. (1981). Palaeoenvironmental implications of fixed dune systems in southern Africa. *Palaeogeography, Palaeoclimatology, Palaeoecology*, **33**, 327–46.

Livingstone, I. (1988). New models for the formation of linear sand dunes. *Geography*, **73**, 105–15.

Mabbutt, J. A. (1971). The Australian arid zone as a prehistoric environment. In D. J. Mulvaney and T. Golson (eds) *Aboriginal Man and Environment*, ANU Press, Canberra, 66–79.

Madigan, C. T. (1936). The Australian sand-ridge deserts. *Geographical Review*, **26**, 205–27.

McKee, E. D. (1979). Introduction to a study of global sand seas. In E. D. McKee (ed.), *A study of global sand seas*. US Geological Survey Professional Paper 1052, 1–19.

Marshall, J. K. (1970). Assessing the protective role of shrub-dominated rangeland vegetation against soil erosion by wind. *Proceedings, Eleventh International Grassland Congress*, 19–23.

Muhs, D. R. (1985). Age and paleoclimatic significance of Holocene dune sands in northeastern Colorado. *Annals, Association of American Geographers*, **75**, 566–82.

Norris, R. M. (1966). Barchan dunes of Imperial Valley, California. *Journal of Geology*, **74**, 292–306.

Purdie, R. (1984). *Land systems of the Simpson Desert*. CSIRO Natural Resources Series 2.

Pye, K. (1982). Thermoluminescence dating of coastal dunes. *Nature*, **299**, 376.

Sarnthein, M. (1978). Sand deserts during glacial maximum and climatic optimum. *Nature*, **272**, 43–6.

Sarnthein, M. and Diester-Hass (1977). Eolian sand turbidites. *Journal of Sedimentary Petrology*, **47**, 868–70.

Sarre, R. D. (1987). Aeolian sand transport. *Progress in Physical Geography*, **11**, 157–82.

Seely, M. (1987). *The Namib*, Shell Oil, Windhoek.

Shields, L. M., C. Mitchell, and F. Drouet (1957). Alga- and lichen stabilized surface crusts as soil nitrogen sources. *American Journal of Botany*, **44**, 489–98.

Thomas, D. S. G. (1984). Ancient ergs of the former arid zones of Zimbabwe, Zambia and Angola. *Institute of British Geographers, Transactions*, **NS9**, 75–88.

Thomas, D. S. G. (1986). Dune pattern statistics applied to the Kalahari Dune Desert, southern Africa. *Zeitschrift für Geomorphologie*, **NF30**, 231–42.

Thomas, D. S. G. (1988a). The geomorphological role of vegetation in the dune systems of the Kalahari. In G. F. Dardis and B. P. Moon (eds), *Geomorphological Studies in Southern Africa*, Balkema, Rotterdam, pp. 145–58.

Thomas, D. S. G. (1988b). The biogeomorphology of arid and semi-arid environments. In H. A. Viles (ed.), *Biogeomorphology*, Blackwell, Oxford, pp. 193–221.

Thomas, D. S. G. (1989a). The nature of arid environments. In D. S. G. Thomas (ed.), *Arid Zone Geomorphology*, Belhaven Press, London, pp. 1–8.

Thomas, D. S. G. (1989b). Aeolian sand deposits. In D. S. G. Thomas (ed.), *Arid Zone Geomorphology*, Belhaven Press, London, pp. 232–61.

Tsoar, H. (1976). Characterization of sand dune environment by their grain-size, mineralogy and surface texture. In D. H. K. Amiran and Y. Ben-Arieh (eds), *Geography in Israel*, Israel I.G.U., pp. 327–43.

Tsoar, H. (1978). *The dynamics of longitudinal dunes*. Final technical report DA-ERO 76-G-072 European Research Office, US Army, London.

Tsoar, H. (1982). Internal structure and surface geometry of longitudinal (seif) dunes. *Journal of Sedimentary Petrology*, **52**, 823–31.

Tsoar, H. (1983). Dynamic processes acting on a longitudinal (seif) dune. *Sedimentology*, **30**, 566–78.

Tsoar, H. (1985). Profiles analysis of sand dunes and their steady state signification. *Geografisker Annaler*, **67A**, 47–59.

Tsoar, H. and J. T. Møller (1986). The role of vegetation in the formation of linear sand dunes. In W. G. Nickling (ed.), *Aeolian Geomorphology*, Allen and Unwin, Boston, pp. 75–95.

Tsoar, H. and K. Pye (1987). Dust transport and the question of desert loess formation. *Sedimentology*, **34**, 139–53.

Twidale, C. R. (1972). Evolution of sand dunes in the Simpson desert, central Australia. *Transactions, Institute of British Geographers*, **56**, 77–106.

Van Den Ancker, J. A. M., P. D. Jungerius, and L. R. Mur (1985). The role of algae in the stabilization of coastal dune blowouts. *Earth Surface Processes and Landforms*, **10**, 189–92.

Wasson, R. J. and R. Hyde (1983). Factors determining desert dune type. *Nature*, **304**, 337–9.

Wasson, R. J. and P. M. Nanninga (1986). Estimating wind transport of sand on vegetated surfaces. *Earth Surface Processes and Landforms*, **11**, 505–14.

Wasson, R. J., S. N. Rajaguru, V. N. Misra, D. P. Agrawal, R. P. Dhir, A. K. Singhvi and K. Kameswara Rao (1983). Geomorphology, late Quaternary stratigraphy and palaeoclimatology of the Thar dunefield. *Zeitschrift für Geomorphologie, Supplementbd*, **45**, 117–51.

Watson, A. (1987). Variations in wind velocity and sand transport on the windward flanks of desert sand dunes. *Sedimentology*, **34**, 511–16.

West, N. E. and J. Skujins (1978). *Nitrogen in desert ecosystems*. US/IBP Synthesis Series 9, Dowden, Hutchinson and Ross, New York.

29 Quaternary Coastal and Vegetation Dynamics in the Palk Strait Region, South Asia — The Evidence and Hypotheses

WALTER ERDELEN
Department of Biogeography, University of the Saarland

and

CHRISTOPH PREU
Department of Physical Geography, University of Augsburg

SUMMARY

A qualitative model is presented for potential exchange of flora and fauna between South India and Sri Lanka during an interglacial/ glacial/interglacial cycle. Late Quaternary data on glaciation at higher northern latitudes, sea levels, and climate allow the reconstruction of land/sea distributions and climatic changes during the cycle. Climatic cyclicity has resulted in alternating phases of range expansion of rainforests and contraction of 'dry' vegetation types viz. semi-deciduous forests and savanna, and vice versa. Time lags most likely facilitated exchange of rainforest species during land-bridge phases over the Palk Strait and might thus offer a new explanation for the similarities in rainforest biota between South India and Sri Lanka. The model is expandable to previous Quaternary cycles and may contribute to the formulation of new hypotheses for the biogeographical evolution of India and Sri Lanka.

Vegetation and Erosion
Edited by J. B. Thornes
©1990 John Wiley & Sons Ltd

INTRODUCTION

During the last decades it has repeatedly been shown that climatic conditions in the tropics have undergone periodic changes during the Quaternary (Flenley, 1979; Prance, 1982; Whitmore, 1987). These considerations are particularly relevant for an understanding of present similarity patterns of flora and fauna between continental islands and the neighbouring mainlands. The magnitude and timing of responses of the flora and fauna might have been species-specific thus causing floral and faunal disequilibria and changes in patterns of species abundance (Davis, 1986). The changes encompassed absolute areas per vegetation type, altitudinal contraction or expansion of plant communities and changes in species distributions and therefore opportunities for speciation and extinction (Walker, 1982).

The larger part of the biogeographical evolution of Sri Lanka and India has taken place in pre-Quaternary times (for details see Mani, 1974; Ashton and Gunatilleke, 1987). However, as documented in Deraniyagala (1958), the extant composition of the flora and fauna of Sri Lanka has been profoundly affected by Quaternary dynamics of climate and sea level. The latter were of major importance in the Palk Strait Region where already small-scale regressions led to the formation of a land-bridge between Sri Lanka and India.

Only Moore (1960), in his study on South Asian squirrels, has presented a model for the Quaternary to explain floral and faunal similarity patterns between India and Sri Lanka. However, his model is based on the classical 'pluvial theory' which assumes (1) contemporaneous changes between sea level and climate, and (2) wet climatic conditions during glacials and drier conditions during interglacials, respectively. Accordingly, Moore (1960) postulates rainforests on the glacial land-bridges and concomitant faunal exchange between India and Sri Lanka and separate evolution during interglacial transgression phases. As modern views of Quaternary climatic dynamics (e.g. Imbrie and Imbrie, 1985) do not support Moore's assumptions, his model cannot explain floral and faunal relations between India and Sri Lanka.

In this chapter we outline in short the present physical geography and biogeography of the Palk Strait Region, summarize Quaternary changes in climate and sea levels of Sri Lanka, and present a qualitative model for potential exchange of flora and fauna between South India and Sri Lanka (over the Palk Strait) during an interglacial/glacial/interglacial cycle.

THE STUDY UNIT: PALK STRAIT REGION

Physical Geography

The shallow (maximum depth 15 m) Palk Strait with a northeast–southwest extension of some 150 km connects the Bay of Bengal and the Gulf of Mannar and separates the continental island Sri Lanka from the southern tip of Peninsula India (Figure 29.1). The minimum distance between them is 35 km. The geology and geomorphology of the areas adjoining the Palk Strait Region, which has been tectonically stable during the Quaternary (Preu, 1988), is summarized in Bremer (1981), Cooray (1967), Mani (1974), Vitanage (1972), and Wadia (1976).

The rainfall patterns of the Palk Strait Region are part of the large-scale weather phenomena in South Asia for which the seasonal shifts of the ITCZ (the Intertropical Convergence Zone) form the most important component. ITCZ shifts are responses to large-scale heating-pressure patterns on the Tibetan Plateau and cause the predominance of seasonally different wind systems in South Asia, the monsoons (Flohn, 1952, 1953, 1958; Suppiah, 1988). The relief is responsible for spatial differences in precipitation (Legris and Viart, 1961; Domrös, 1974).

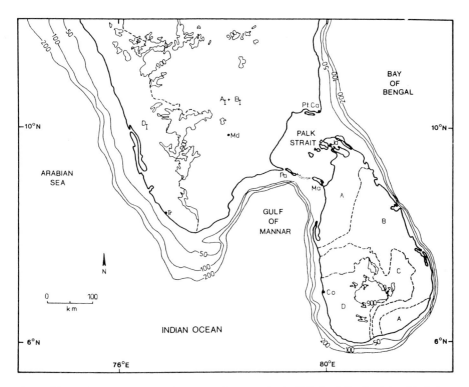

Figure 29.1 Bathymetry and major vegetation types in the Palk Strait Region (simplified after Legris, 1963; Mueller-Dombois, 1968; Pascal, 1984). A, A$_I$ = Monsoon scrub jungle; B, B$_I$ = Semi-evergreen forest; C = Intermediate forest (not shown for India); D, D$_I$ = Lowland rainforest. Above 900 m: Montane and cloud forests. Locations: Co = Colombo, Ja = Jaffna, Ma = Mannar, Md = Madurai, Pa = Pamban Island, Pt.Ca = Point Calimere, Tr = Trivandrum. Area D in Sri Lanka more or less congruent with the Wet Zone, remaining areas belong to the Dry Zone. For further explanation see text

More specifically, during northern summers the ITCZ reaches latitudes 28°–30°N (Himalayan Region, Bay of Bengal) which results in the predominance of southwesterly winds from May to September. Due to the high humidity and instability of these air masses, heavy convectional rains occur in the southwest of Sri Lanka and along the western slopes of the South Indian hills. In northern winters the ITCZ reaches latitudes 10°–12°S. During this season (December to February) the relatively dry and stable air masses of the northeast trade winds predominate and lead to rainfall maxima in the northern and eastern parts of Sri Lanka and South India. Based on these seasonal and spatial rainfall patterns, Kularatnam (1968), Fernando (1968), and Domrös (1974) subdivide Sri Lanka into two hygroclimatic zones, viz. the Wet Zone (mean annual rainfall higher than 1904 mm) and Dry Zone (mean annual rainfall less than 1904 mm; see Figure 29.1. These terms have not been adopted for India although the southwestern slopes of the Deccan (Figure 29.1, D_I) have rainfall patterns similar to the Wet Zone, and the area $A_I + B_I$ has patterns similar to the Dry Zone of Sri Lanka.

Average temperatures vary little from month to month. Mean annual temperatures are about 27 °C at sea level, 24 °C at 500 m, and 15 °C at 1900 m (Domrös, 1974; Legris and Viart, 1961). Humidity is generally high throughout the year; annual averages are between 80 and 85 per cent.

Vegetation and Biogeography

The spatial rainfall patterns reflect the distribution of the major vegetational communities. However, small-scale spatio-temporal differences in amounts of precipitation are important for the classification of the vegetation (Fernando, 1968; Gaussen et al., 1964, 1968; Koelmeyer, 1957, 1958). For this study we adopted the subdivision into four zones (cf. Mueller-Dombois, 1968; see Figure 29.1):

A—Monsoon scrub jungle
B—Semi-evergreen forest
C—Intermediate forest
D—Rainforest and grassland

Whereas zones A to C are comparatively uniform in floral composition zone D is differentiated (not shown in Figure 29.1) into lowland forests (sea level to approx. 900 m), montane forests (900–150 m), and cloud forests (above 1500 m). Nomenclatures for the Wet Zone and hill forests are given in Werner (1984). The distribution and origin of the grasslands in zone D we do not treat further here (for details see Pemadasa, 1984).

The patterns of similarity between South India and Sri Lanka consist of two corresponding pairs of biogeographic regions (e.g. Bhimachar, 1945), viz. (1) the

Malabar Tract of South India and the Wet Zone lowlands and mountains of Sri Lanka (rainforest communities zone D; see above) and (2) the southeastern part of the Indian peninsula, the Carnatic Tract, and the Dry Zone of Sri Lanka ('dry' vegetation types; zones A, B, and C). The greater number of closely related species occurs in the rainforest regions on both sides of the Palk Strait (Erdelen, in press). The fossil record documents that the onset of the Quaternary marks a new period in the biogeographical evolution in South India and Sri Lanka (Deraniyagala, 1958). Nevertheless the similarities between the rainforest regions in South India and Sri Lanka which are at present far apart from each other has not been explained satisfactorily (for an overview see Erdelen, in press). McKay (1984), in his study of mammal biogeography, assumed that climatically induced habitat changes might have been the most important aspects for an explanation of the similarity patterns or the disjunctions between South India and Sri Lanka. Following McKay's argument and those of others (e:g. Moore, 1960; Silas, 1955), the Quaternary, with its repeated 'short-term' cycles (compared with preceding geological periods) is very likely to have profoundly affected extant floral and faunal relations between the two regions. However, on the one hand the exchange of typical rainforest species over land-bridges during the Quaternary seems necessary to explain the present similarity patterns, but on the other hand, recent studies, for instance of the last glacial maximum (18 000 yr B.P.) suggest drier climatic conditions for South India and Sri Lanka (e.g. Van Campo, 1986).

PRESENT VIEWS ON QUATERNARY CHANGES IN CLIMATE AND SEA LEVELS

Climate

Data available on Quaternary climatic fluctuations in South Asia are mostly from the last glacial and the Holocene. From Flenley's (1979) studies in the tropical African and Malesian region it follows that the transition phase between the last interglacial (125 000 yr B.P.) and the glacial maximum (18 000 yr B.P.) is characterized by a rather marked climatic change no sooner than at 30 000 yr B.P.

For the last glacial maximum, Kuhle (1987) postulates a maximum lowering in temperature of 11°C in Tibet and lowering of the snow line of 1100–1500 m and extended glaciation in the Himalayan region. The concomitant significant changes in the large-scale heating-pressure patterns on the Tibetan Plateau prevented the ITCZ reaching the northern extension it had during the previous interglacial. Aridity in Sri Lanka increased due to the weakening of the southwest airflow and trade winds prevailed. From the calculations of Gates (1976a, 1976b) it follows that air temperatures in the South Indian and Sri Lankan region were about 6°C below present values and that precipitation rates during the last glacial maximum (18 000 B.P.) must have been markedly below recent values for the whole island. From Gates's calculations, Verstappen (1980) estimates precipitation to have been approximately 30 per cent below recent values. Tropical dry climate (TD) predominated even in the Wet Zone of Sri Lanka (Späth, 1981) and the corresponding zones in South India (see Figure 29.1). This is also documented by the fossil record of Sri Lanka (Deraniyagala, 1981). The general decrease in precipitation is also corroborated by analyses of drilling profiles and pollen studies from the Arabian Sea for the last glacial maximum (Kolla and Biscaye, 1977; Van Campo, 1986). Contrary to the findings of Fairbridge (1961) and Verstappen (1980), sea surface temperatures (SST) of the Northern Indian Ocean increased by 1°C and decreased by 7–8°C in the Southern Indian Ocean (Prell, 1984; Williams and Johnson, 1975).

After the last glacial maximum the climate in South Asia gradually became warmer, the Himalayan glaciers started melting, and humidity increased. For the study area precipitation reached a maximum at 11 000 to 10 000 yr B.P. (Van Campo, 1986) and interglacial climatic conditions, here called tropical wet climate (TW), prevailed. Studies from other tropical areas (Ericson and Wollin, 1956a, b; Conolly, 1967; Verstappen, 1974) corroborate this value. This higher precipitation resulted from a northward shift of the ITCZ in South Asia, coupled with an increase in importance of the southwest monsoon. Post-glacial lake level maxima in North and East Africa have been radiocarbon dated between 12 450 yr B.P. and 9 650 yr B.P. (Hillaire-Marcel, Carro and Casanova, 1986; Nicholson, 1981). It follows that precipitation increased in short time after the last glacial maximum at 18 000 yr B.P. Warming-up occurred at lower latitudes, reaching the Holocene temperature maximum at approximately 8 000 yr B.P. (Flenley, 1979). To account for such lake level maxima the position of the ITCZ must have reached latitudes higher than at present. Following this phase of increase in precipitation and temperature there was a gradual transition of precipitation, temperature, and position of the ITCZ towards present conditions.

No data on monsoonal dynamics before the last glacial maximum are available for South Asia. However, studies of global climatic cycles during the Quaternary (e.g. Shackleton and Updyke, 1976; Denton and Hughes, 1983) suggest that the climatic changes just outlined, i.e. the rainfall patterns and temperatures, are typical for a glacial/interglacial cycle and thus have been similar during earlier glacial/interglacial cycles. This generalization applies to the qualitative characteristics of the changes and their temporal aspects.

Sea Levels

Eustatic sea level changes repeatedly affected the South Asian coasts in the Quaternary. For South India, only Chatterjee (1961) published a rather general approach on Quaternary sea levels. Therefore, we discuss data from Sri Lanka only.

Sedimentological and geomorphological studies of beach levels presented evidence for three transgression phases only (Preu, 1988; Preu and Weerakkody, 1987). These transgression maxima were at 5–6 m, 3 m, and 1–1.5 m above present mean sea level. [14]C datings of shells and corals suggest ages of 6200–5100 yr B.P. for the 3 m, and 3200–2300 yr B.P. for the 1–1.5 m level (Katupotha and Fujiwara, 1987; Verstappen, 1987). The 5–6 m maximum is of last interglacial origin (Preu, 1988). Its sediments contain no ^{14}C datable material. Terrestrial sediments of the coastal hinterland of Sri Lanka (e.g. gravel remains, Pisolith deposits) show that additional maxima must have existed, but there are no corresponding marine deposits which might allow more specific statements. Altimetrical interpretations of 'hanging valleys' which Swan (1983) and Bremer (1981) had used to obtain their estimates of 82 m and approximately 100 m, respectively, cannot provide evidence for these previous Pleistocene marine transgressions (Preu, 1988). Furthermore, other studies (Haile, 1971; Batchelor, 1979; Radtke and Ratusny, 1987) also support our view that these earlier Quaternary sea level maxima in Sri Lanka most probably were of similar scale to the values for the last interglacial and the mid-Holocene.

The evaluation and dating of the glacial regression phases is much more complicated compared with the transgressions. The seismic surveys of Sarathchandra *et al.* (1986), Wickremaratne *et al.* (1986), and Wijeyananda *et al.* (1986) show that the shelf of Sri Lanka consists of a tropical inselberg relief covered with three separate sediment bodies. This inselberg relief is of Tertiary age (Preu, 1985). The sediment bodies are of terrestrial and Quaternary origin, the sediments having mainly been transported while the shelf was dry (Preu, 1988). As these sediments are at a depth of 70 m at the outer fringe of the shelf, sea level lowering was at least of the same magnitude. Other data available (e.g. Fairbridge, 1961; Gates, 1976a, 1976b) support this value. The three-layered composition of the sediment body substantiates three regression phases. Most likely, however, more regression phases (with sea level minima ≤ -70 m) may have occurred, as shown by global climatic and sea level curves (Fairbridge, 1961; Shackleton and Updyke, 1976). No datings for the sediment bodies are available. Generally, the onset of the global drop in sea level during the last interglacial/glacial transition phase took place a short time after the 125 000 yr B.P. mark (Denton and Hughes, 1983) and reached its lowest level for the Indian Ocean at approximately 15 000 yr B.P. (Emery, Niino and Sullivan, 1971). The sea level of the Indian Ocean began to rise again at approximately 13 000 yr B.P. (after Emery, Niino and Sullivan, 1971; Verstappen, 1980).

ENVIRONMENTAL AND VEGETATIONAL CYCLES—A MODEL

In the following, we present a qualitative model for potential exchange of flora and fauna between South India and Sri Lanka (over the Palk Strait) during an interglacial/glacial/interglacial cycle (Figure 29.2). Climatic and sea level data for the Late Quaternary cycle form the basis of the model, which is expandable to encompass also previous Quaternary cycles.

The data input into the model consists of changes in (1) glaciation at higher northern latitudes, (2) sea levels in Sri Lanka, and (3) climate. Only qualitative

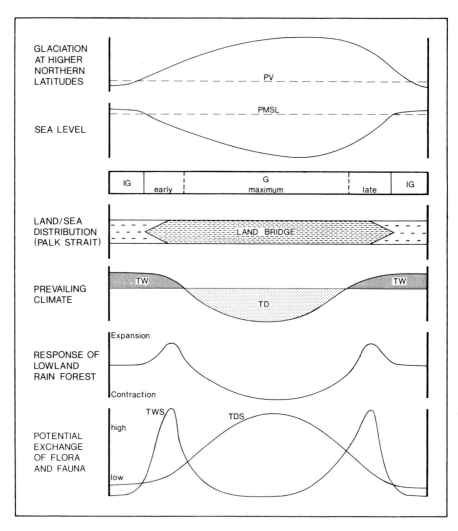

Figure 29.2 Qualitative model for potential exchange of flora and fauna between South India and Sri Lanka over the Palk Strait during an interglacial/glacial/interglacial cycle. Arbitrary scales on abscissa (time) and ordinate. PV = Present value, PMSL = Present mean sea level, TW = Tropical wet climate, TD = Tropical dry climate, IG = Interglacial (only partly shown), G = Glacial, TDS = species (both plant and animal) of tropical dry communities, TWS = species of tropical wet communities

trends are shown in the model. From (2) we derive the land/sea distribution and from (3) we derive climatic conditions (TW, TD) in the Palk Strait during one interglacial/glacial/interglacial cycle. Moreover, we qualitatively describe response of lowland rainforest and potential exchange of flora and fauna across the Palk Strait.

Four assumptions for the Quaternary underlie our model:

1. Present topographical conditions in Sri Lanka, South India and the Palk Strait.
2. Tectonic stability. Geomorphological studies (Bremer, 1981; Preu, 1988) have not corroborated neotectonic movements as postulated for Sri Lanka by Vitanage (1970, 1972).
3. No major shifts in ecological requirements of most of the extant taxa. This should apply despite the facts that species compositions and abundances have not remained constant (see above; and Deraniyagala, 1958).
4. Time lags during interglacial/glacial/interglacial cycles. Two types of time lags occur during one cycle, viz. (a) between the onset of changes in sea level and climate in the transitional phases between interglacial/glacial and glacial/interglacial (see above), and (b) between prevailing climate and vegetation type. We assume that these time lags have been short enough for the establishment of rainforest on the land-bridge.

In the following, we outline the stages of one interglacial/glacial/interglacial cycle (see Figure 29.2) and, for clarity, illustrate it with data for the Late Quaternary.

Parallel to the increase in glaciation at higher northern latitudes, the sea level fell. When the sea level fell some 5 m after the interglacial (shortly after 125 000 yr B.P.; see above) two small land-bridges emerged, one between Jaffna peninsula and Point Calimere, the other between Mannar and Pamban Island (see Figure 29.1). A lowering of 10 m resulted in 50 per cent of the Palk Strait area falling dry. As the climatic change started time-delayed (at approximately 30 000 yr B.P.; see above), it seems very likely that humid and warm conditions prevailed on the land-bridge already existing in the early glacial phase (Figure 29.2). This facilitated primary succession with lowland rainforest as the dominating community on the land-bridge. In this stage a narrow but continuous rainforest belt existed between South India and Sri Lanka. Potential exchange rates were high for the plant and animal species of these forests (TWS in Figure 29.2). When the climate became drier (at least at 30 000 yr B.P.) and thus less favourable, the lowland rainforests contracted in their range and became restricted to areas in the periphery of the Central Hills in Sri Lanka (see fossil data in Deraniyagala, 1981) and the West Ghats of South India. This climatic change induced secondary succession and led to the dominance of tropical dry vegetation on the land-bridge and the adjoining regions. The potential exchange rates for the respective plant and animal species of this community increased (TDS curve in Figure 29.2). During the glacial maximum (18 000 yr B.P.) lowland rainforests had their minimum geographic extension and the land-bridge over the Palk Strait was predominantly covered by tropical dry vegetation (semi-evergreen forests or savanna-like vegetation) and the TDS curve had reached its peak. The glacial maximum coincided with horizontal shrinkage in range of the lowland rainforests and depression of the vegetation zones. Based on Flenley (1979), Ashton and Gunatilleke (1987) estimated a general lowering of 800 m for the vegetational zones in South Asia during the last glacial maximum. Adopting this value for Sri Lanka, cloud forests (now at altitudes >1500 m; see above) would have had their lower limit at 700 m, and montane forests (now between 900 and 1500 m) would have covered an altitudinal range between 100 m and 700 m. However, data from Sumatra and New Guinea (see Flenley, 1979, for details) indicate that the vegetational zones, particularly with regard to montane and cloud forests, were not only depressed but also contracted. Moreover, cloud forests might even have disappeared and only their characteristic elements might have survived as rare components of the upper fringes of the montane forests. To date no evidence has been presented to evaluate whether processes similar or even identical to these should also be postulated for the Palk Strait Region.

In the glacial/interglacial transition phase the onset of the climatic change precedes the onset of the sea level rise (at approx. 13 000 yr B.P.). In this late glacial phase, TW climatic conditions established themselves for the second time (at approximately 10 000 yr B.P.), and secondary succession again replaced the tropical dry communities by tropical rainforests. The TWS curve had reached its second peak and declined steeply when the rising sea level covered the land-bridge. Interglacial conditions prevailed. During the subsequent transition phase towards the present, vegetational dynamics in South India and Sri Lanka were more or less independent of each other.

From the evidence as given above we suggest that, for the repeated Quaternary cycles, most likely a pulse system existed in distribution of the major vegetation types, similar to those documented for the South American, African and Malesian regions (Flenley, 1979; Prance, 1982; Whitmore, 1987), in the Palk Strait Region.

CRITICAL OUTLOOK

The new model describes the characteristics of one Quaternary interglacial/glacial/interglacial cycle with its climatic and vegetational dynamics. The time lags documented for the last Quaternary cycle, which we also suggest for similar previous cycles, offer new opportunities for a biogeographical analysis of extant floral and faunal similarity patterns in the Palk Strait Region. This new approach may contribute to a solution for the seeming paradox of similarities between rainforest plant and animal species between South India and Sri Lanka. However, there are several shortcomings.

1. Generally there is a substantial lack of data on the Quaternary history of the Palk Strait although oil exploration drillings have been carried out (Cantwell, Brown and Mathews, 1978). We would require sediment data for more specific statements on the dating of the land-bridge phases and morphodynamics of the Palk Strait. In addition, pollen data might provide evidence on the multiple climatic changes and information on vegetation cover and its spatio-temporal dynamics. The pollen data available need to be calibrated against present pollen rain and do not yet allow detailed conclusions (cf. Flenley, 1979). Palynological evidence would be particularly necessary to estimate changes in altitudinal zonation of vegetation. Furthermore, detailed new palaeontological studies might provide insight into the Quaternary processes described here.

2. We need more specific estimates for the vertical changes in vegetation zones. We do not know the exact patterns of vegetational changes from the Palk Strait Region itself, particularly when and where both processes, viz. depression and contraction, resulted in equilibrium with the dry and cool climate. The prevalence of dry and cool climatic conditions during extended periods of time suggests that the ranges of the rainforests changed drastically within short periods of time (with regard to the land-bridge dynamics) when compared with monsoon forests or tree savannas. The impact of this temporal bottleneck for evolutionary processes in rainforest communities and possible changes in biogeographical patterns remains unknown.

3. Our rather gross model needs refinement. Computer simulations, based on new data from the Palk Strait Region, might offer opportunities to discuss differing exchange rates of flora and fauna at the onset and at the end of a glacial stage and to estimate time-lag values critical for such exchange processes.

4. Our knowledge of the biogeography of India *sensu latu* is still very poor. In particular, we should propose new hypotheses for an understanding of the altitudinal amplitude of the Sri Lankan and South Indian tree floras and the high degree and geographic localization of endemicity in the Indo-Sri Lankan Region.

ACKNOWLEDGEMENTS

We thank H. Crusz, J. Jacobs, and F. Wieneke for critical reviews of earlier drafts of the manuscript. M. Lagger helped preparing the figures and T. Saks commented on style. The Deutsche Forschungsgemeinschaft (German Research Foundation) provided funds for this study.

REFERENCES

Ashton, P. S. and C. V. S. Gunatilleke (1987). New light on the plant geography of Ceylon. I. Historical plant geography. *Journal of Biogeography,* **14**, 249–85.
Batchelor, B. C. (1979). Discontinuously rising late Cenozoic eustatic sea levels, with special reference to Sundaland, Southeast Asia. *Geologie en Mijnbouw*, **58**, 1–20.
Bhimachar, B. S. (1945). Zoogeographical divisions of the Western Ghats as evidenced by the distribution of hill stream fishes. *Current Science*, **14**, 12–16.
Bremer, H. (1981). Reliefformen und reliefbildende Prozesse in Sri Lanka. *Relief, Boden, Paläoklima*, **1**, 7–183.
Cantwell, Th., Th. E. Brown and G. Mathews (1978). Petroleum geology of the northwest off shore area of Sri Lanka. *Seapex Conference, Singapore*, 1–13.
Chatterjee, S.P. (1961). Fluctuations of sea level around the coast of India during the Quaternary period. *Zeitschrift für Geomorphologie, Supplement-Band*, **3**, 48–56.
Conolly, J. R. (1967). Postglacial-glacial change in climate of the Indian Ocean. *Nature (London)*, **214**, 873–5.
Cooray, P. G. (1967). An introduction to the geology of Ceylon. *Spolia Zeylanica*, **31**, 1–324.
Davis, M. B. (1986). Climatic instability, time lags, and community disequilibrium. In J. Diamond and T. J. Case (eds.), *Community Ecology*, Harper and Row, pp. 269–84.
Denton, G. H. and T. J. Hughes (1983). Milankovitch theory of ice ages: Hypothesis of ice-age linkage between regional insolation and global climate. *Quat. Res.*, **20**, 125–44.
Deraniyagala, P. E. P. (1958). *The Pleistocene of Ceylon*, Government Press, Colombo.
Deraniyagala, S. U. (1971). Prehistoric Ceylon—A summary in 1968. *Ancient Ceylon*, **1**, 3–46.
Deraniyagala, S. U. (1981). Was there a Palaeolithic (Old Stone Age) in Lanka? *Ancient Ceylon*, **4**, 143–56.
Domrös, M. (1974). The agroclimate of Ceylon. *Geoecological Research*, **2**, 1–265.
Emery, K. O., H. Niino and B. Sullivan (1971). Postpleistocene levels of the East China Sea. In K. K. Turekian (ed.), *The Late Cenozoic Glacial Ages*, Yale University Press, pp. 381–90.
Erdelen, W. (in press). Aspects of the biogeography of Sri Lanka. In U. Schweinfurth (ed.), *Forschungen auf Ceylon III*, Steiner, Wiesbaden.
Ericson, D. B. and G. Wollin (1956a). Correlation of six cores from the equatorial Atlantic and the Caribbean. *Deep-Sea Research*, **3**, 104–25.
Ericson, D. B. and G. Wollin (1956b). Micropaleontological and isotopic determinations of Pleistocene climates. *Micropaleontology*, **2**, 257–70.
Fairbridge, R. W. (1961). Eustatic changes in sea level. *Physics and Chemistry of the Earth*, **4**, 99–185.
Fernando, S. N. U. (1968). *The Natural Vegetation of Ceylon*, Swabasha, Colombo.
Flenley, J. R. (1979). *The Equatorial Rain Forest: A Geological History*, Butterworths, London.
Flohn, H. (1952). Allgemeine atmosphärische Zirkulation und Paläoklimatologie. *Geologische Rundschau*, **40**, 153–78.
Flohn, H. (1953). Studien über die atmosphärische Zirkulation in der letzten Eiszeit. *Erdkunde*, **7**, 266–75.
Flohn, H. (1958). Beiträge zur Klimakunde von Hochasien. *Erdkunde*, **12**, 294–308.
Gates, W. L. (1976a). Modeling the ice-age climate. *Science*, **191**, 1138–44.
Gates, W. L. (1976b). The numerical simulation of ice-age climate with a global general circulation model. *Journal of the Atmospheric Sciences*, **33**, 1844–73.

Gaussen, H., P. Legris, M. Viart and L. Labroue (1964). *International Map of Vegetation, Ceylon*, Ceylon Survey Department, Colombo.

Gaussen, H., P. Legris, M. Viart and L. Labroue (1968). *Explanatory Notes on the Vegetation Map of Ceylon*, Government Press, Colombo.

Haile, N. S. (1971). Quaternary shorelines in W. Malaysia and adjacent parts of the Sunda shelf. *Quaternaria*, 15, 338–43.

Hillaire-Marcel, C., O. Carro and J. Casanova (1986). ^{14}C and Th/U dating of Pleistocene and Holocene stromatolithes from East African paleolakes. *Quat. Res.*, 25, 312–29.

Imbrie, J. and K. P. Imbrie (1985). *Ice Ages*, Harvard University Press, Cambridge, Mass.

Katupotha, J. and K. Fujiwara (1987). Holocene sea level change of the southwest and south coast in Sri Lanka. In *Final Field Symposium on Late Quaternary sea level correlations and applications*, NATO Advanced Study Institute.

Koelmeyer K. O. (1957). Climatic classification and the distribution of vegetation in Ceylon. *Ceylon Forester*, 3, 144–63.

Koelmeyer, K. O. (1958). Climatic classification and the distribution of vegetation in Ceylon. *Ceylon Forester*, 3, 265–288.

Kolla, V. and P. E. Biscaye (1977). Distribution and origin of quartz in the sediments of the Indian Ocean. *Journal of Sedimentary Petrology*, 47, 642–9.

Kuhle, M. (1987). Subtropical mountain- and highland glaciation as the ice-age trigger and the waning of the glacial periods in the Pleistocene. *Geo Journal*, 14, 393–421.

Kularatnam, K. (1968). Ceylon. In S. P. Chatterjee (ed.), *Developing Countries of the World*, Calcutta, pp. 308–17.

Legris, P. (1963). *La végétation de l'Inde: Écologie et flore*, Vol. 6, Travaux de la Section Sciences et Techniques, Institut Français de Pondichéry.

Legris, P. and M. Viart (1961). *Bioclimates of South India and Ceylon*, Vol. 3, Travaux de la Section Scientifique et Technique, Institut Français de Pondichéry.

McKay, G. M. (1984). Ecology and biogeography of mammals. In C. H. Fernando (ed.), *Ecology and Biogeography in Sri Lanka*, Martinus Nijhoff, pp. 413–29.

Mani, M. S. (ed.) (1974). *Ecology and Biogeography in India*, Junk Publishers, The Hague.

Moore, J. C. (1960). Squirrel geography of the Indian subregion. *Syst. Zool.*, 9, 1–17.

Mueller-Dombois, D. (1968). Ecogeographic analysis of a climate map of Ceylon with particular reference to vegetation. *Ceylon Forester*, 8, 39–58.

Nicholson, S. E. (1981). The historical climatology of Africa. In T. M. Wigley, M. J. Ingram and G. Farmer (eds), *Climate and History*, Cambridge University Press, pp. 249–70.

Pascal, J. P. (1984). *Les Forêts denses humides sempervirentes des Ghâts Occidentaux de l'Inde*, Vol. 20, Traveaux de la Section Scientifique et Technique, Institut Français de Pondichéry.

Pemadasa, M. A. (1984). Grasslands. In C. H. Fernando (ed.) *Ecology and Biogeography in Sri Lanka*, Martinus Nijhoff, pp. 99–131.

Prance, G. T. (ed.) (1982). *Biological Diversification in the Tropics*, Columbia University Press, New York.

Prell, W. L. (1984). Monsoon climate of the Arabian Sea during the Late Quaternary. A response to changing solar radiation. In A. Berger (ed.), *Milankovitch and Climate*, Part I, Wiley, pp. 349–66.

Preu, Ch. (1985). Erste Forschungsergebnisse quartärmorphologischer Untersuchungen an den Küsten Sri Lankas. *Kieler Geographische Schriften*, 62, 115–25.

Preu, Ch. (1988). *Zur Küstenentwicklung Sri Lankas im Quartär. Untersuchung der Steuerungsmechanismen und ihrer Dynamik im Quartär zur Ableitung eines Modells der polygenetischen Küstenentwicklung einer Insel in den wechselfeuchten Tropen*, Habilitationsschrift, University of Augsburg.

Preu, Ch. and U. Weerakkoddy (1987). Mapping of geomorphology of estuarine coasts using remote sensing techniques. A case study from Sri Lanka. *Berliner Geographische Schriften*, **25**, 389–401.

Radtke, U. and A. Ratusny (1987). Pleistozäne Meeresspiegelschwankungen. Forschungsgeschichtlicher Rückblick und aktuelle Perspektiven. *Berliner Geographische Schriften*, **25**, 9–33.

Sarathchandra, M. J., W. S. Wickremeratne, N. P. Wijeyananda and N. G. Ranatunga (1986). *Preliminary results of seismic reflection and bathymetric studies off Dondra and Matara*. Unpublished manuscript.

Shackleton, N. J. and N. D. Updyke (1976). Oxygen-isotope and paleomagnetic stratigraphy of Pacific core, V28–239, late Pliocene to latest Pleistocene. *Mem. Geol. Soc. Amer.*, **145**, 449–64.

Silas, E. G. (1955). Speciation among the freshwater fishes of Ceylon. *Bull. Nat. Inst. Sci. India*, **7**, 248–59.

Späth, H. (1981). Bodenbildung und Reliefentwicklung in Sri Lanka. *Relief, Boden, Paläoklima*, **1**, 185–238.

Suppiah, R. (1988). Atmospheric circulation and Sri Lankan rainfall. *Science Reports of the Institute of Geoscience*, **9**, 75–142. University of Tsukuba, Japan.

Swan, S. B. St. C. (1983). *An Introduction to the Coastal Geomorphology of Sri Lanka*, Government Press, Colombo.

Van Campo, E. (1986). Monsoon fluctuations in two 20,000-Yr B.P. oxygen-isotope/ pollen records off southwest India. *Quaternary Research*, **26**, 376–88.

Verstappen, H. Th. (1974). On palaeo climates and landform development in Malesia. *Modern Quaternary Research in SE Asia*, **1**, 3–35.

Verstappen, H. Th. (1980). Quaternary climatic changes and natural environment in Southeast Asia. *Geo Journal*, **4**, 45–54.

Verstappen, H. Th. (1987). Geomorphologic Studies on Sri Lanka with special emphasis on the northwest coast. *ITC Journal*, **1**, 1–17.

Vitanage, P. W. (1970). A study of the geomorphology and the morphotectonics of Ceylon. *Proceedings of the Second Seminar on Geochemical Prospecting Methods and Techniques*, 391–405.

Vitanage, P. W. (1972). Post-Precambrian uplifts and regional neotectonic movements in Ceylon. *Proceedings of the 24th International Geological Congress, Section 3*, 642–54.

Wadia, D. N. (1976). *Geology of India*, McGraw-Hill, New Delhi.

Walker. D. (1982). Speculations on the origin and evolution of Sunda-Sahul rain forests. In G. T. Prance (ed.) *Biological Diversification in the Tropics*, Columbia University Press, New York, pp. 554–75.

Werner, W. L. (1984). *Die Höhen- und Nebelwälder auf der Insel Ceylon (Sri Lanka)*, Steiner, Wiesbaden.

Whitmore, T. C. (ed.) (1987). *Biogeographical Evolution of the Malay Archipelago*, Clarendon Press, Oxford.

Wickremaratne, W. S., N. P. Wijeyananda, M. J. Sarathchandra and N. G. Ranatunga (1986). *Quaternary geological research of the continental shelf and slope off Panadura*. Unpublished manuscript.

Wijeyananda, N. P., W. S. Wickremeratne and M. J. Sarathchandra (1986). *Geophysical, sedimentological studies around the Great Basses Ridge—Southeast Coast of Sri Lanka*. Unpublished manuscript.

Williams, D. F. and W. C. Johnson (1975). Diversity of recent planktonic foraminifera in the Southern Indian Ocean and Late Pleistocene paleotemperatures. *Quat. Res.*, **5**, 237–50.

Author Index

Subject Index

aerial photography 74
algae 10, 11, 16, **479**
alluvial fans **77**
asymmetry **25, 398**
Atlantic period 293
attractor 49

basal control of erosion 132
basin scale response 61
beach rock 16
biocrusts 479
biogeographical evolution 491
biogeomorphology 6
biological weathering 8
biomass
    heathland 188, 193, 194
    mediterranean
        measured 368
        modelled 37, 45
    seasonal variations 33
    wind erosion 93
bioturbation 15
boreal forest 13
Bowen ratio method 28
Bronze Age 19

calcrete 8, 16
cave deposits 16
channel adjustments
    aggradation **151**
    climatic change 146
    vegetation recovery 157
chaparral 14, **270**
characteristic forms 17
chemical interactions
    fertilizer application rates 163
    metal ores **240**
    nitrate 175
    nitrogen **161**
    phosphorus **161**
chemical sedimentation 17
colluvial hollows 76

competition 42
conservation practise 62
cracking soils 151, 205, 389
crop
    residues 99
    resistance to wind 89–90
    rotations 57
crops
    barley 95, 163, 164
    corn 57
    cotton 57
    kale 163, 170
    sugar beet 95
    wheat 203

debris avalanche **67**
debris torrents 71
dendrochronology 76, 151, 442, 463
deposition 15–17
    bank accretion 131, 148
    biogenic 17
    channel bars 146, 229
    coastal rates 119
    metals **240**
desertification 363
desert varnish 16
disturbance 13, 452
drag coefficient for vegetation 89
drainage
    bank stability 137
dunes 15, 16, **471**

earthworms 7
EPIC model 60
erosion 12–17, **42**, 56
    control 107
    fluvial 133
    mats 100
    modelled 31, 47
    overland flow **326**
    slope steepness 101
    splash 100, 457
    wind **86**